LAMBORGHINI

**with Italy
for Italy**

Letizia Battaglia	**Palermo**
Stefano Guindani	**Sicilia**
Davide De Martis	**Sardegna**
Guido Taroni	**Calabria**
Gabriele Micalizzi	**Puglia**
Camilla Ferrari	**Basilicata**
Marco Casino	**Campania**
Roselena Ramistella	**Molise**
Valentina Sommariva	**Abruzzo**
Anna Di Prospero	**Lazio**
Wolfango Spaccarelli	**Marche**
Alessandro Cinque	**Umbria**
Gabriele Galimberti	**Toscana**
Piero Gemelli	**Emilia-Romagna**
Marco Valmarana	**Veneto**
Mattia Balsamini	**Friuli Venezia Giulia**
Simone Bramante	**Trentino-Alto Adige**
Vincenzo Grillo	**Lombardia**
Chiara Mirelli	**Piemonte**
Alberto Selvestrel	**Liguria**
Fulvio Bugani	**Valle d'Aosta**

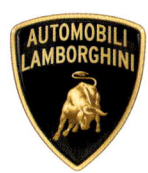

LAMBORGHINI
with Italy for Italy

21 Views for a New Drive

SKIRA

Davide Rampello

Johann Wolfgang von Goethe arrived in Bologna in October 1786. Accustomed to seeking out a lofty position for an overview of where he was visiting, on the evening of 18 October, he climbed the Torre degli Asinelli: "What a splendid view! To the north you can see the hills of Padua, then the Swiss, Tyrolean and Friulian Alps … To the west, an unblocked horizon, from which only the towers of Modena emerge. To the east, a uniform plain reaching to the Adriatic … To the south, the first hills of the Apennines, cultivated and luxuriant to their peaks, populated by churches, palaces and villas, like the hills of the Vicentino." The clear, unpolluted air would have made it possible to observe an immense landscape and distinguish towers, church steeples and villages.
Sant'Agata Bolognese, with its convents, palaces, and the Torre del Barbarossa lay to the west, in the autumnal countryside furrowed by silvered canals.
This is where "With Italy, for Italy" could start: with the travel notes of one of the stars of the Grand Tour age. Today, however, the star of this new journey on which we are about to embark is a car that "guides" the paths of twenty interpreting gazes: a formidable car capable of inspiring glances, feelings and visions.
An overture precedes the start of our story: a prologue dedicated to the capital of the Mediterranean – Palermo, a city that is "all port" (Pánormos = παν-όρμος), a city of welcome, of shared differences, a city of the flavours, scents and feelings cultivated by the deep vitality of the people who live here.
Although partially disfigured by criminal speculation during the post-war period, in the depths of its soul Palermo retains the luminous splendour that inspired the Phoenician people to call it a "flower" ("zyz").
Letizia Battaglia had the job of narrating this city with a child's "innocent gaze". The very young "woman" with the copper-coloured hair looking enigmatic and crowned with white roses, a young Saint Rosalie, is the "daughter" of the Saint portrayed from behind by Anthony van Dyck in his *Madonna of the Rosary* in 1625, the year after the great plague of 1624.

A coincidence and disturbing circumstance that the powerful energy of the yellow Aventador SVJ in the background leads back to a reassuring calm. The images that Letizia has captured, the bursting vitality of the innocent and fragile young "women" from the Vucciria district, portrayed together with this symbol of power, strength and reliability, create an "Italian" landscape with the secret dimension of dream. In his book *Antropologia del paesaggio* (Anthropology of Landscape), Eugenio Turri points out that "the environment implies the being there, the living there", whereas landscape is "the sense-based manifestation of the environment, spatial reality seen and felt". This subtle distinction elicits a reflection on the quality and intensity of our "being" and "living" a place, on our ability to perceive its spatial reality. The brief we gave our photographers was to interpret Italy's twenty regions with an exceptional companion, a "subject" that had to appear in the image and in the story. The beautiful car would have to be placed, "set" within the chosen space to become part of the landscape, unlike on traditional fashion shoots, where the starring object, the dress, degrades the most beautiful landscape to mere "location". The chosen place's capacity for and intensity of feeling, its reality, alienates any instrumental indifference.
"With Italy, for Italy" is a story of relationships, one that adds the meditated breathing space of a rest stop to the metaphor of the journey. Every stop becomes a chance "to be" and "to live" that place. The splendid Lamborghinis become a thing among things, a shape among shapes, a colour among colours, seeking out a detail in the interplay of distance, finding wonder in the chromatic concordances of a sunset.
In a word: Landscapes.
Landscape as re-invention and re-discovery, revealing the infinite souls that the "memory" of a place preserves and protects.
The underlying theme "With Italy, for Italy" suggests that a never-ending quest for quality and perfection, abandoning ourselves to listening and to beauty, can reveal a renewed Italy to our eyes.

Stefano Guindani

To restore Italy to its rightful place, to exalt its ethos, beauty, uniqueness and excellence through the art of photography… this was the brief for which Lamborghini hired me to be Artistic Director at the start of this project. "With Italy, for Italy" is an initiative of great cultural and social significance. This Italian brand, one of the most prestigious in the world, wanted to go beyond mere business to pursue a social role in a modern interpretation of Renaissance patronage, restoring culture and art to their highest and deepest value as instruments independent of power, a never-ending gateway to communicating beauty, to aspiring to beauty: indeed, to transforming society by breaking through old modes of expression and mapping out another future.

The result is this anthem to the typically-Italian creativity that has made this country so great in the world. It is one company's act of love for its country, one company keen to offer its support as we face up to the new challenges of the Covid-19 pandemic, as we set off to seek a new competitive identity.

It is an attempt to bring out the richness of the landscape and architecture that make Italy, as Piovene writes in his *Viaggio in Italia* (Journey across Italy), a distillate of the world.

Lamborghini and I approached twenty talented contemporary Italian photographers, many established and some emerging, and invited them to interpret one of Italy's twenty regions in their own style. Almost exclusively, we chose artists from outside the sector who could provide a narrative of Italy with an *auteur*'s eye. We gave them *carte blanche* to experiment with new stylistic approaches, to go beyond the typical self-referentiality of the automotive world. The result was a whirlwind photographic journey that took in the entire Italian peninsula, seeking dialogue and ongoing harmonies between the beauty of this land, its people and Lamborghini's super sports cars, all of which share the same DNA.

The team of photographers comes from the world of fashion and lifestyle, from which I myself hail (Piero Gemelli, Guido Taroni and Vincenzo Grillo); photojournalists (Gabriele Micalizzi, Fulvio Bugani and Alessandro Cinque); documentary photographers (Gabriele Galimberti, Marco Casino and Camilla Ferrari); social photographers (Mattia Balsamini and Roselena Ramistella); architecture and interiors photographers (Valentina Sommariva); staged and art photographers (Anna Di Prospero); car photographers (Wolfango Spaccarelli and Davide De Martis); a photographer from the music industry (Chiara Mirelli); a creative director (Simone Bramante); an Instagram content creator (Marco Valmarana); and a very young and talented landscape photographer (Alberto Selvestrel), showing how invested Lamborghini is in future generations.

This image-based narrative encompasses Italy's twenty regions and twenty Lamborghinis on a journey that wends its way from South to North – a journey in reverse, because changes always require a change of direction and a new perspective.

The book starts with a photography big gun: Letizia Battaglia, one of Italy's most authoritative international photographers, known for her reportage on Palermo, a champion of civil battles and the first European woman to win the Eugene Smith Award in New York in 1985, commemorating the famous *Life* photographer. Battaglia makes a special homage to her Palermo, her images part of an allegorical narrative: the dream of a future free from injustice. Countercurrent as this choice may be, it shows how much Lamborghini cares about its status as a brand with a strong social responsibility, sharing the values and commitment this photographer has always expressed through her works.

This project puts our values at its core, repositioning the role of artistic research. Ultimately, this book crystallizes this approach page after page, telling a story, exciting, awakening and above all restoring the pride of what it is to be Italian.

This book is about us.

Thank you Lamborghini. Thank you Italy.

Cover
Anna Di Prospero, Sperlonga, detail

Design
Marcello Francone

Editorial Coordination
Emma Cavazzini

Copy Editor
Carlotta Santuccio

Layout
Paola Ranzini Pallavicini

Translations
Adam Victor for Scriptum, Rome

First published in Italy in 2020 by
Skira editore S.p.A.
Palazzo Casati Stampa
via Torino 61
20123 Milano
Italy
www.skira.net

© 2020 Automobili Lamborghini
© 2020 Skira editore

All rights reserved under international copyright conventions. No part of this book may be reproduced or utilized in any form or by any means, electronic or mechanical, including photocopying, recording, or any information storage and retrieval system, without permission in writing from the publisher.

Printed and bound in Italy. First edition

ISBN: 978-88-572-4494-5

Distributed in USA, Canada, Central & South America by ARTBOOK | D.A.P. 75, Broad Street Suite 630, New York, NY 10004, USA.
Distributed elsewhere in the world by Thames and Hudson Ltd., 181A High Holborn, London WC1V 7QX, United Kingdom.

The "Automobili Lamborghini" and "Automobili Lamborghini Bull and Shield" trademarks, copyrights, designs and models are used under licence from Automobili Lamborghini S.p.A., Italy.

Contents

Letizia Battaglia	11	**Palermo**
Stefano Guindani	24	**Sicilia**
Davide De Martis	38	**Sardegna**
Guido Taroni	50	**Calabria**
Gabriele Micalizzi	62	**Puglia**
Camilla Ferrari	74	**Basilicata**
Marco Casino	84	**Campania**
Roselena Ramistella	94	**Molise**
Valentina Sommariva	106	**Abruzzo**
Anna Di Prospero	120	**Lazio**
Wolfango Spaccarelli	130	**Marche**
Alessandro Cinque	142	**Umbria**
Gabriele Galimberti	152	**Toscana**
Piero Gemelli	164	**Emilia-Romagna**
Marco Valmarana	178	**Veneto**
Mattia Balsamini	188	**Friuli Venezia Giulia**
Simone Bramante	198	**Trentino-Alto Adige**
Vincenzo Grillo	202	**Lombardia**
Chiara Mirelli	210	**Piemonte**
Alberto Selvestrel	222	**Liguria**
Fulvio Bugani	232	**Valle d'Aosta**
	244	The photographers
	247	With Italy, for Italy: Lamborghini models
	248	Acknowledgements

I chose my hometown Palermo as the location for this project.

I used the Aventador SVJ to highlight a street corner, a church, a small port, the sea. I took the Lamborghini to the Vucciria district, surrounded by festive folk.

I chose to portray little girls because Palermo, for me, is all about this: a little girl with the innocent look of someone who has hopes and dreams as she grows up. The very same way I dream that Palermo, Italy, our world will all become places where human justice prevails.

Such has been my creed and my mission as a photojournalist, in the firm belief that photographs have the power to boost culture, identity and morality.

This is the future I hope for.

Letizia Battaglia

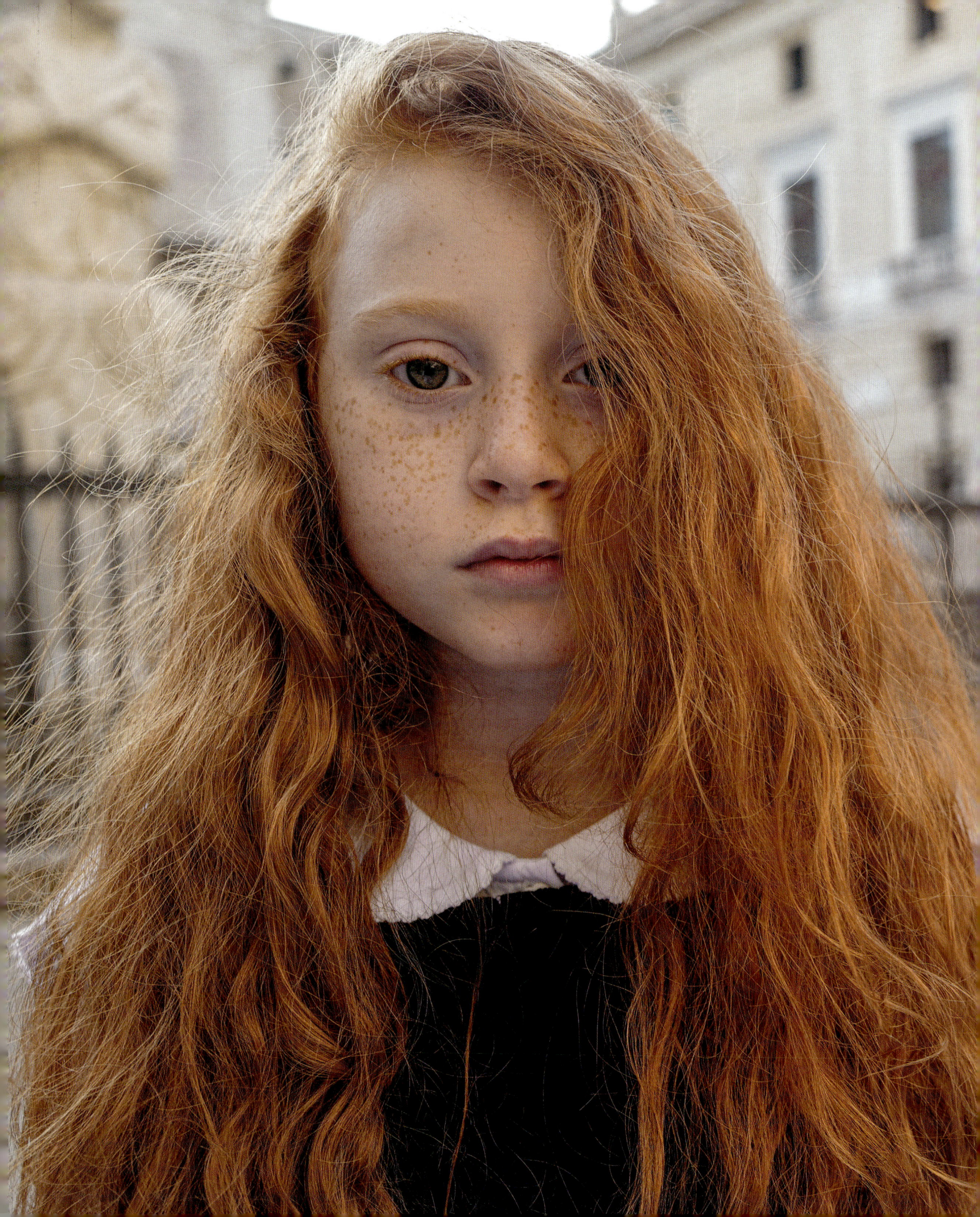

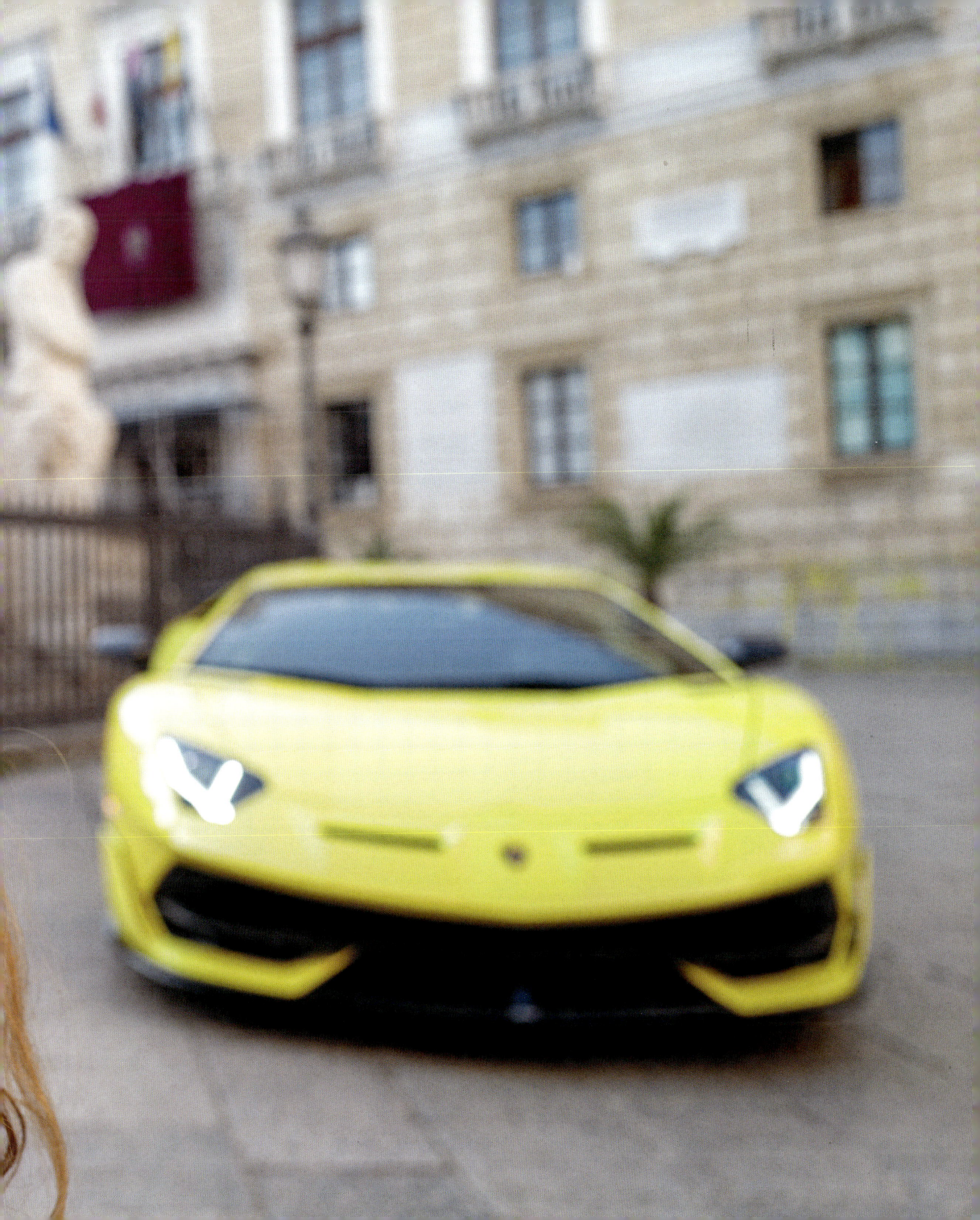

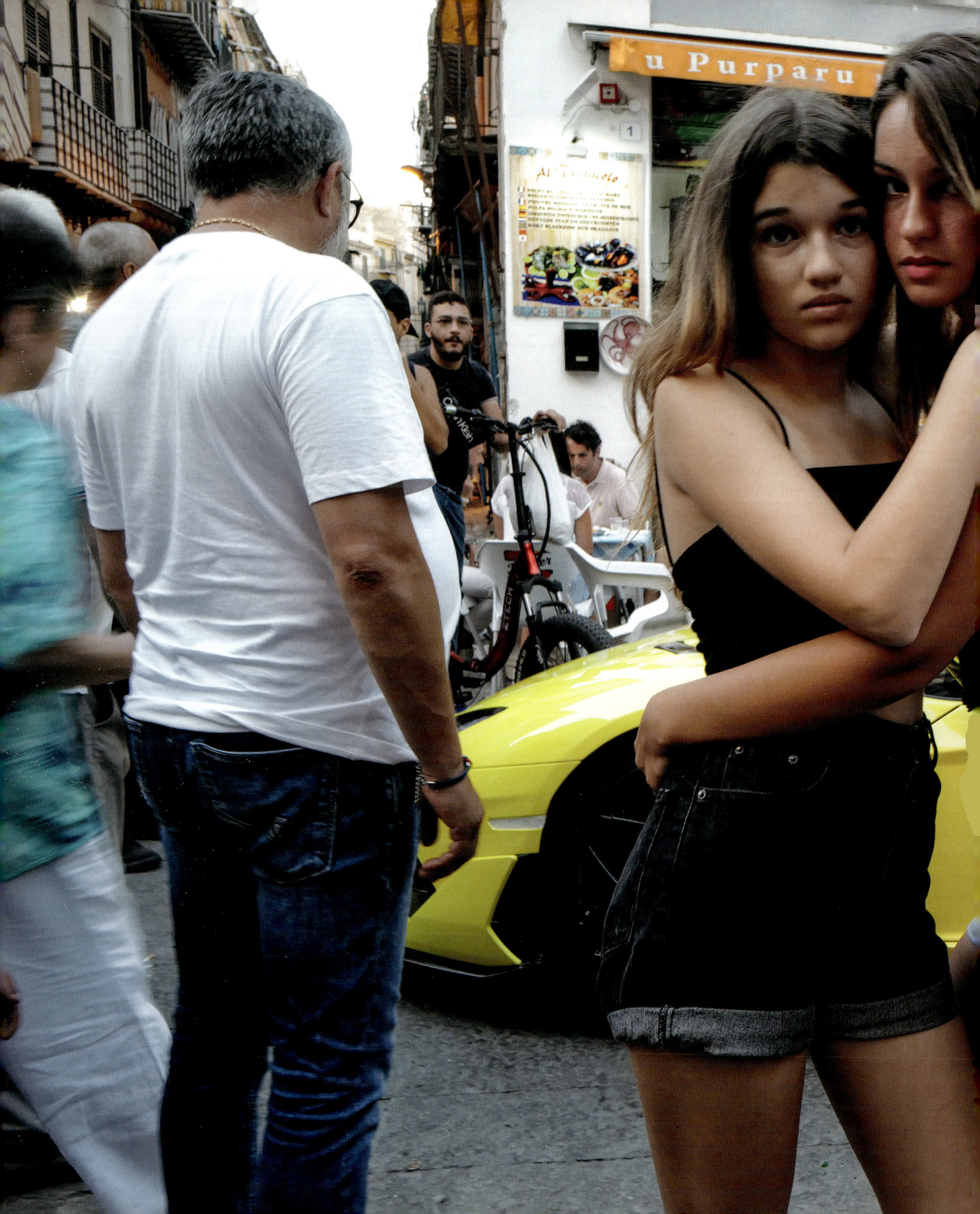

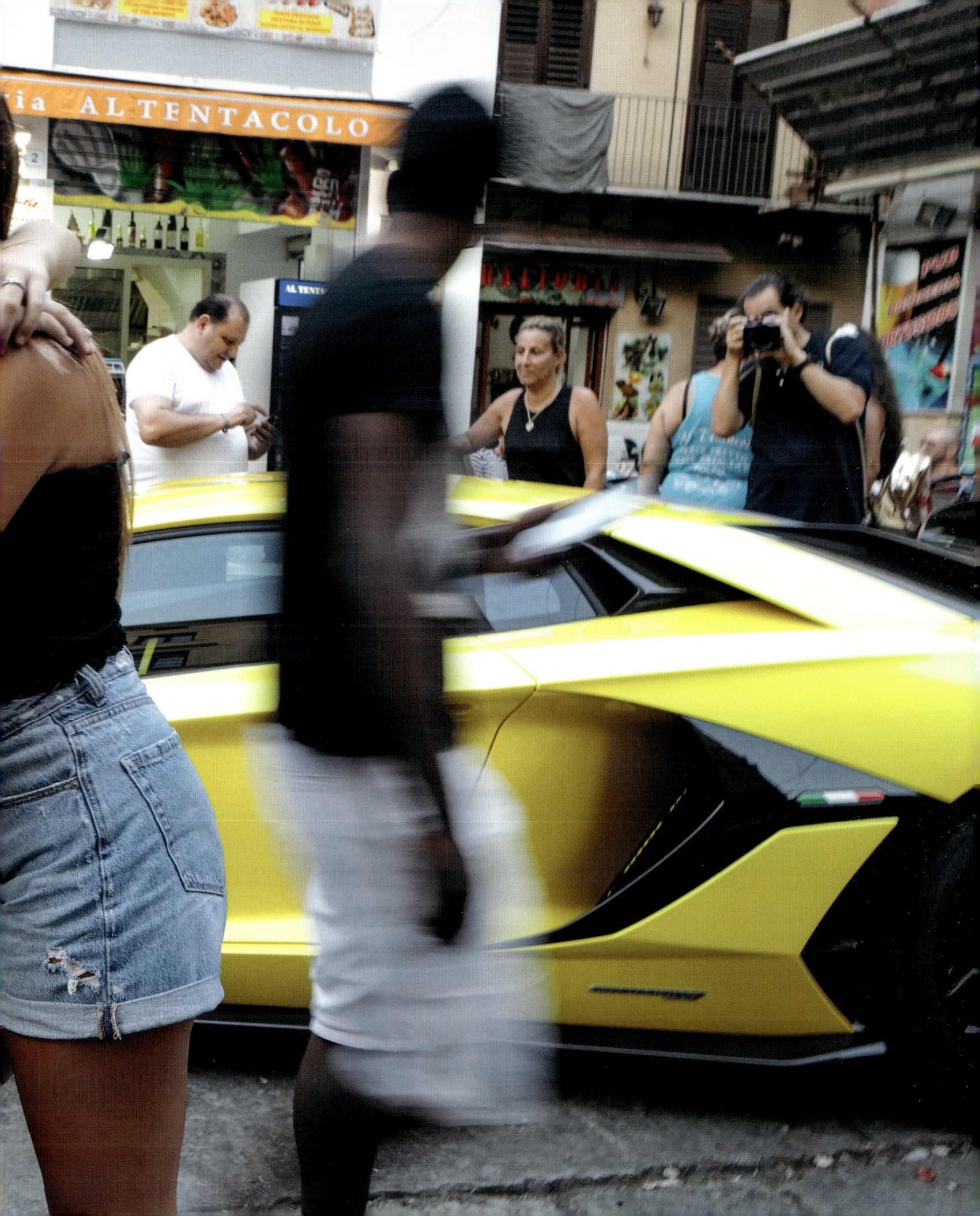

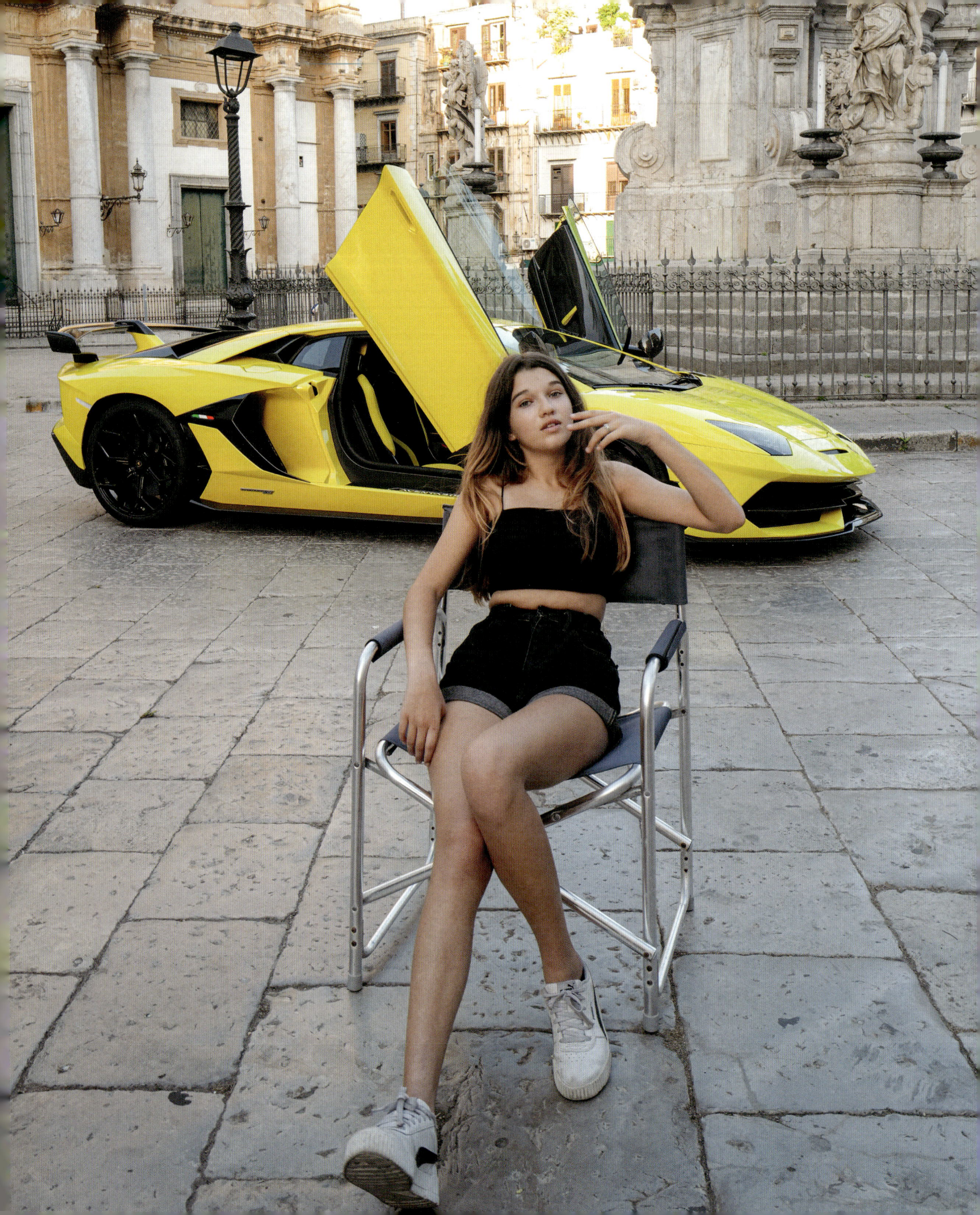

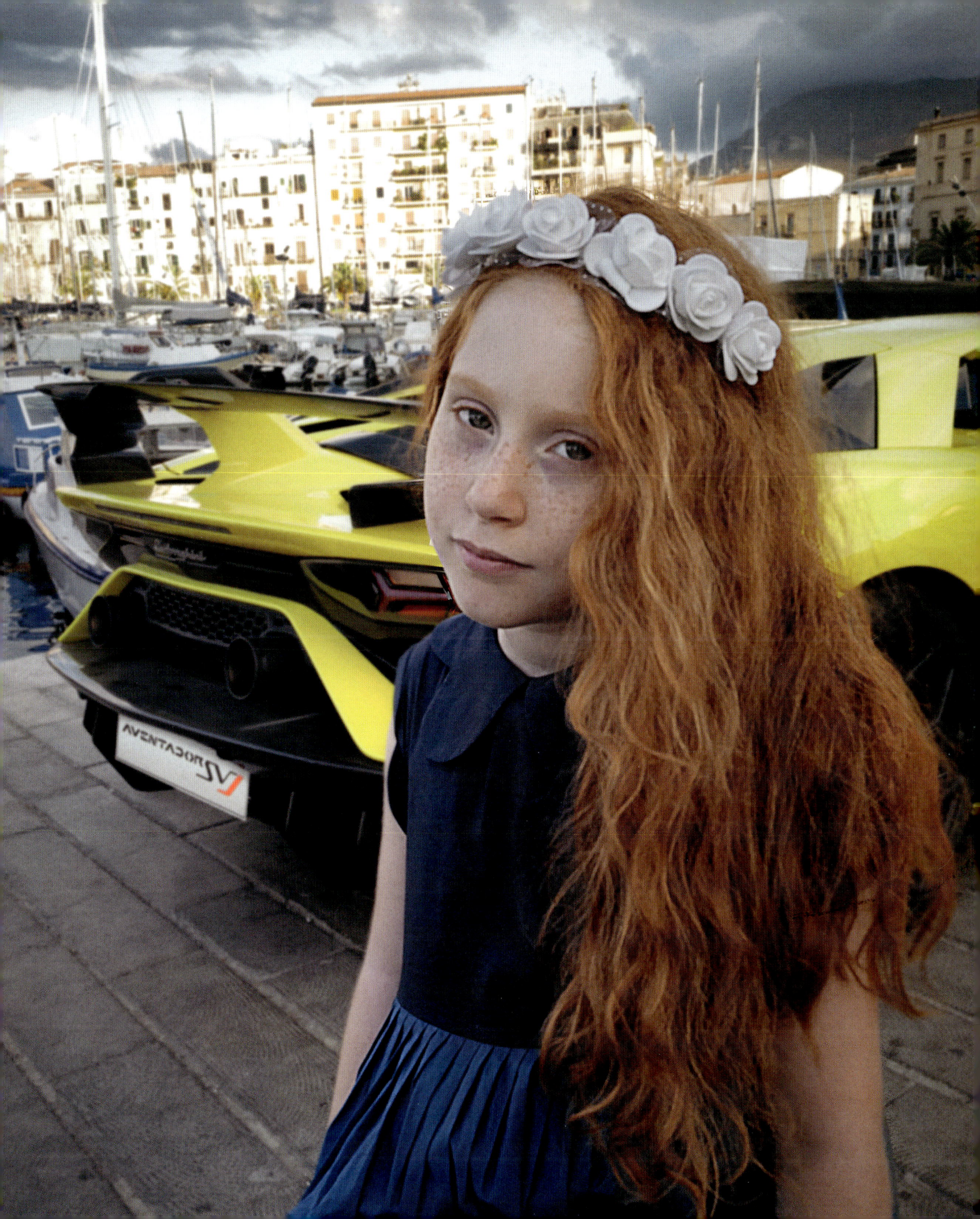

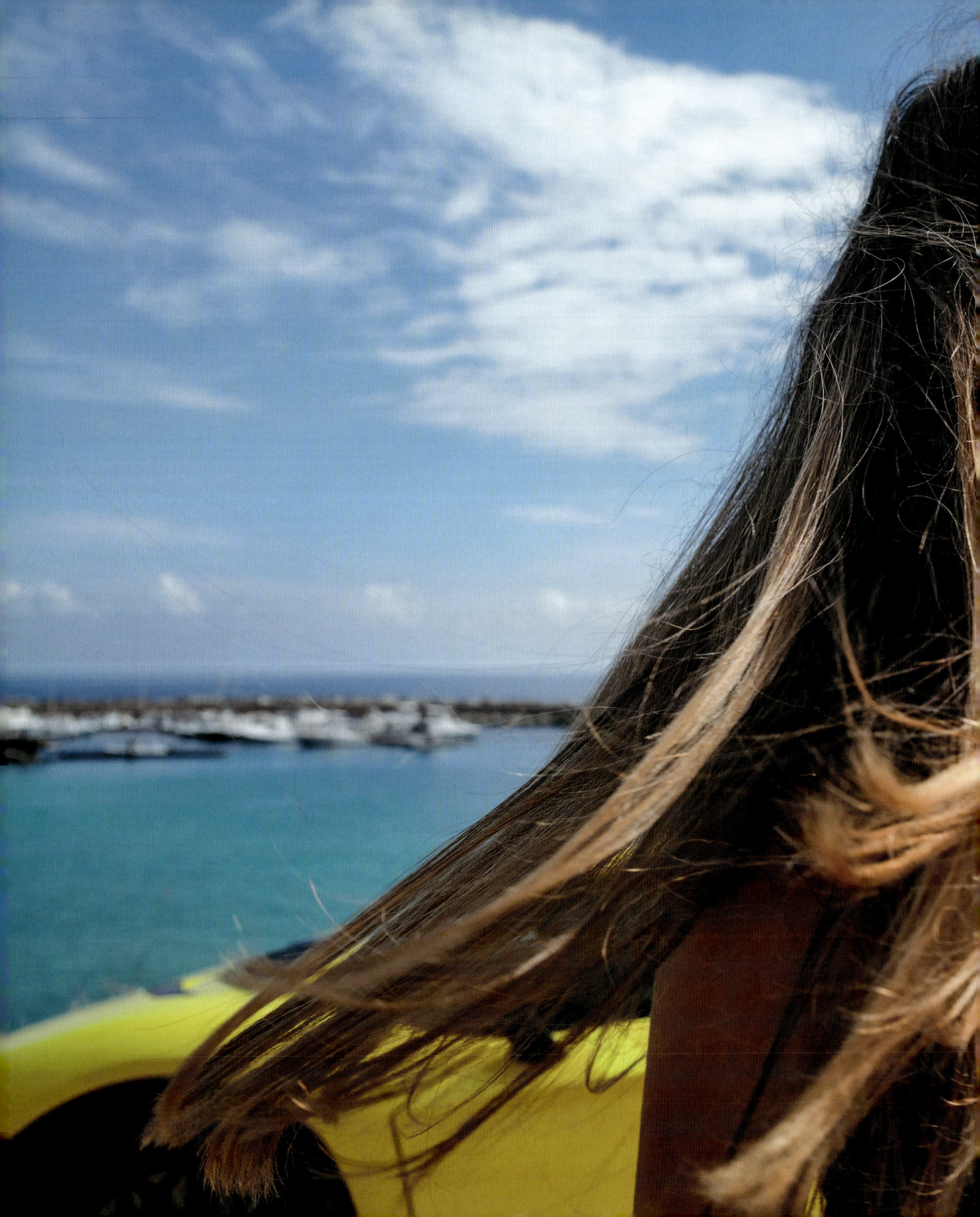

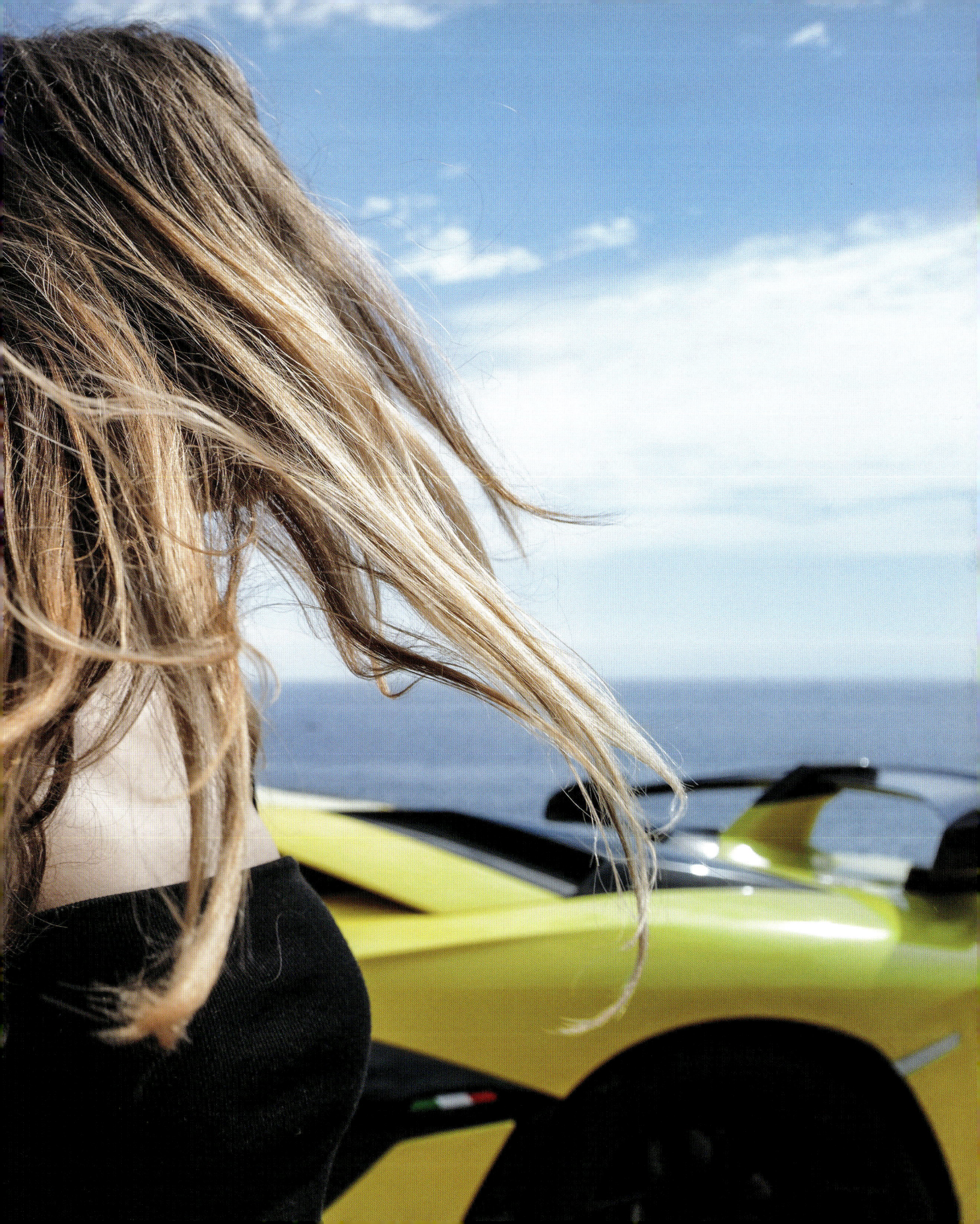

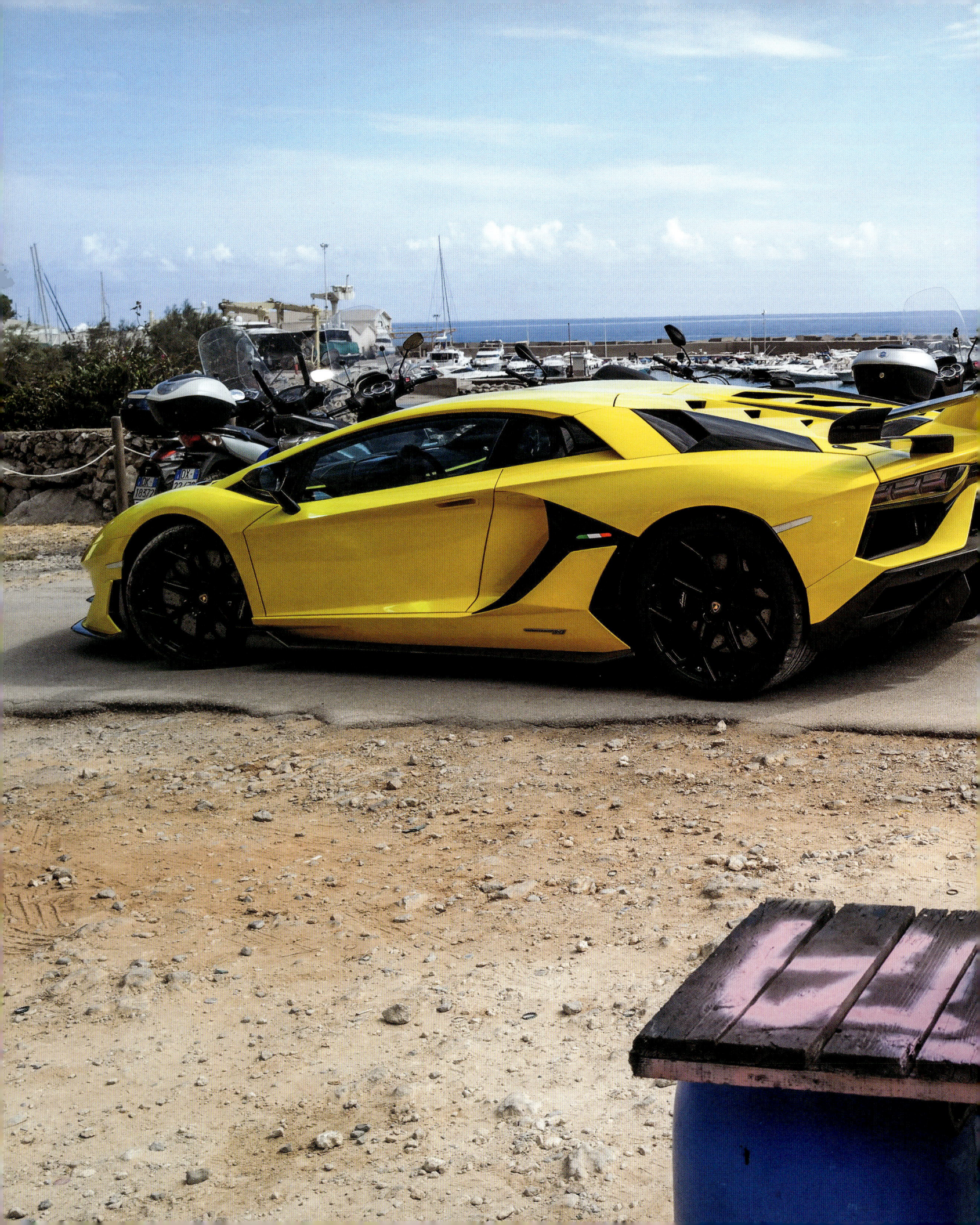

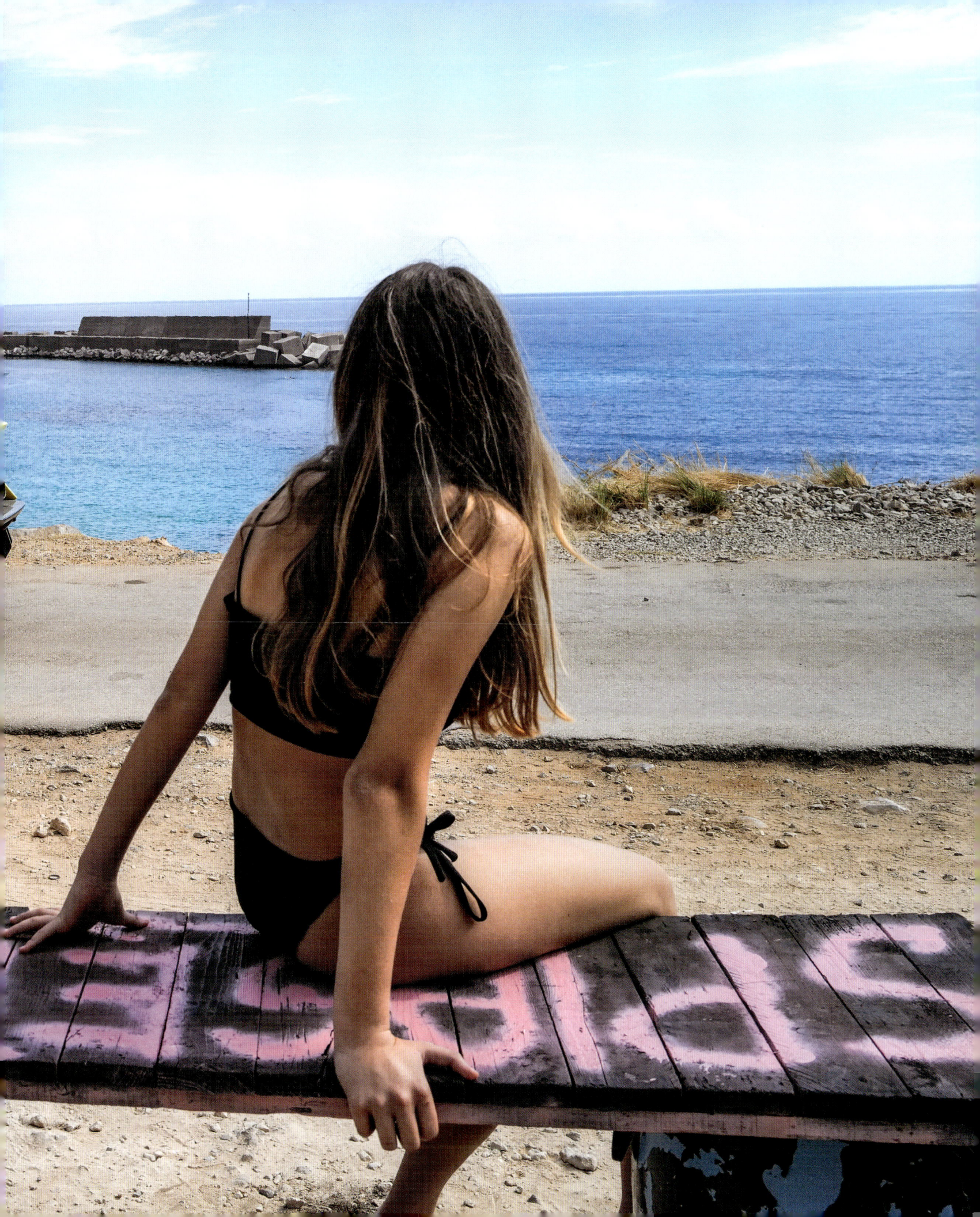

"I love your splendid seas and sublime Alps, I love your solemn monuments and immortal memories;
I love your glory and your beauty"

Edmondo De Amicis

STEFANO GUINDANI
SICILIA

Sicily is a unique region, capable of stimulating all our senses. It overflows with proud, disruptive beauty, its colours and scents interwoven through thousands of years of history and encounters with different peoples and cultures. On my trip with the Lamborghini Urus, I wanted to interpret this region through a lifestyle and street photography approach, creating an image in which the car appears and disappears within the frame of local territory. I wanted to celebrate Sicily in all its natural elegance and poignant beauty.

San Giorgio Cathedral, Modica

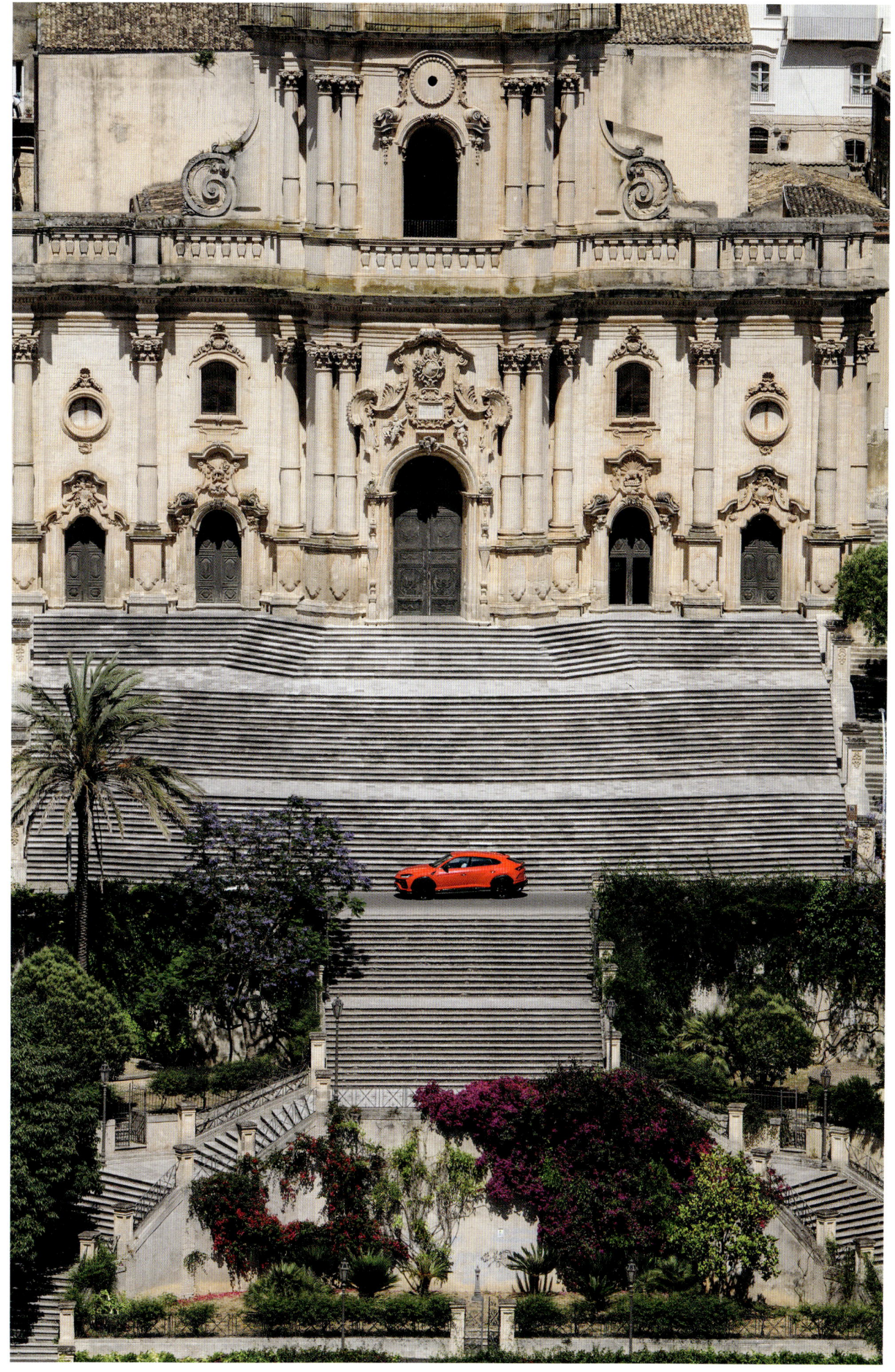

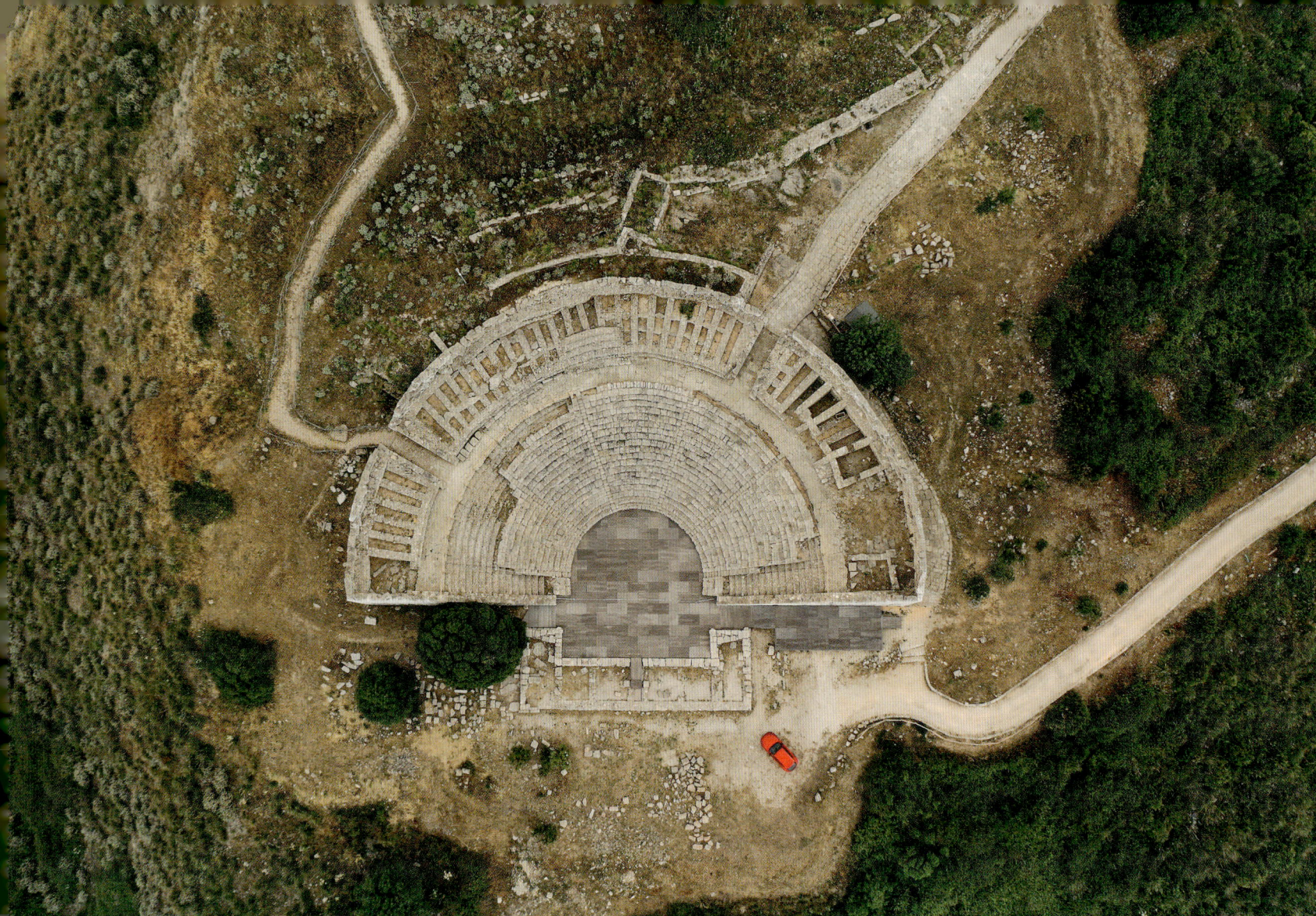

Segesta Archaeological Park, Trapani

"I wanted to celebrate Sicily in all its natural elegance and poignant beauty"

Pages 28–29 Church of San Bartolomeo, Scicli, Ragusa

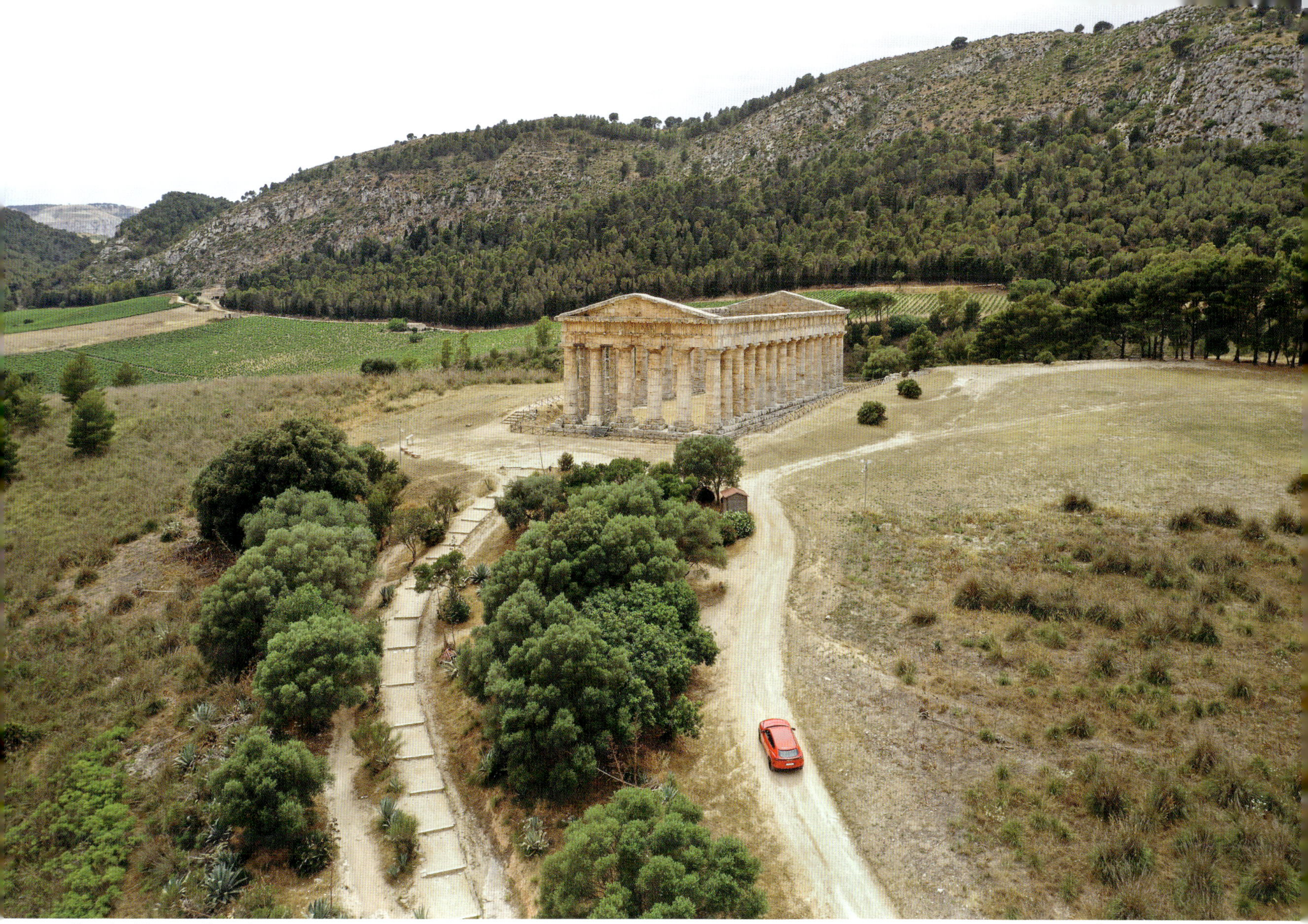

Temple of Segesta, Trapani

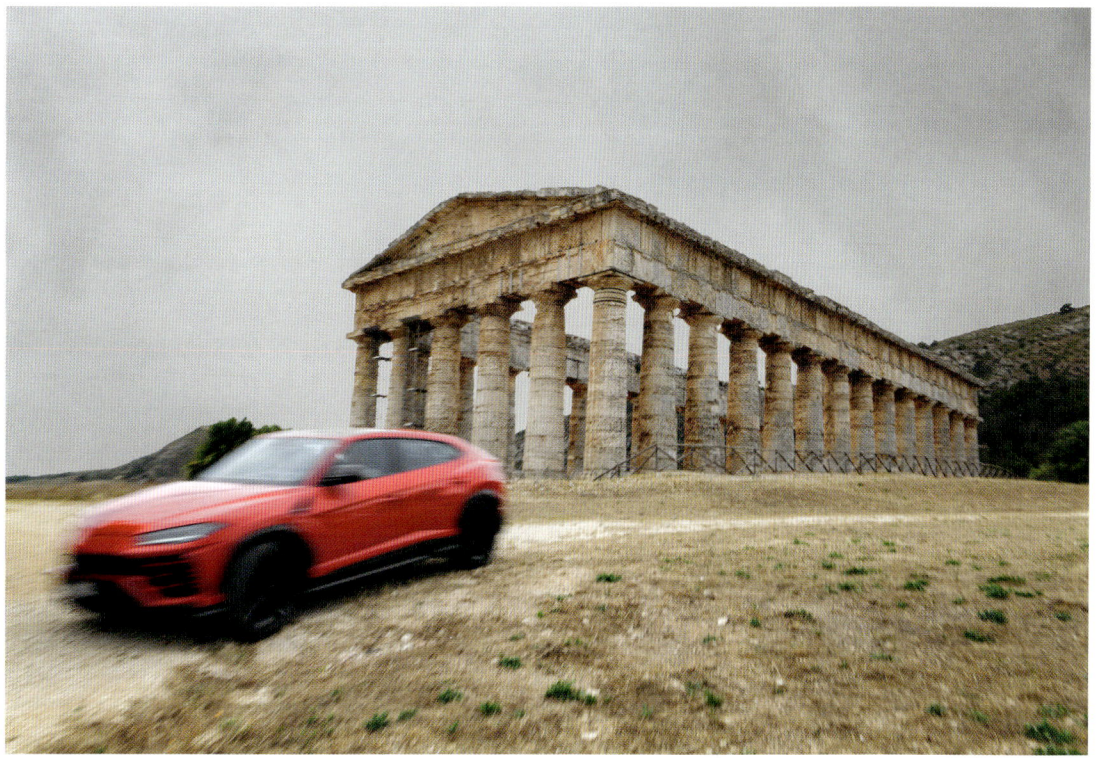

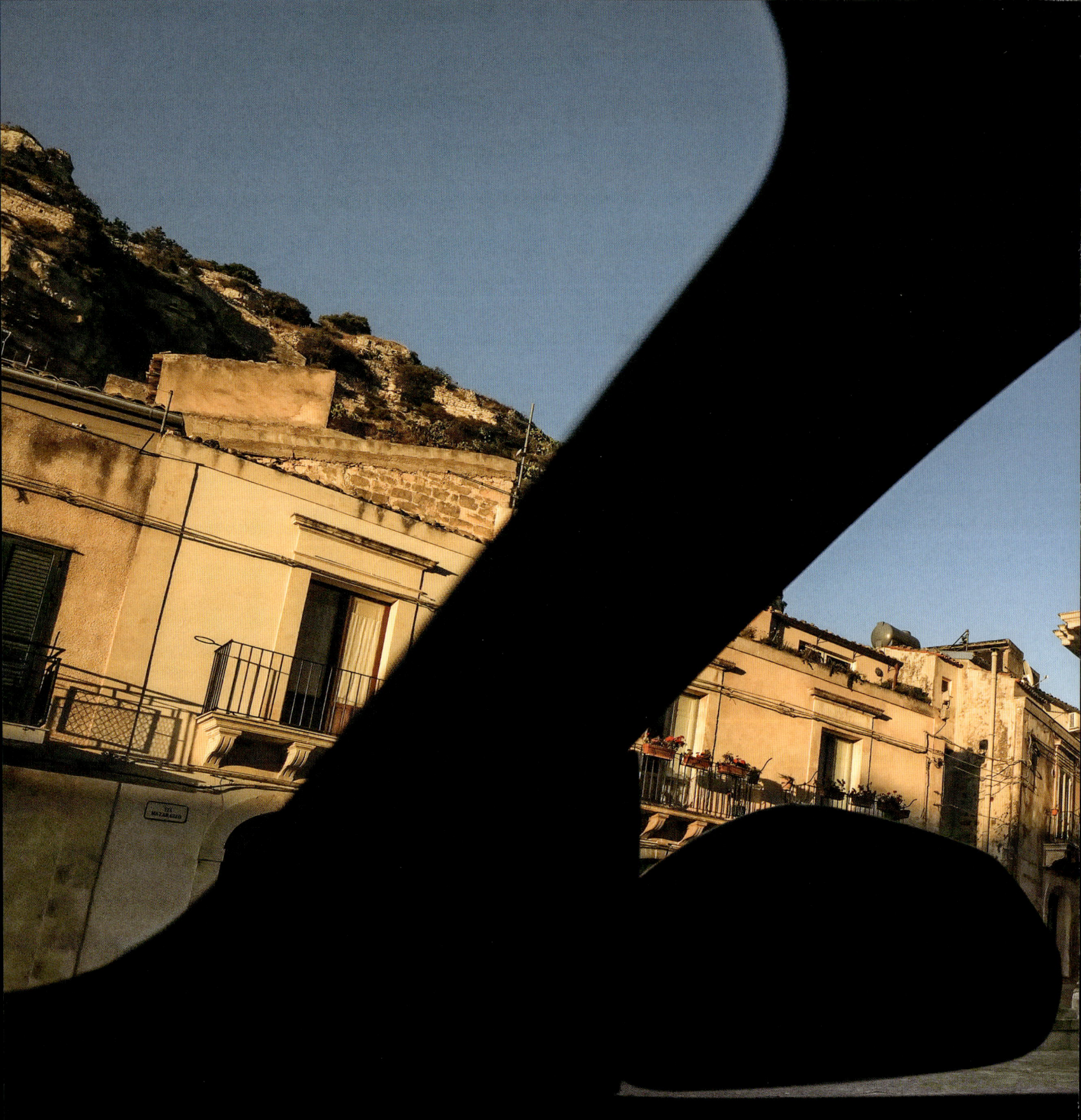

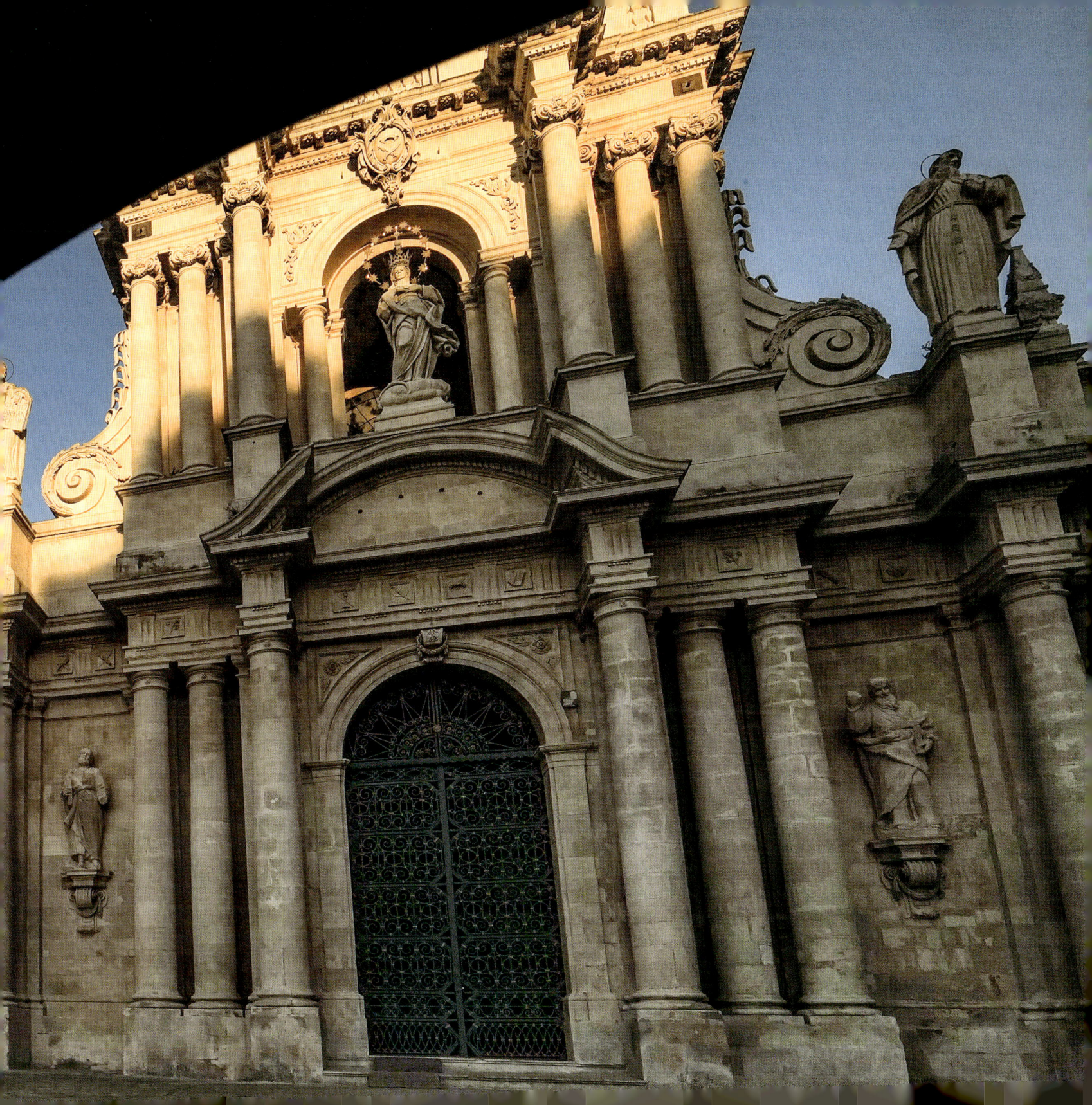

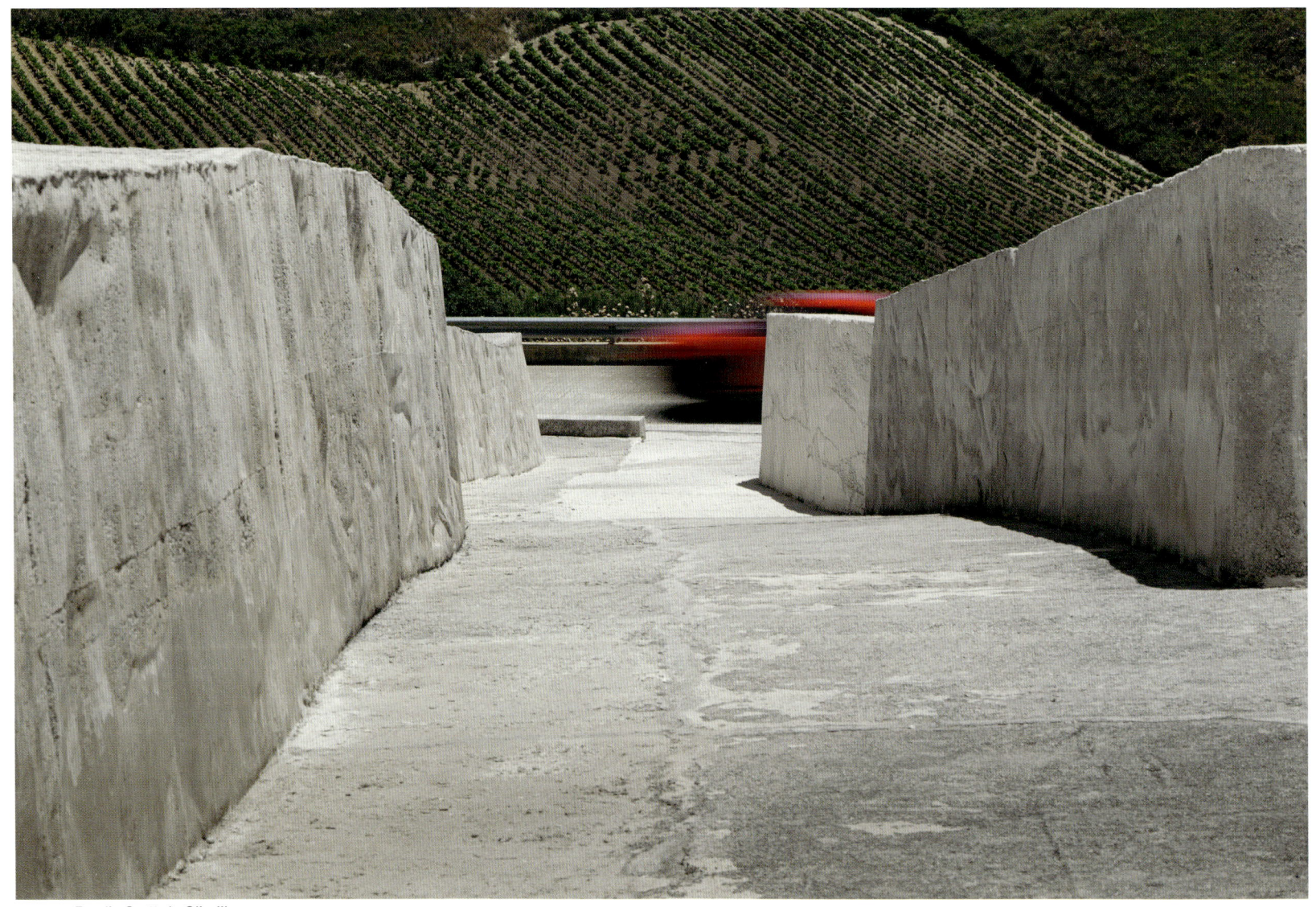
Burri's *Cretto* in Gibellina

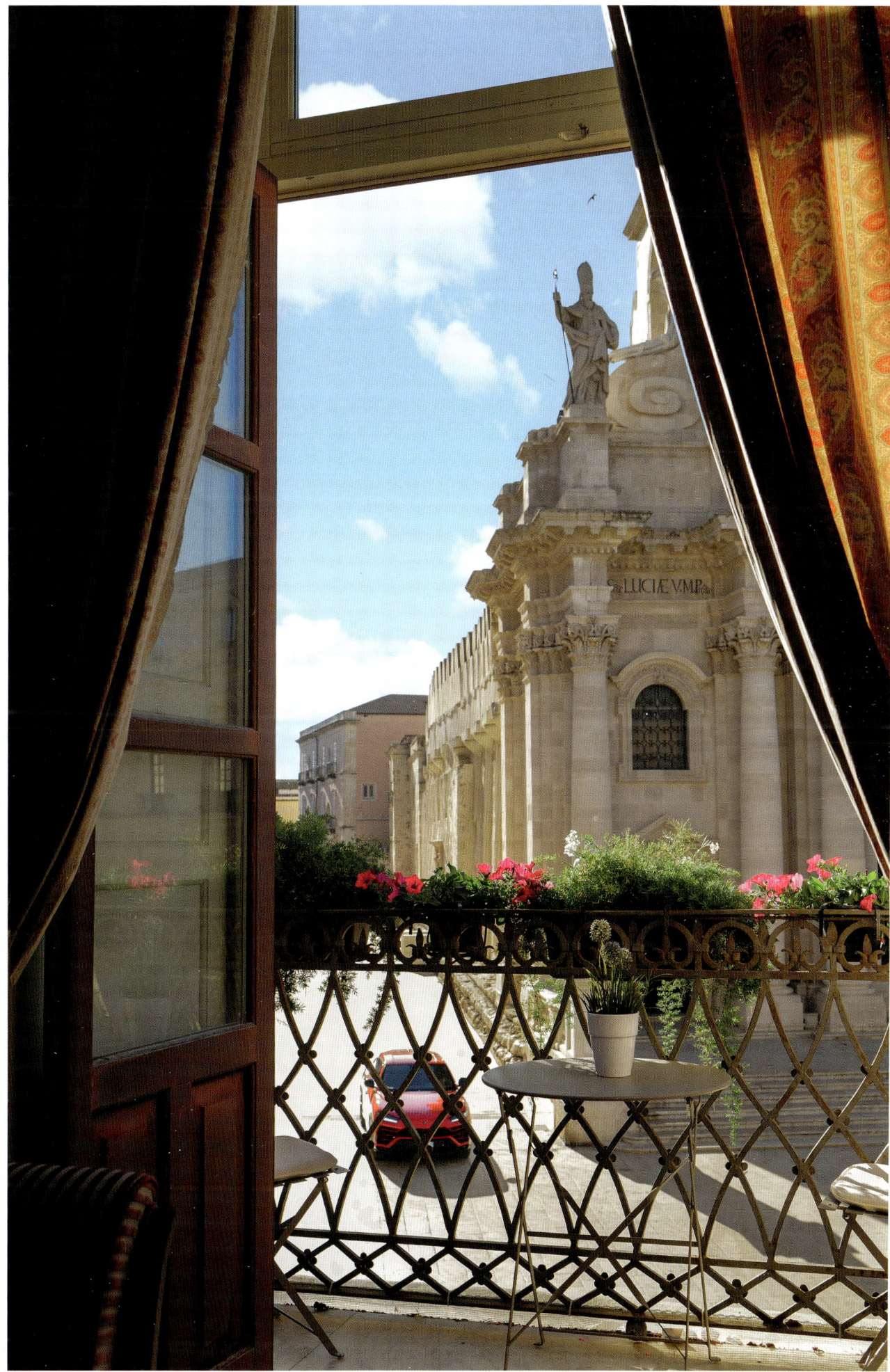

Ortigia Cathedral, Syracuse

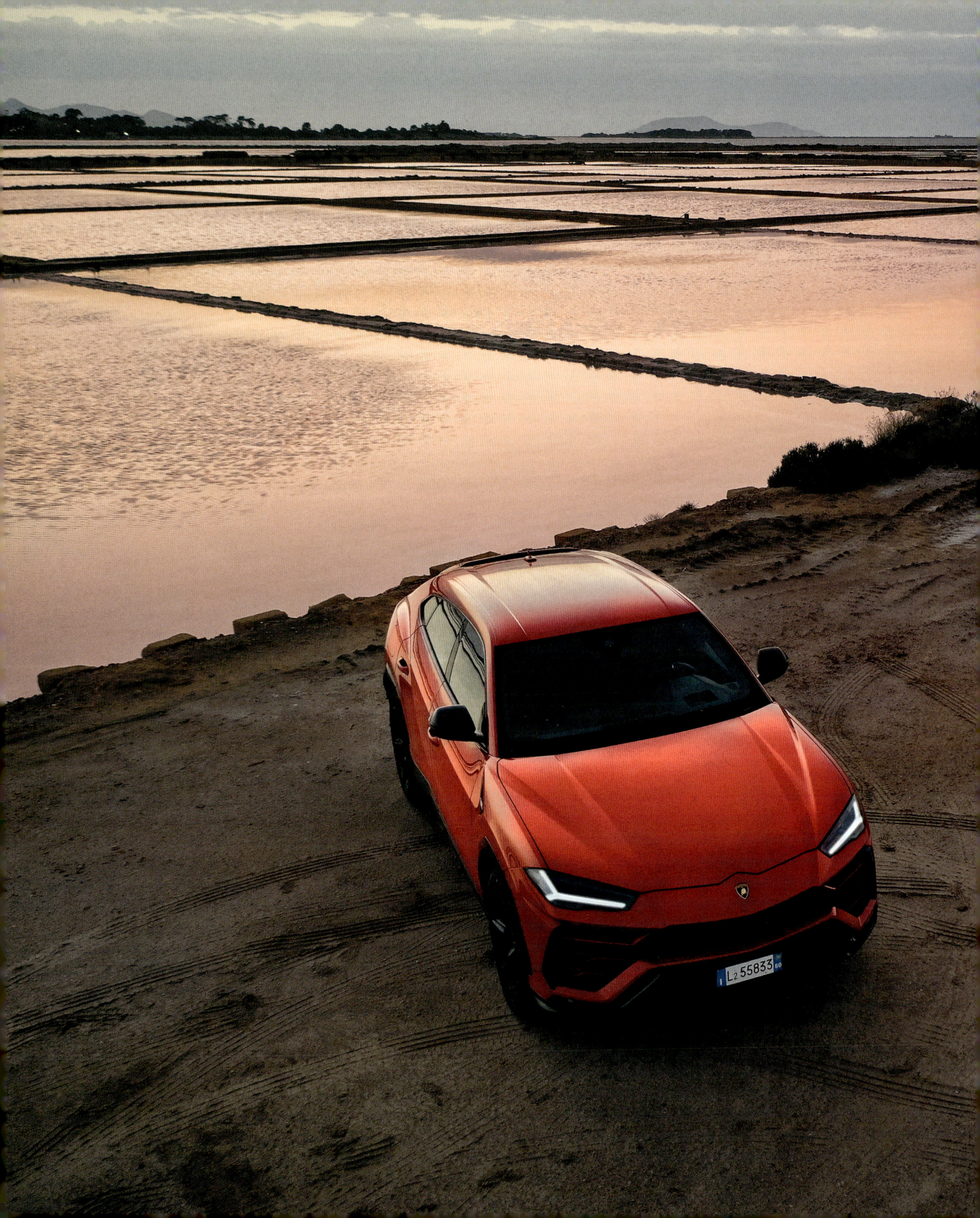

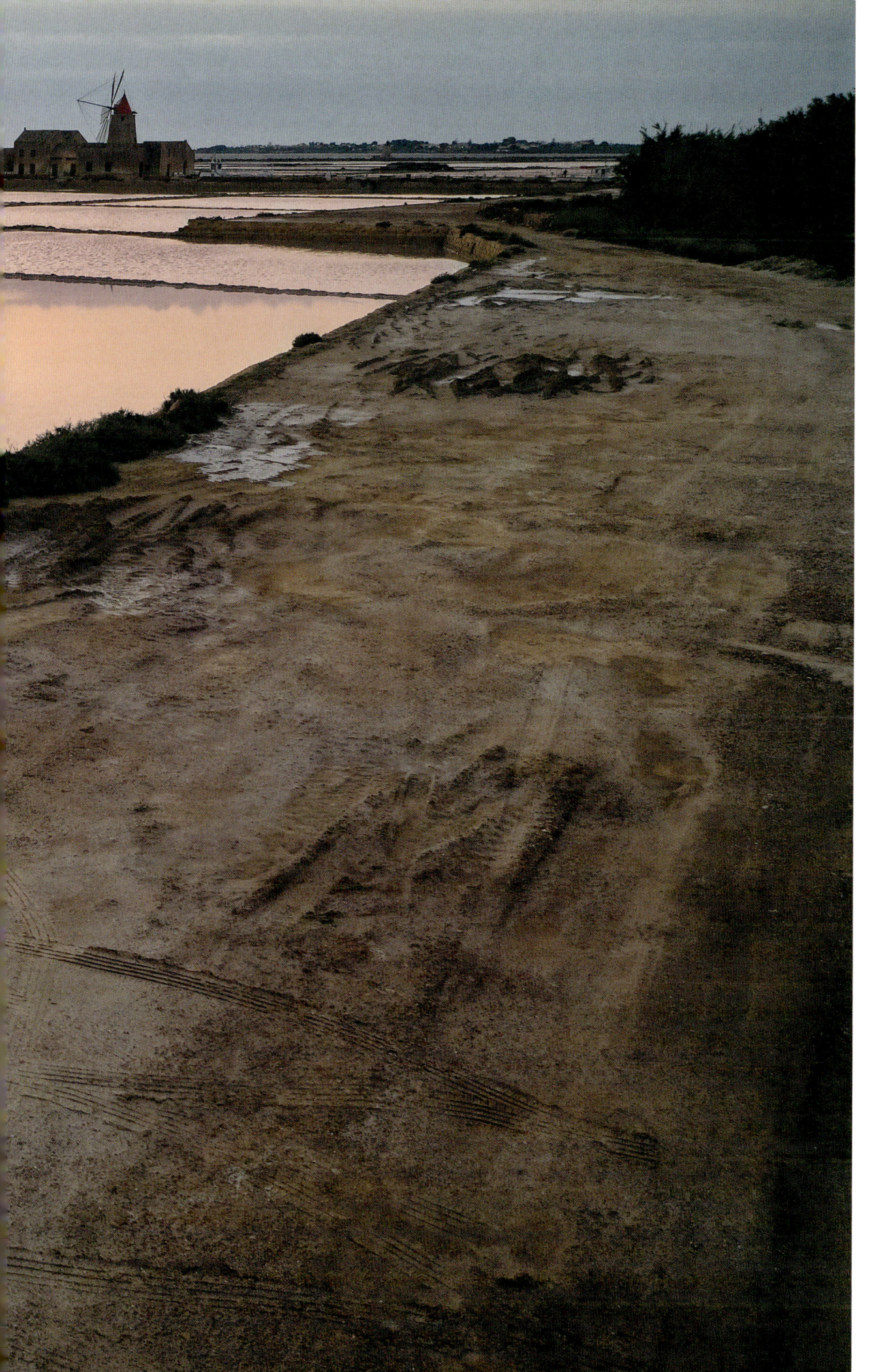

Saline Ettore
and Infersa, Marsala

Pages 34 and 35
Mount Etna
at 3,000 meters
above sea level

Pages 36–37
Scala dei Turchi,
Realmonte, Agrigento

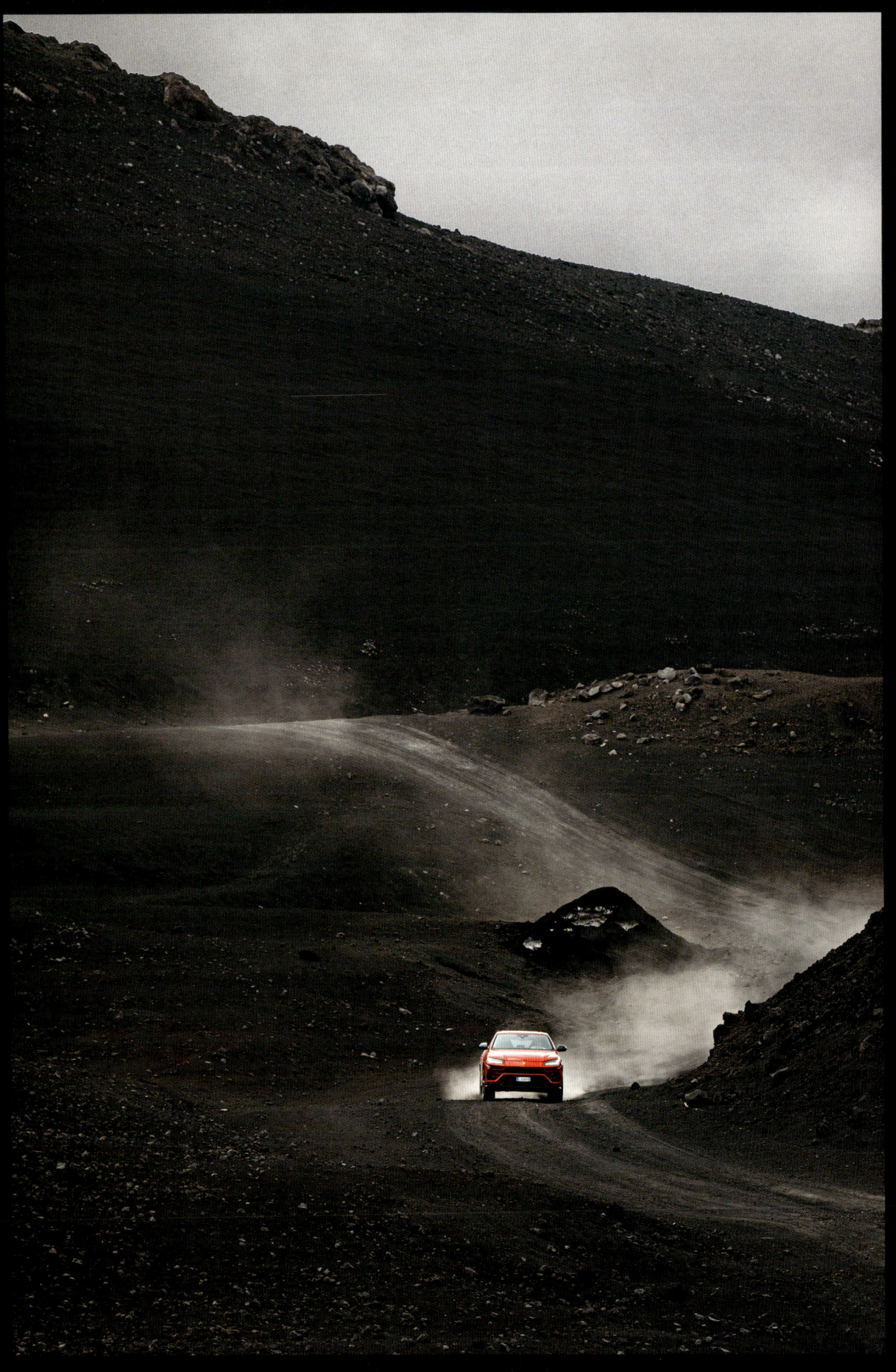

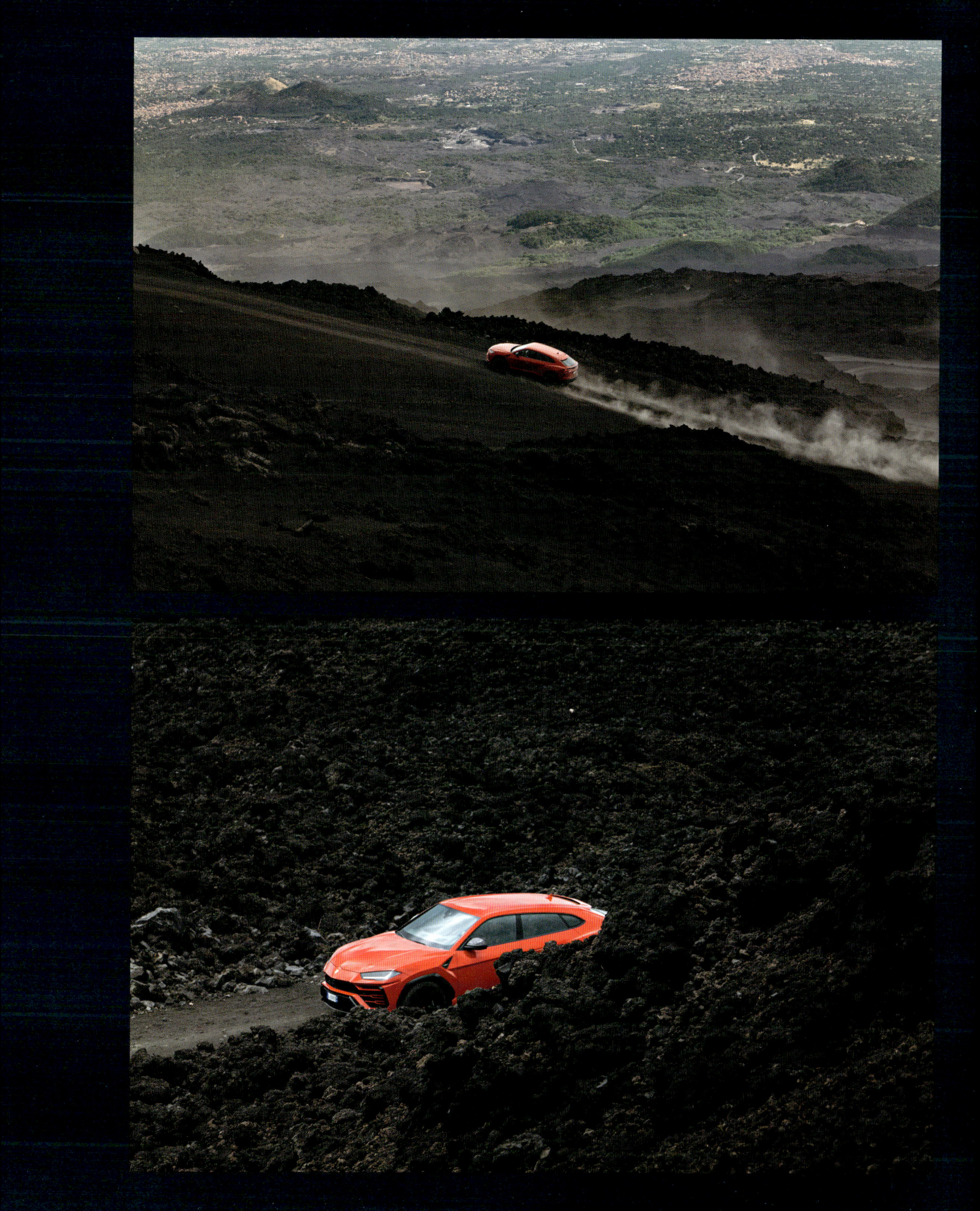

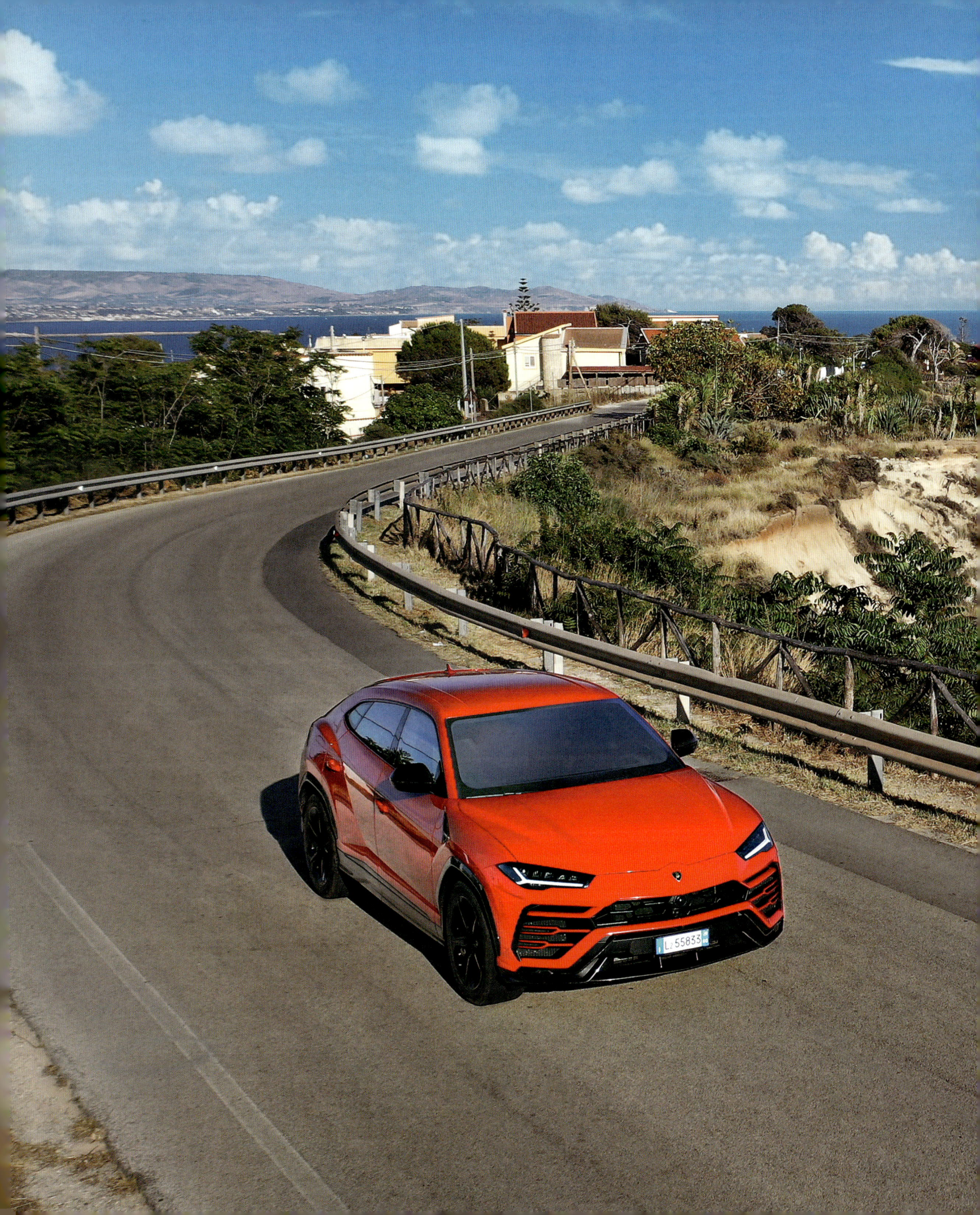

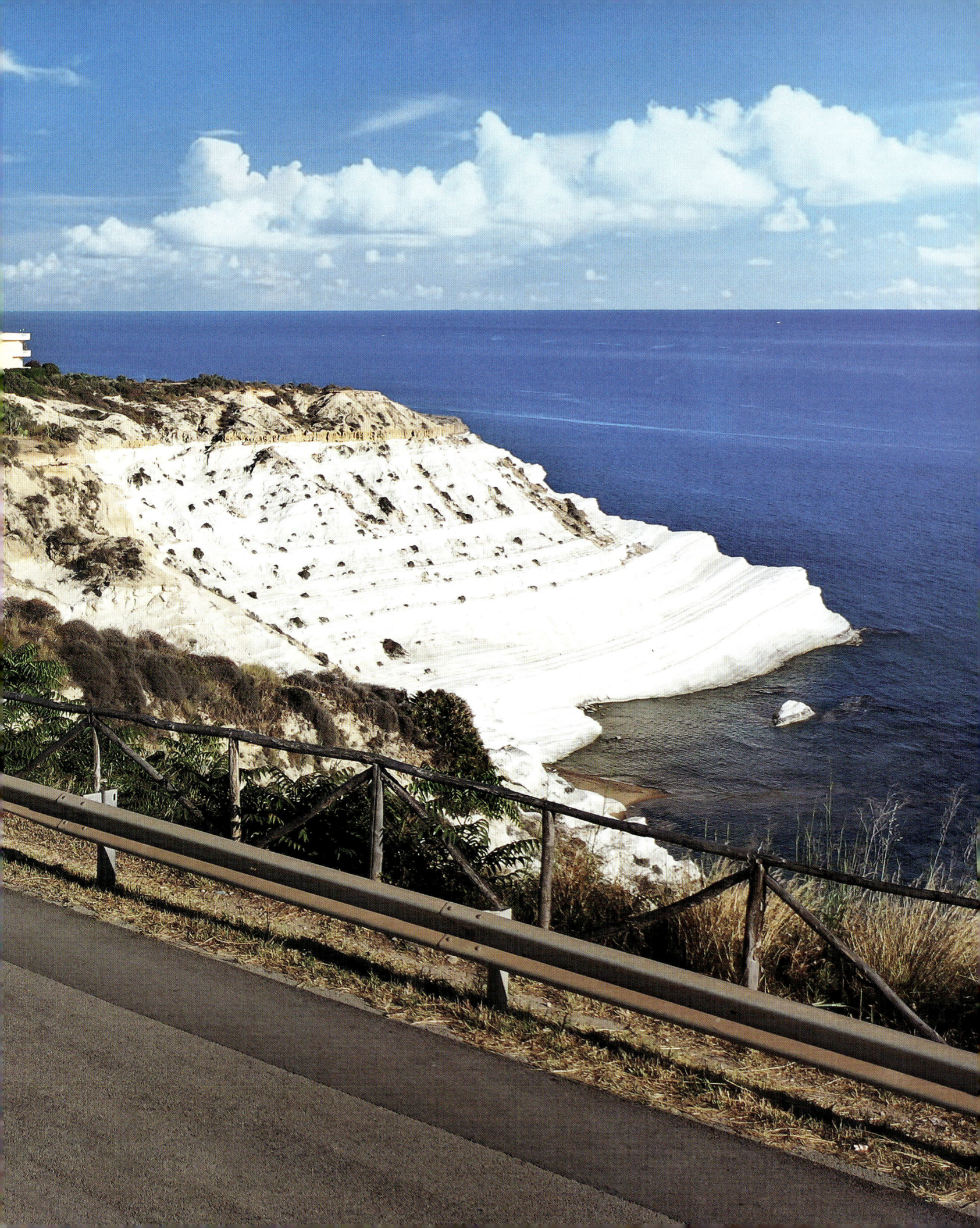

DAVIDE DE MARTIS
SARDEGNA

On this visual path, I wanted to reinstate the most authentic image of Sardinia, the place where I was born and grew up. This ancient land, forged by wind and sea, so generous in granting itself and yet so good at hiding itself away, where time is marked by the slowness of everyday life and the value of traditions… This narrative was reinforced by the other star of the story, a Lamborghini Miura P400 S in metallic gold with a Cagliari licence plate which, fifty years on, returns to its city from Sant'Agata Bolognese, blending in silently with the colours and shapes of the places it passes through on this nostalgic journey home.

Old bridge, Bosa old town, Oristano

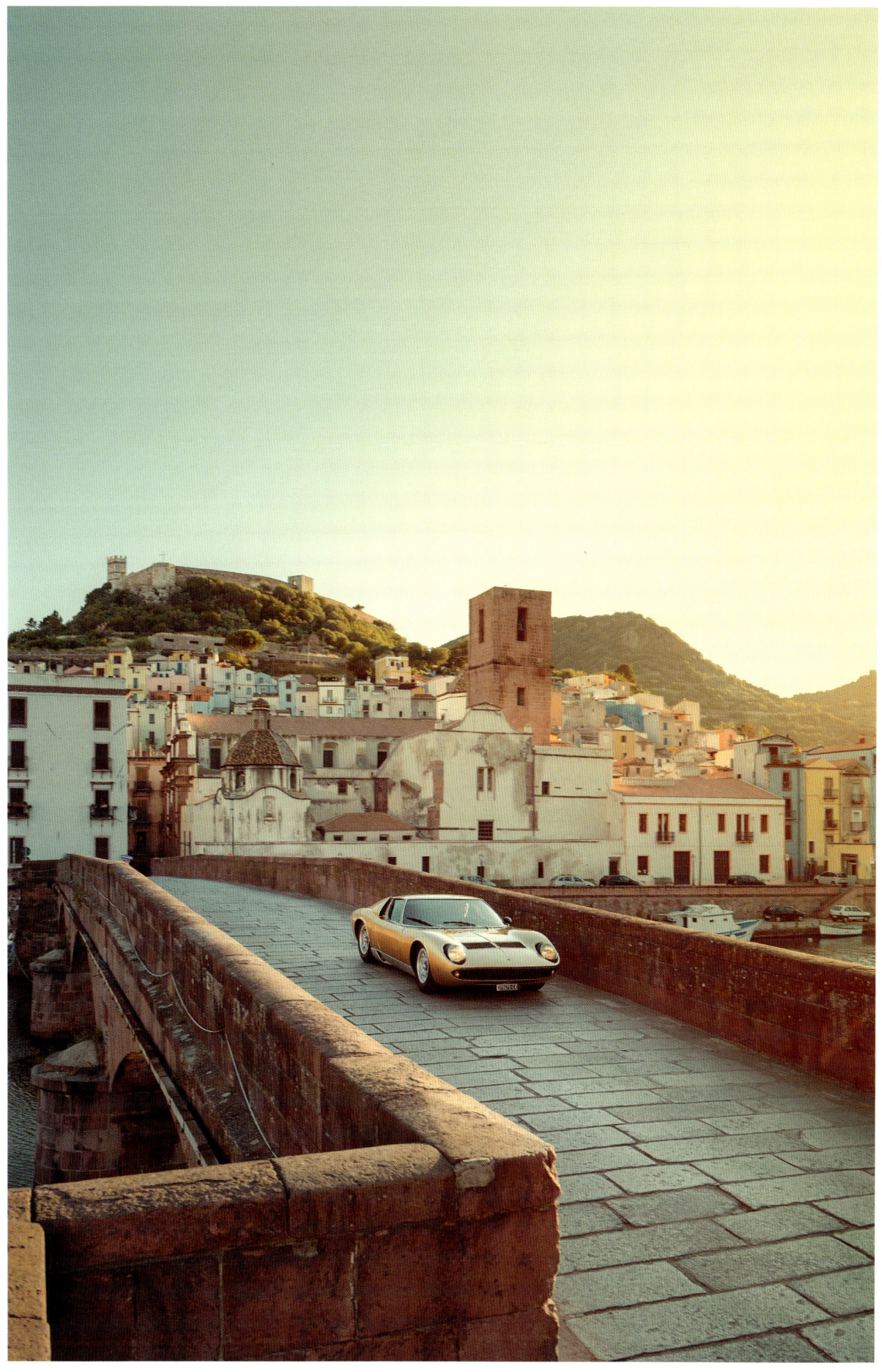

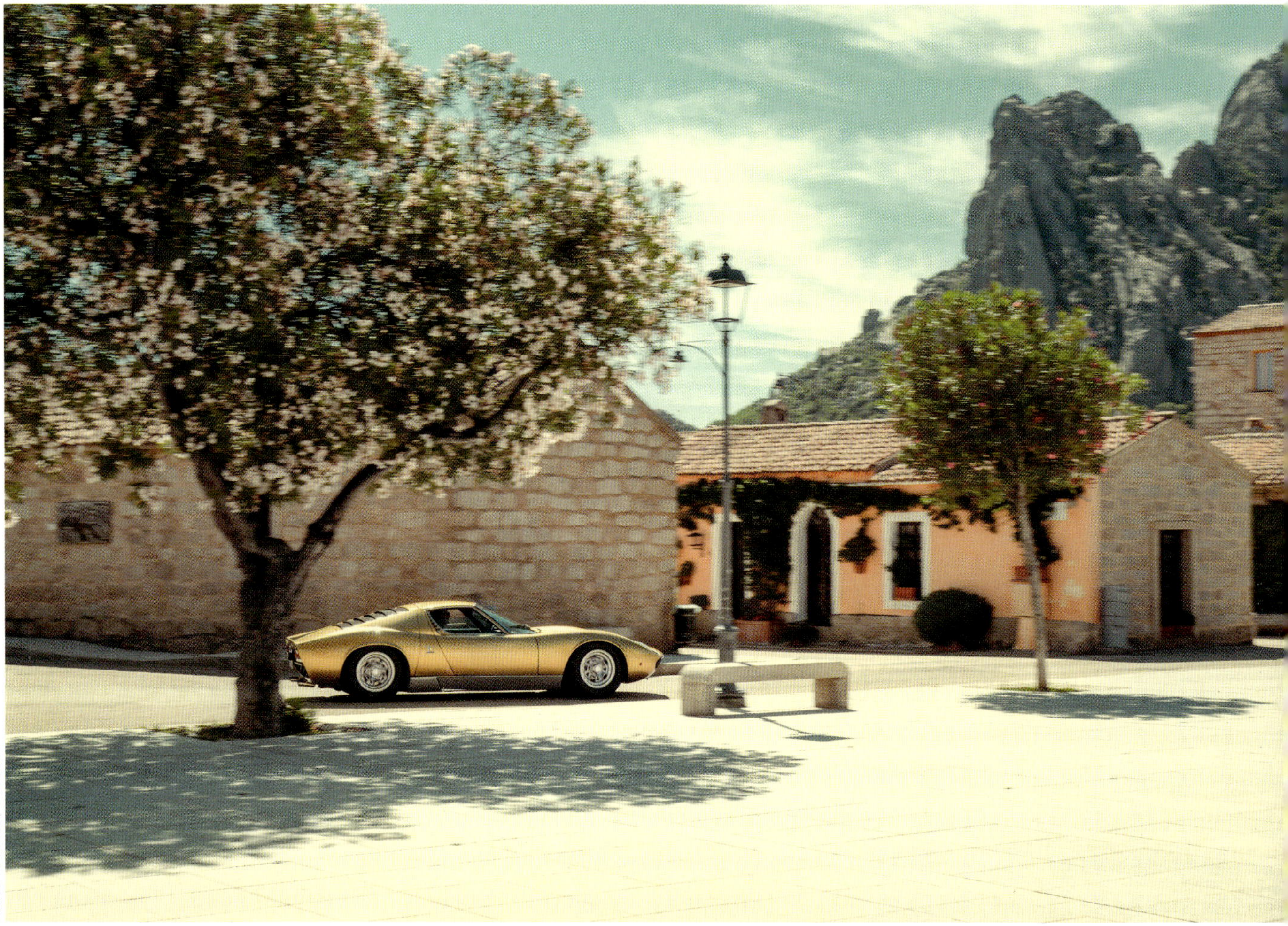

San Pantaleo, Sassari

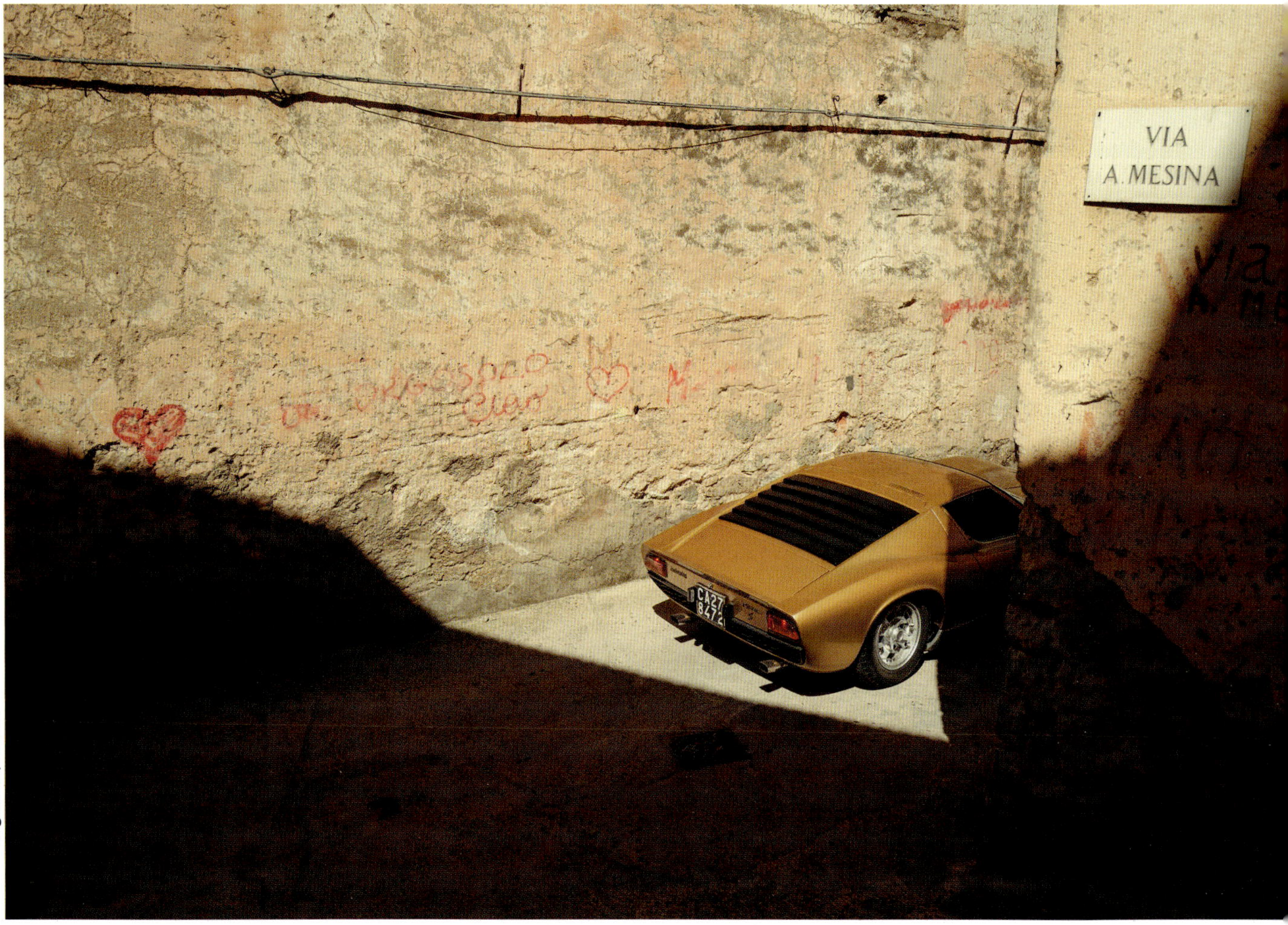

Orgosolo, Nuoro

VIA
A. MESINA

40

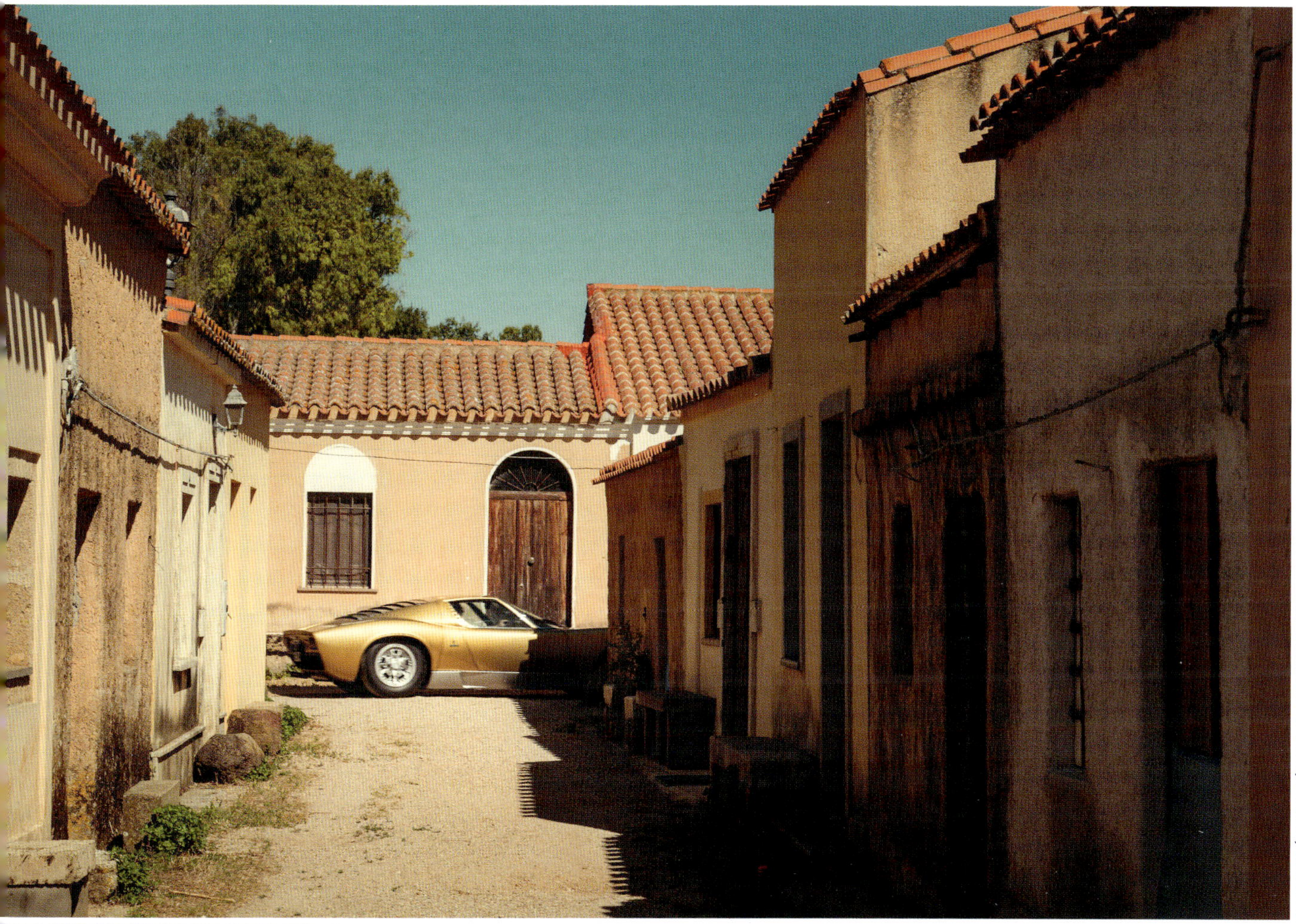

San Salvatore di Sinis, Cabras, Oristano

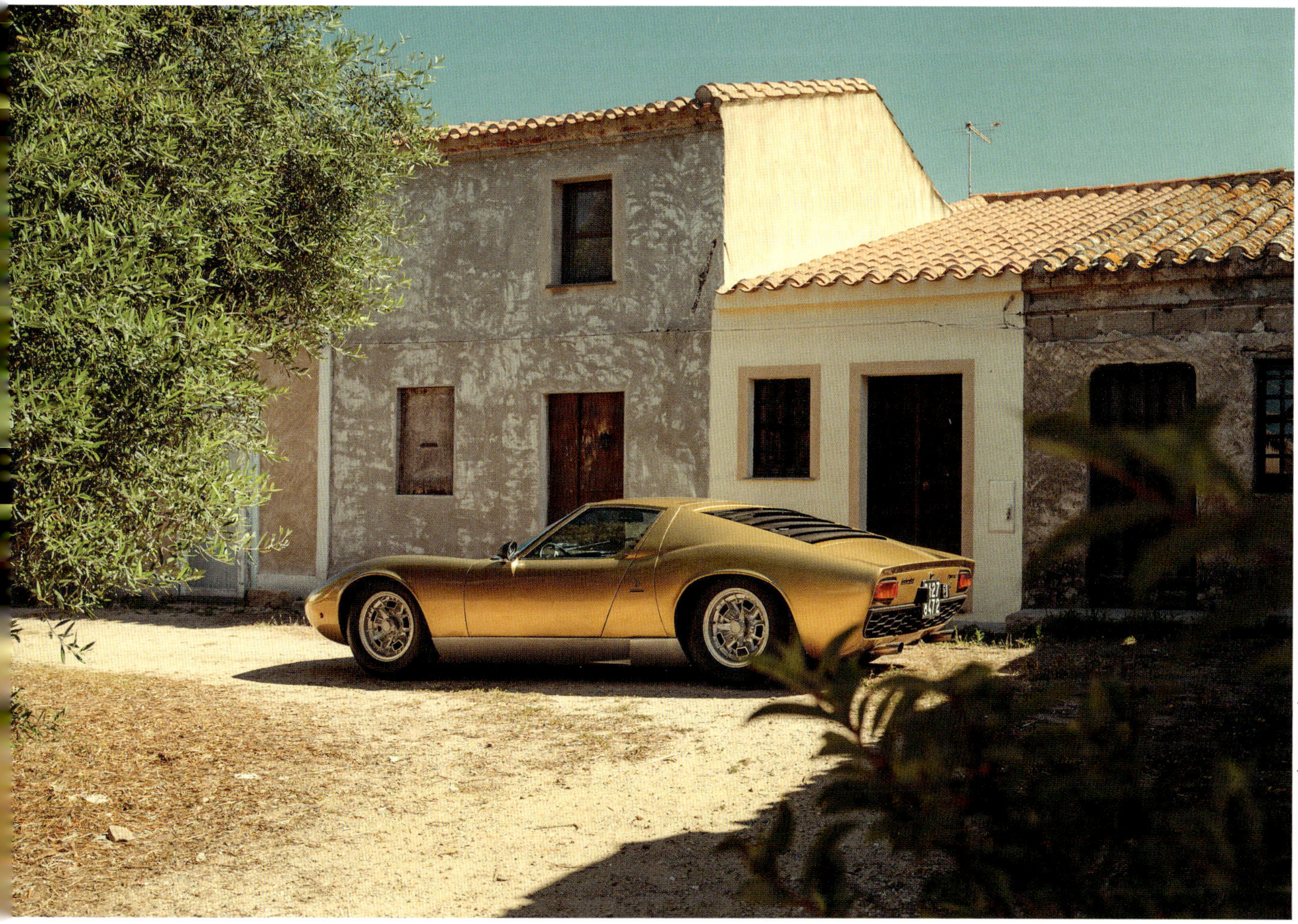

San Salvatore di Sinis, Cabras, Oristano

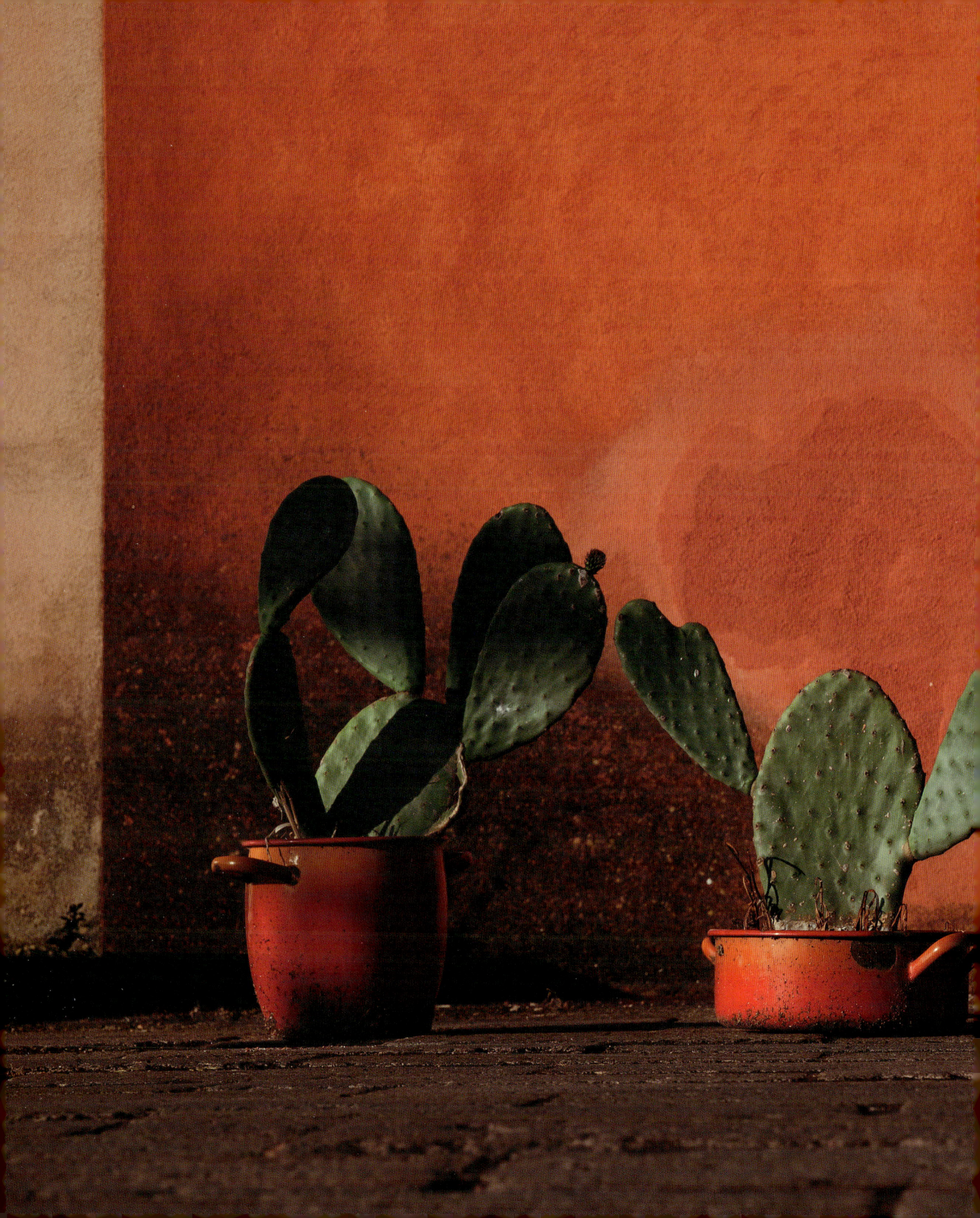

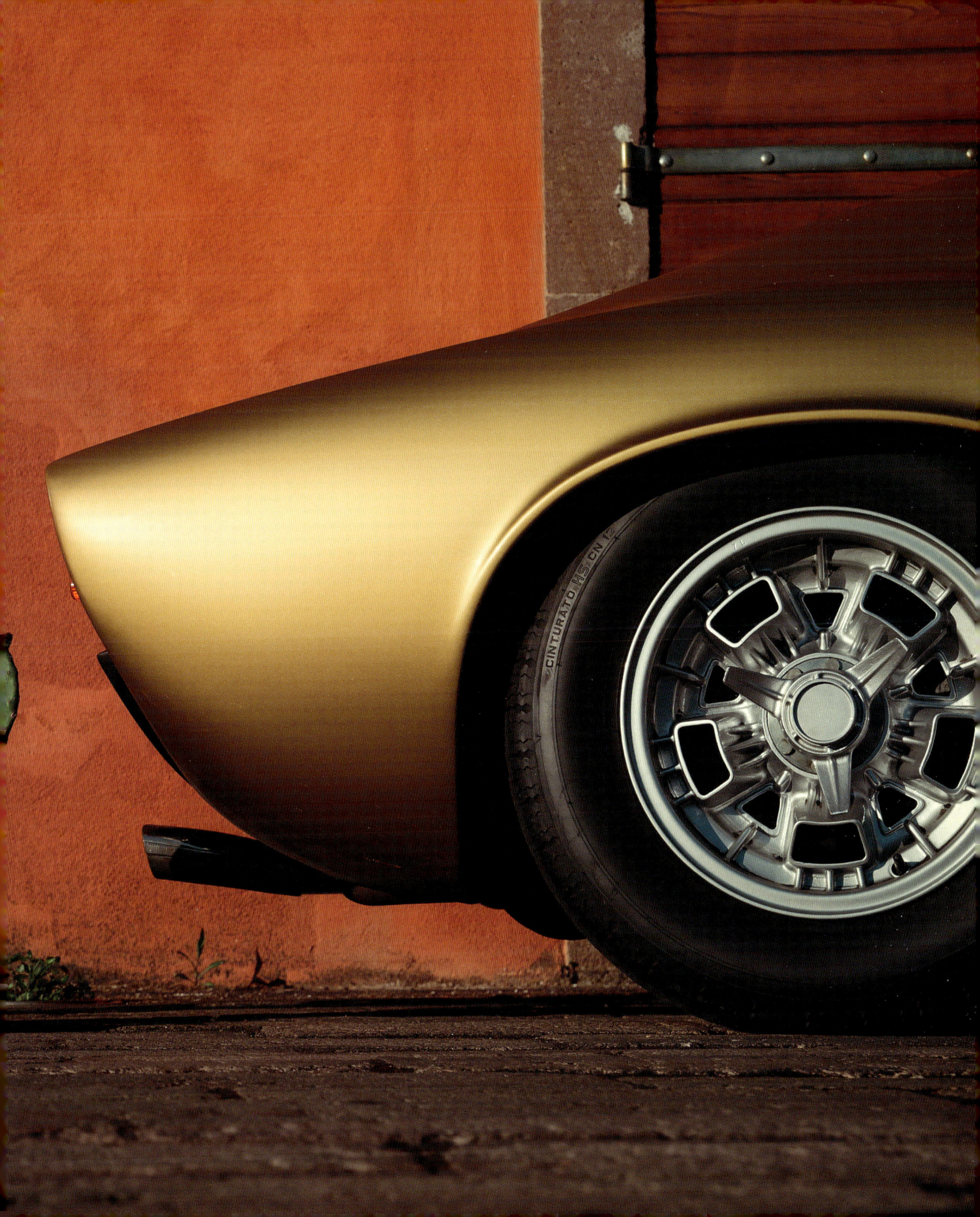

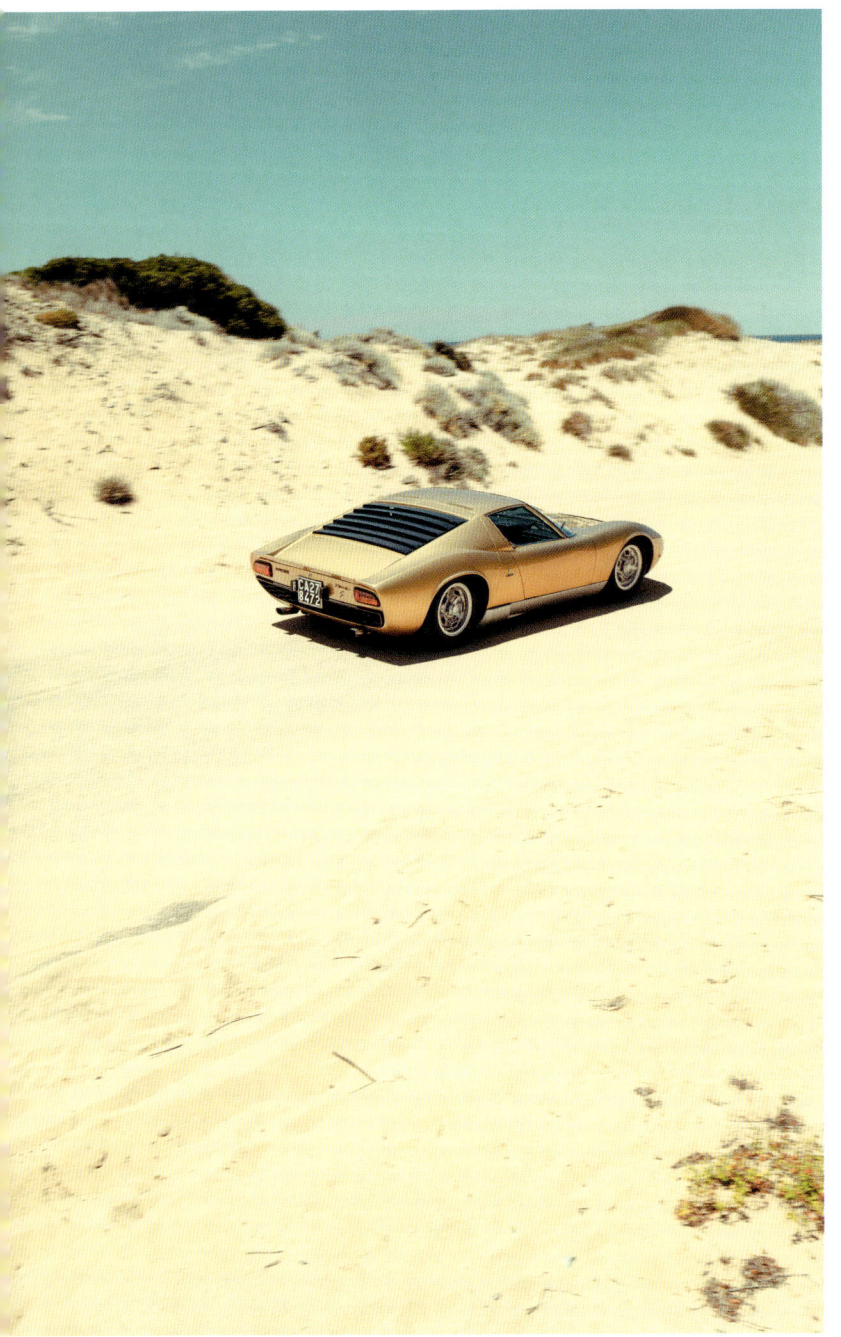

Piscinas, Arbus

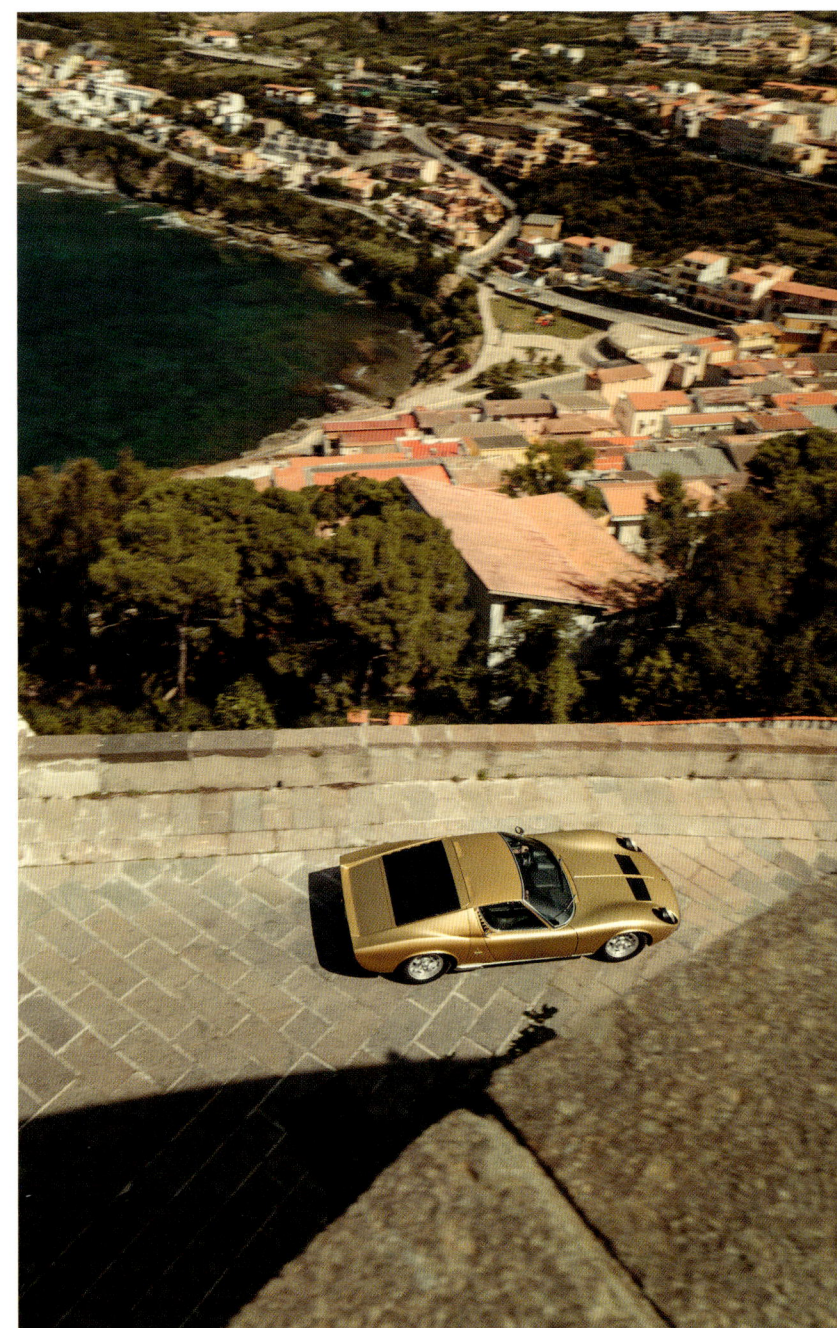

View from the Castelsardo Castle, Sassari

"I wanted to reinstate the most authentic image of Sardinia, the place where I was born and grew up"

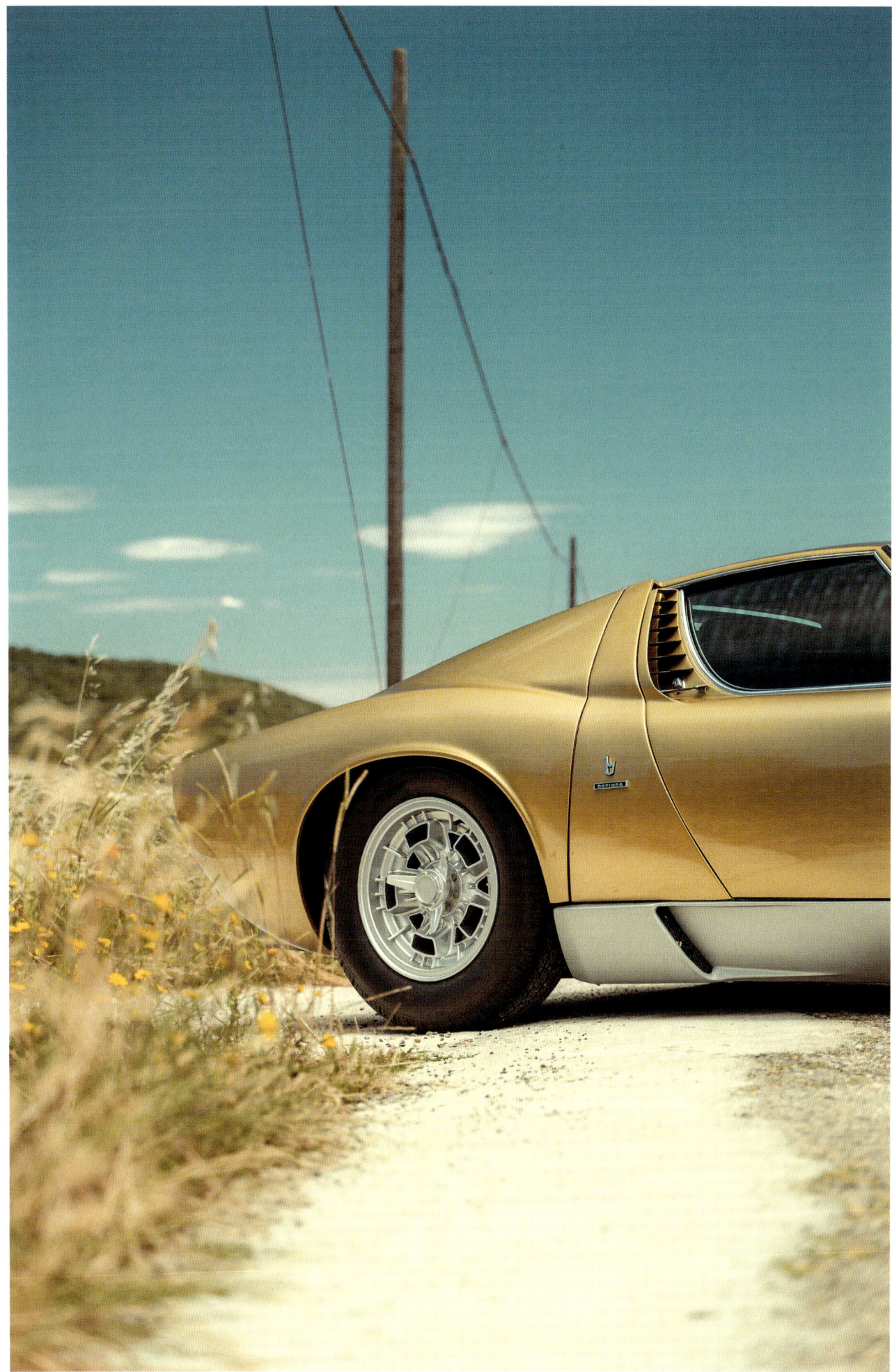

Upper Gallura

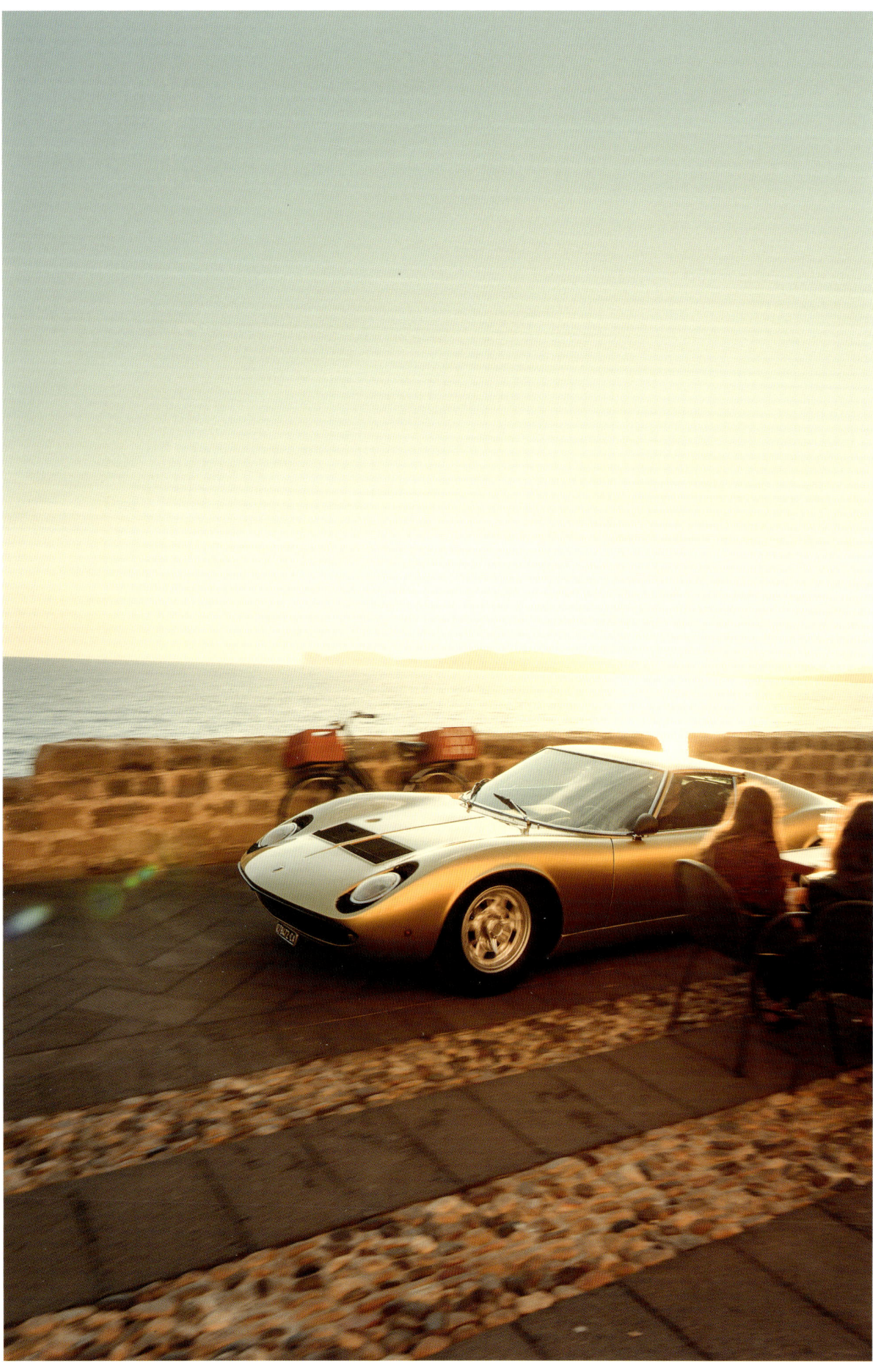

The ramparts at Alghero

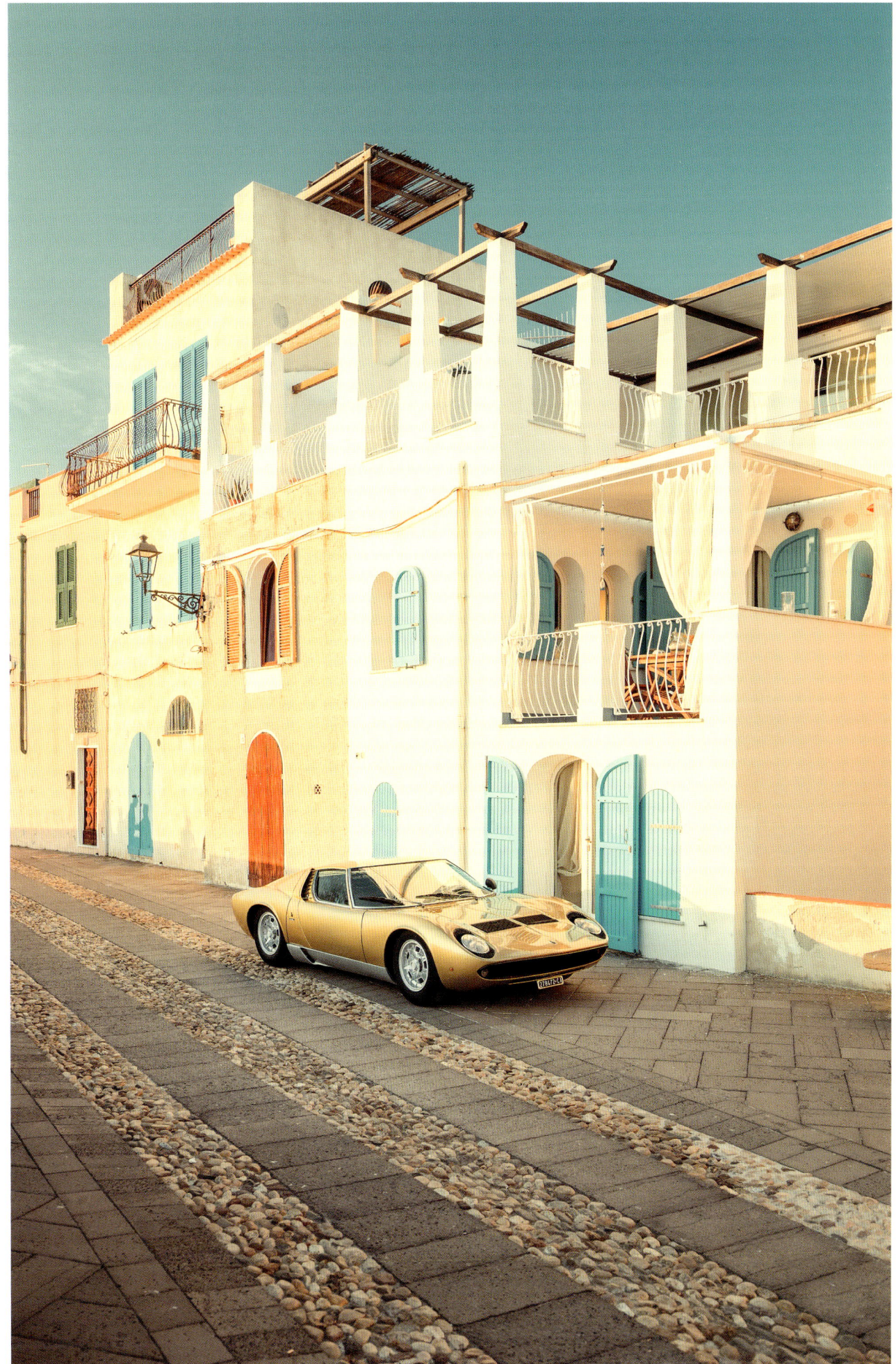

Alghero

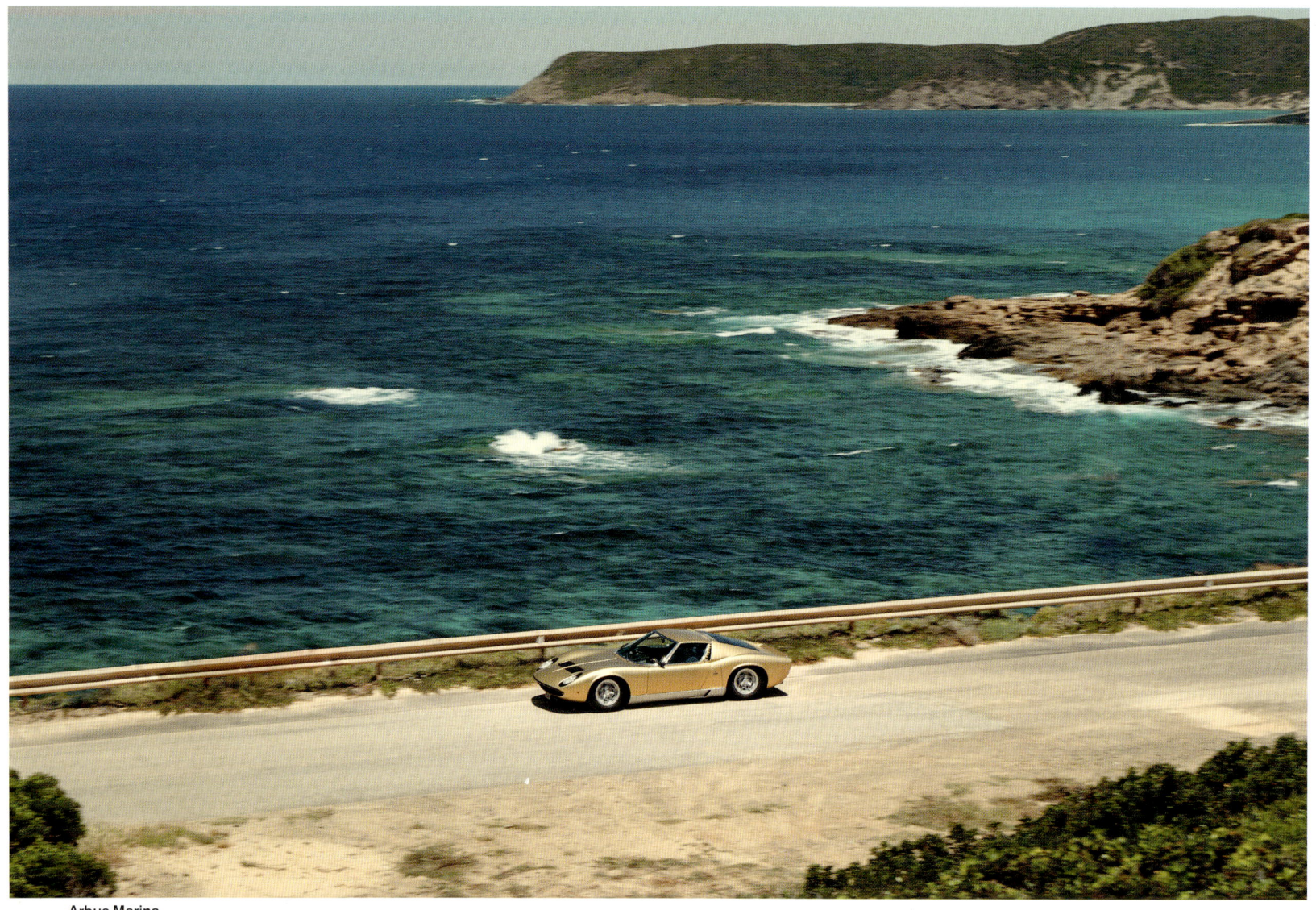

Arbus Marina

San Giovanni Cave, Domusnovas

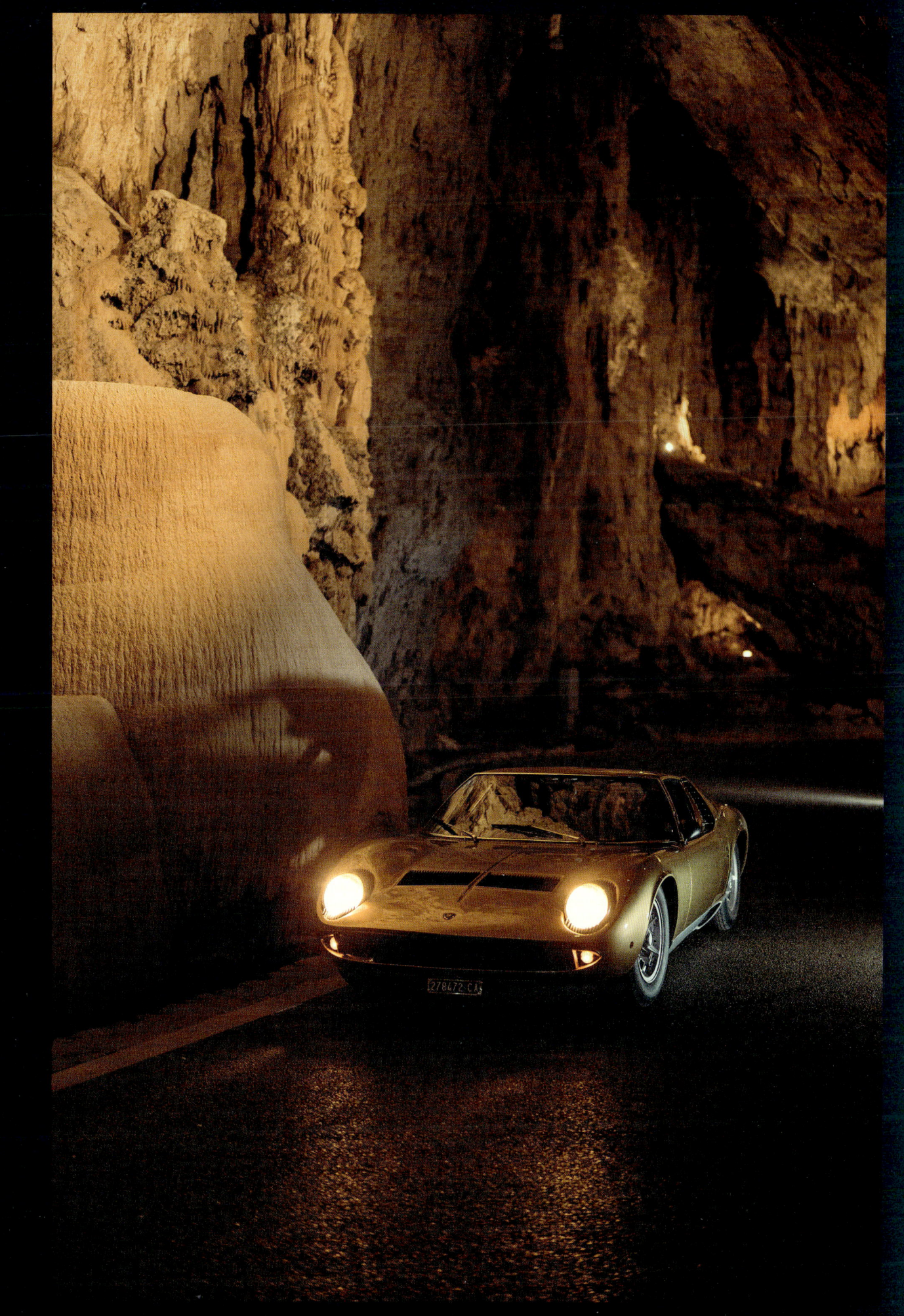

GUIDO TARONI
CALABRIA

Calabria's intensity is expressed through its continuous dialogue

between opposing land types: verdant mountain areas,

the vegetation lush and wild, alternate with sea views of pristine,

fine sand beaches, headlands and a sea of countless shades of blue.

The cradle of Magna Graecia and a land of ancient settlements,

Calabria showcases its traditions in the villages and monuments.

My trip along the tip of Italy's boot has been a journey through

the rich landscapes of this extraordinary region, in which the

Lamborghini Urus, the very same blue as the Calabrian sea,

first hides in nature, only slowly to reveal itself, becoming part

of the daily life of the locals and their traditions, infused

with the irony that is part and parcel of my photographic spirit.

Praialonga cliff, Island of Capo Rizzuto, Crotone

Pages 52–53 Pentedattilo Village, Reggio Calabria

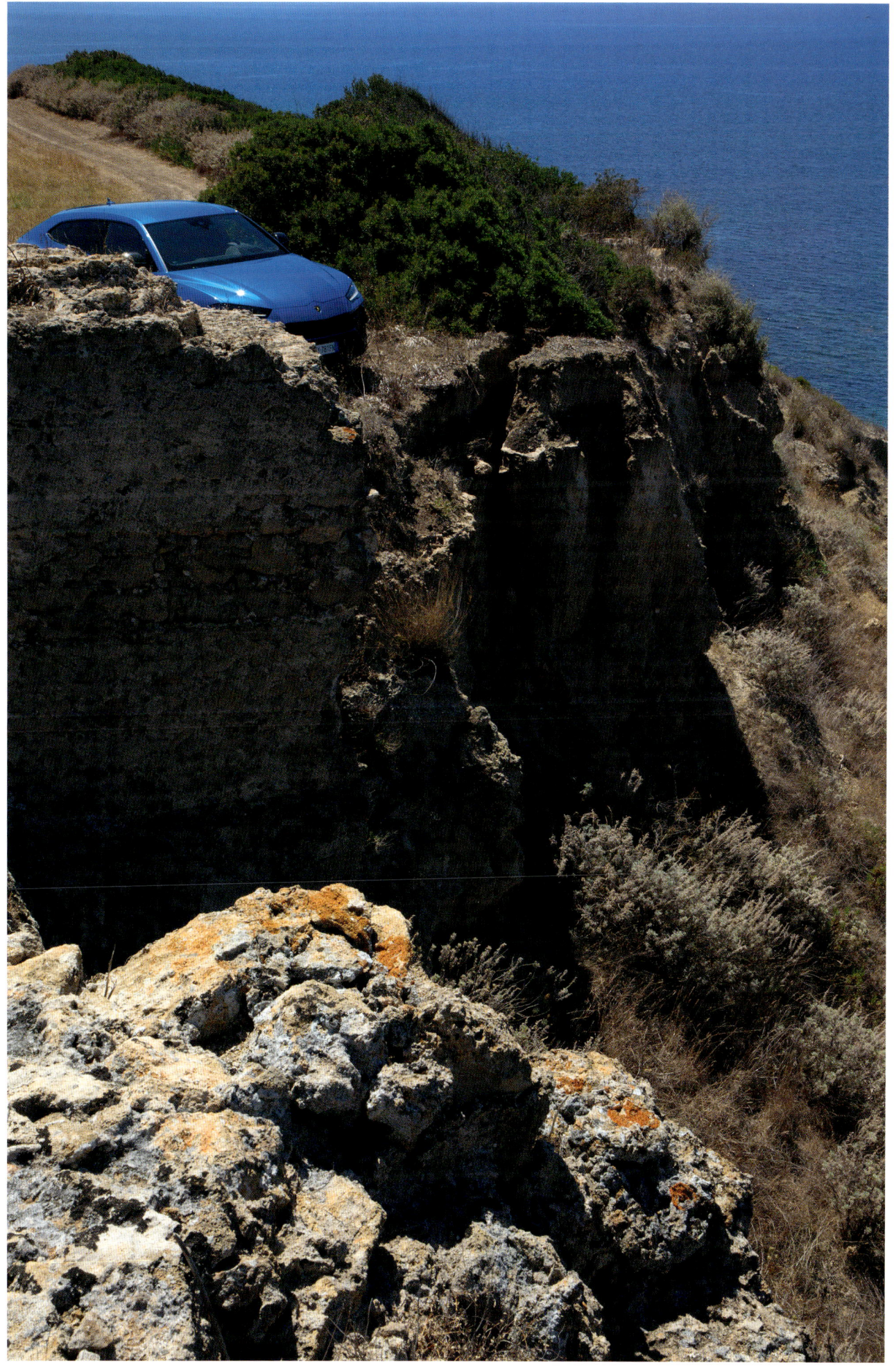

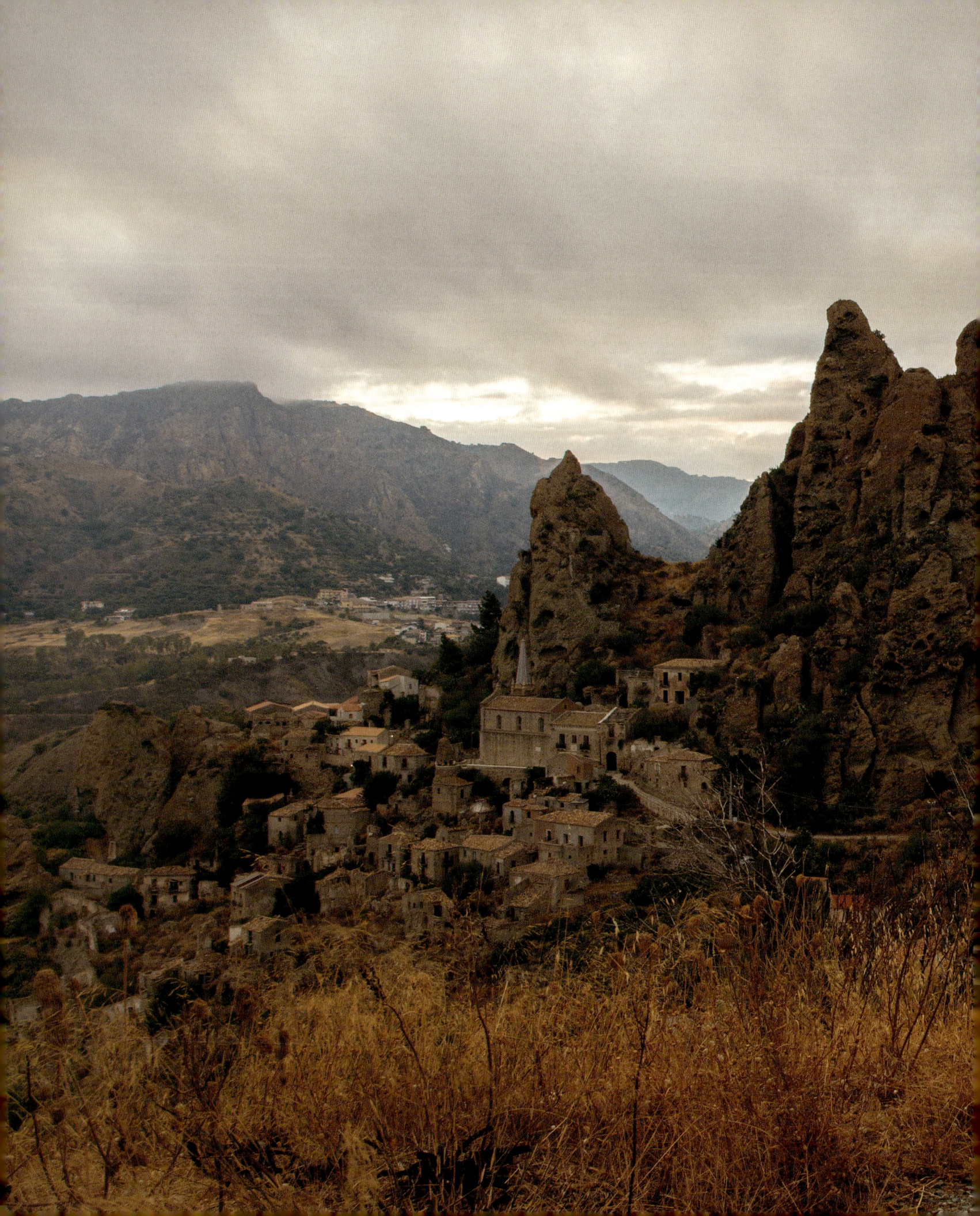

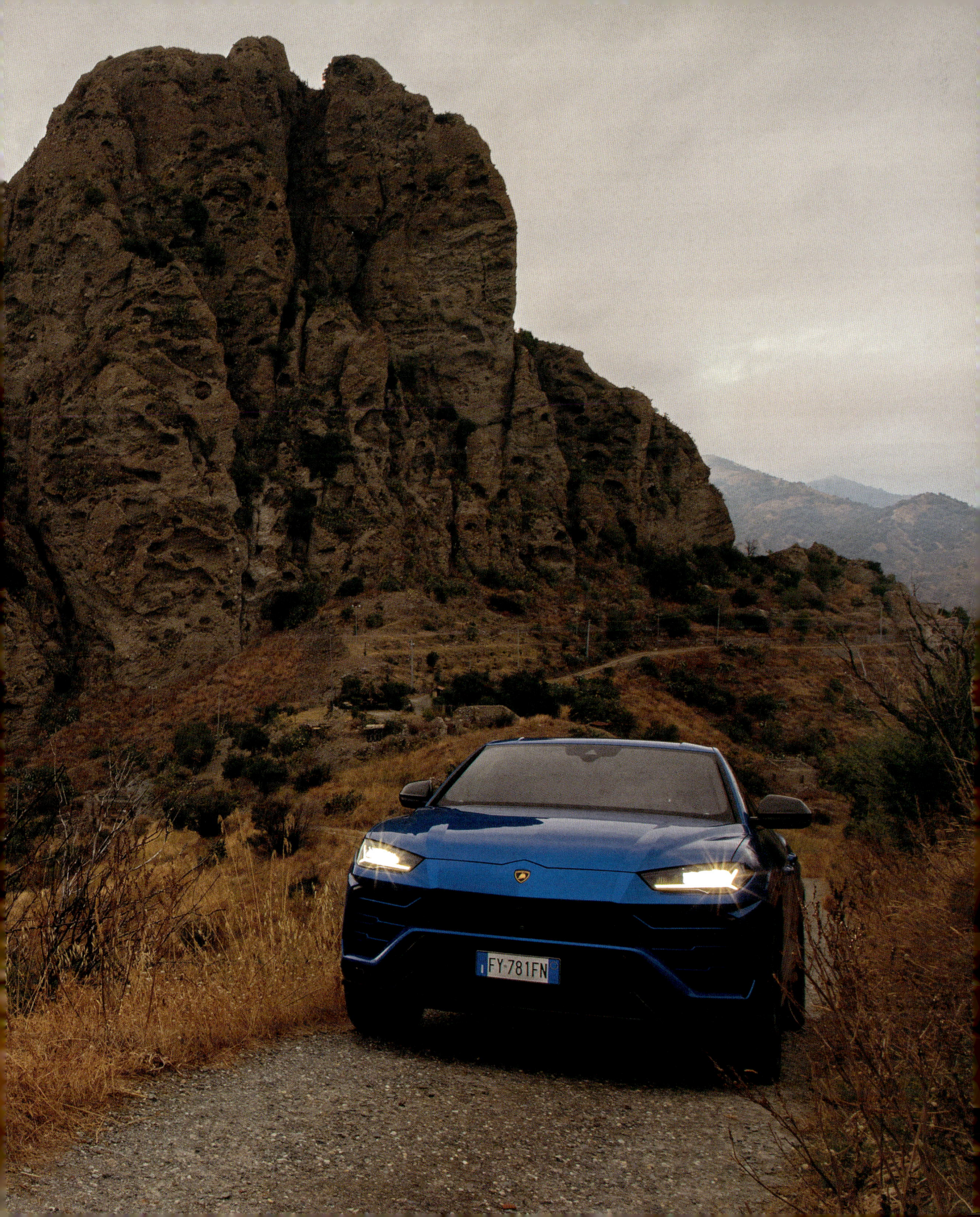

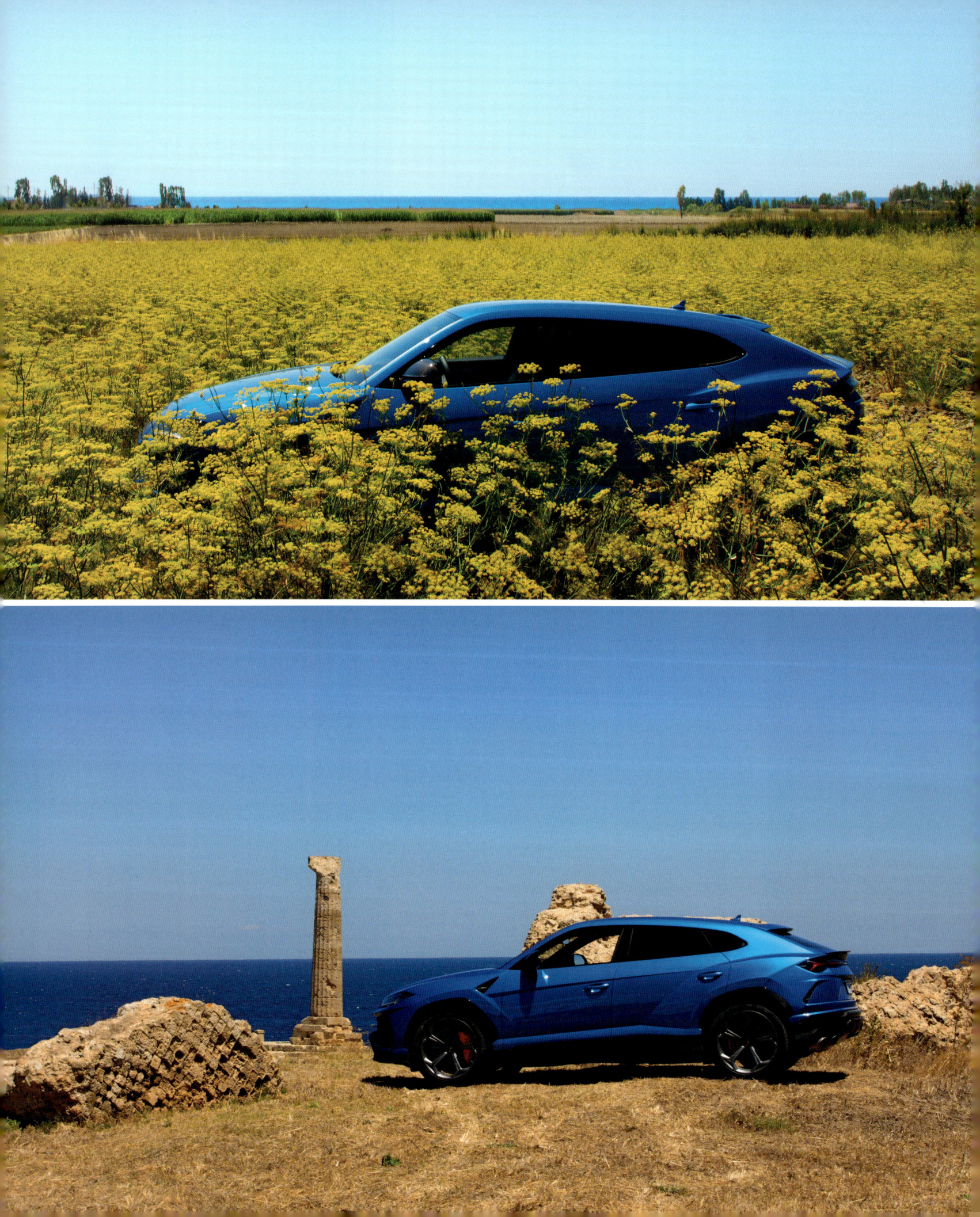

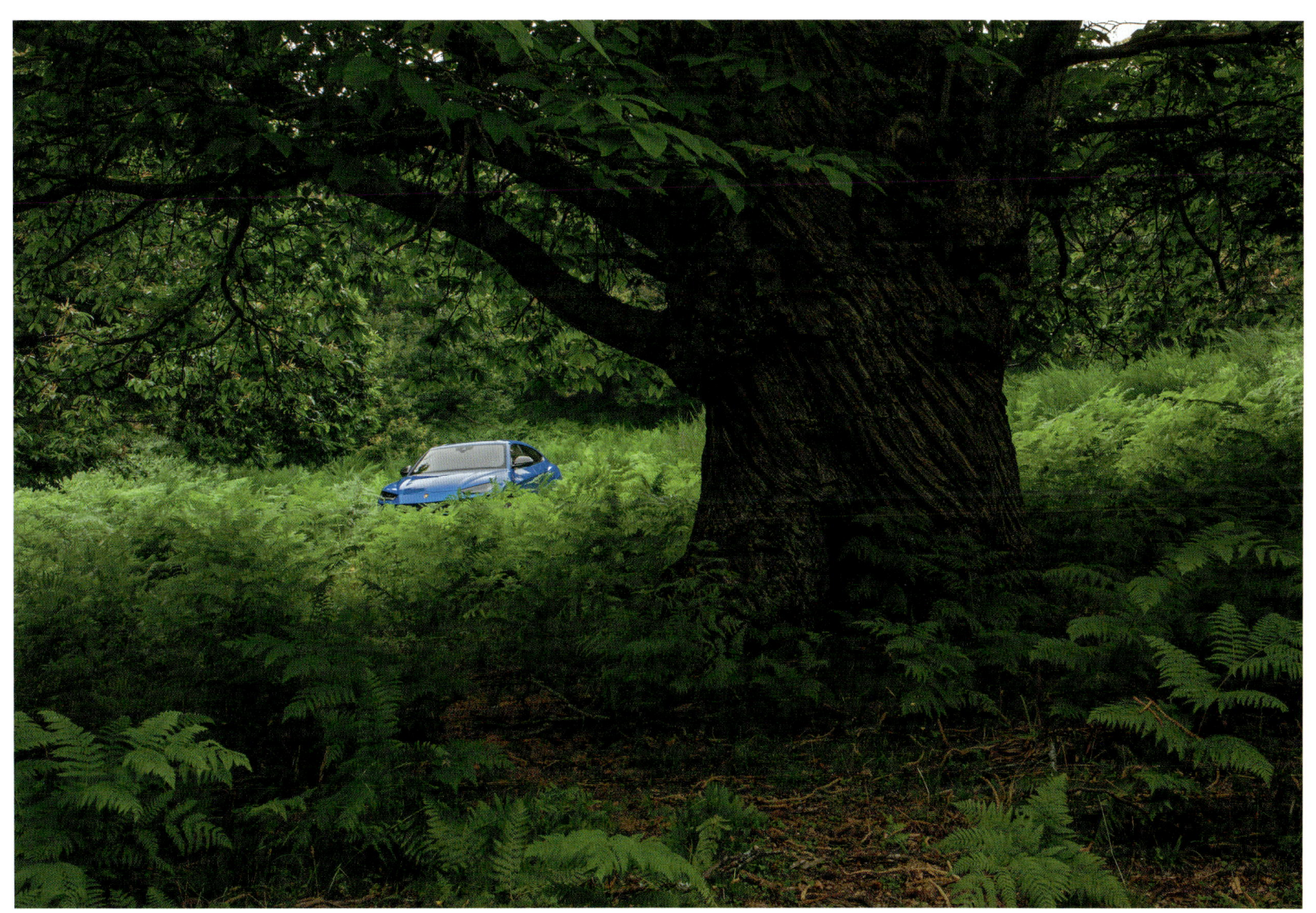
Centuries-old tree, Gigante Melitano, Cerva, Catanzaro

Wild fennel in bloom, Ionian Coast, Botricello, Catanzaro
Remains of the Temple of Hera Lacinia, Capo Colonna, Crotone

Pages 56–57 Pietra Cappa, Aspromonte

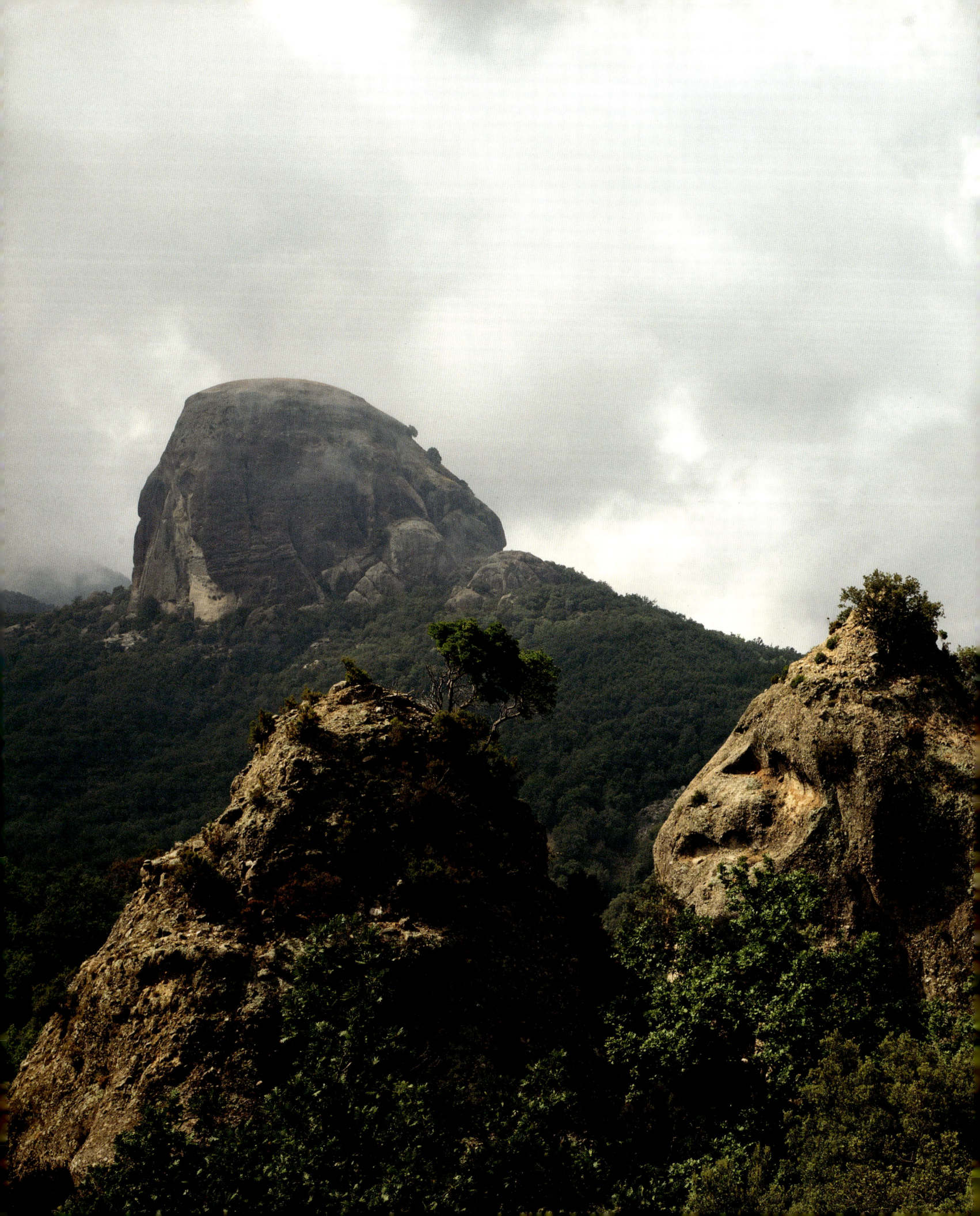

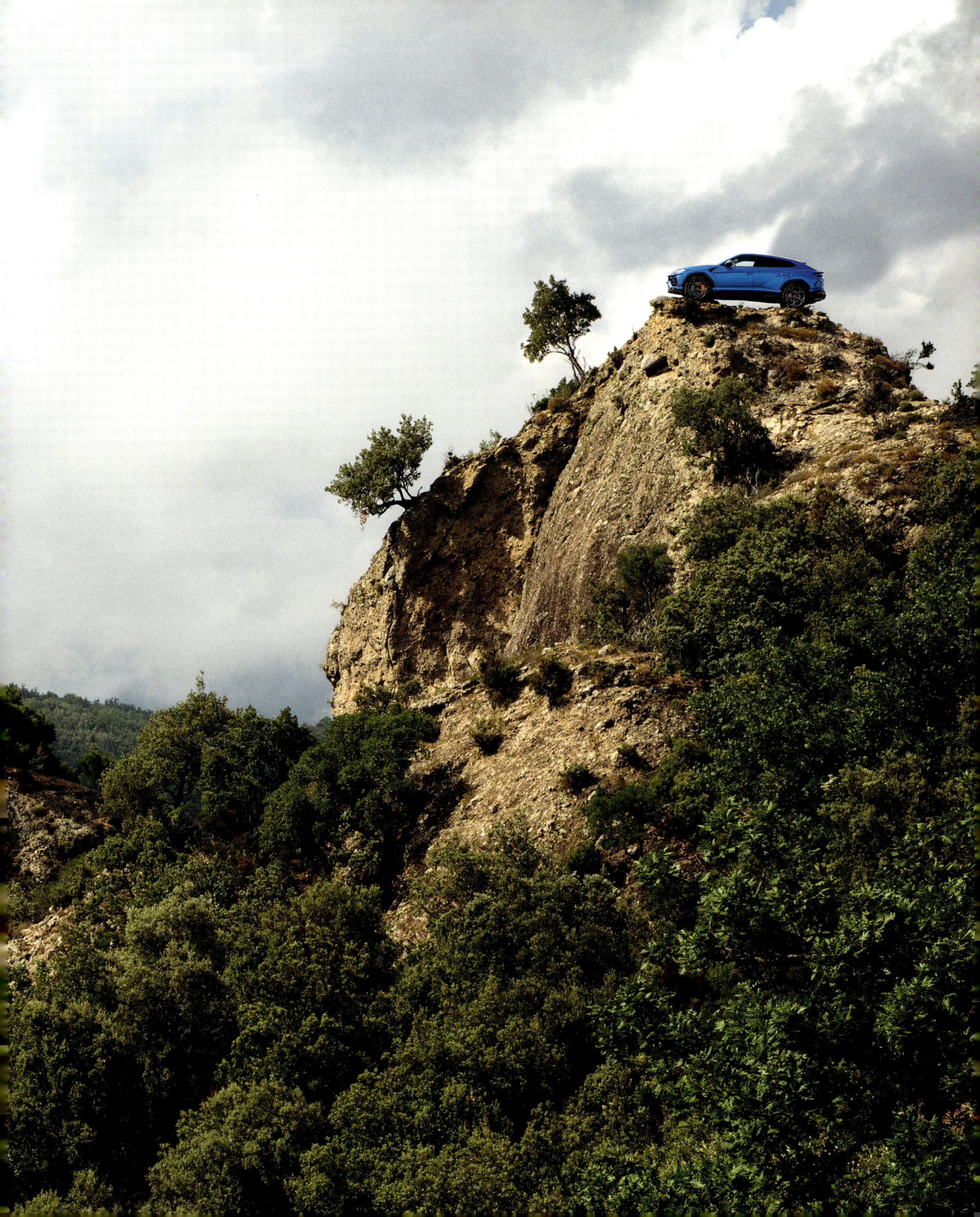

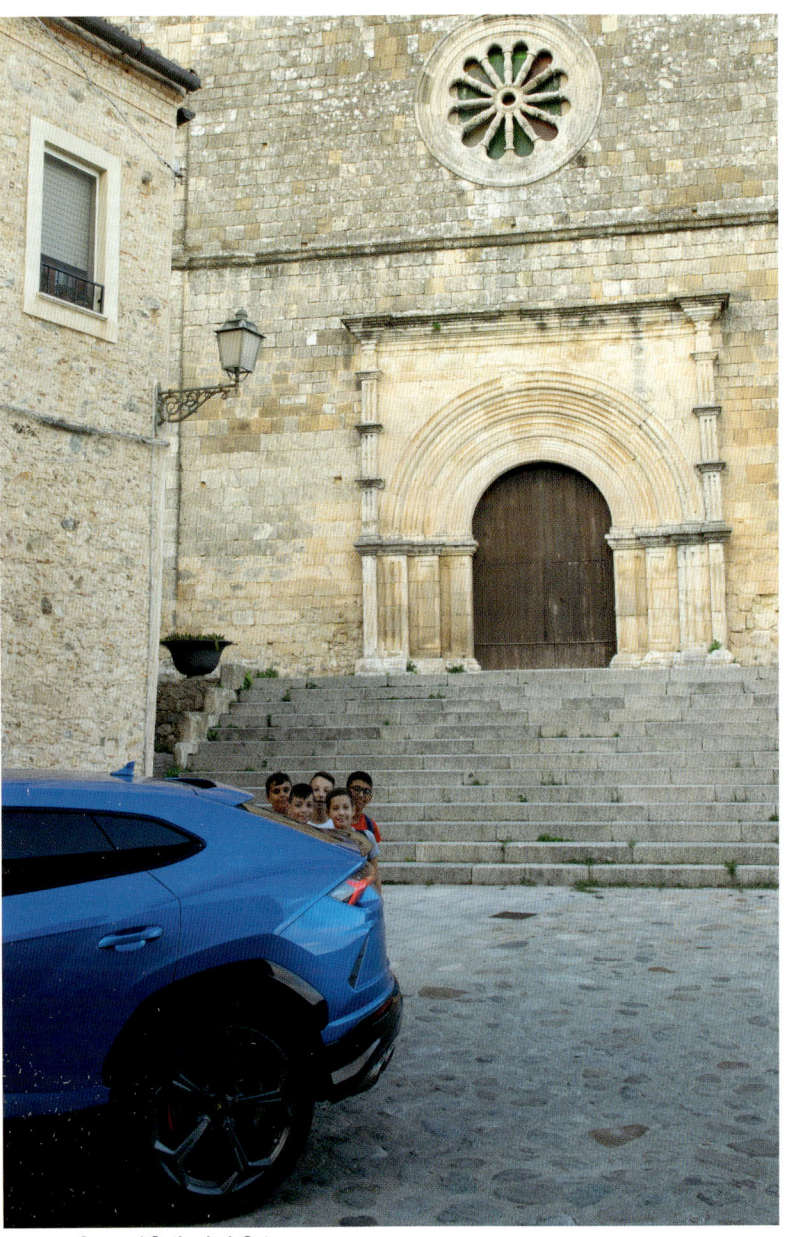

Cropani Cathedral, Catanzaro

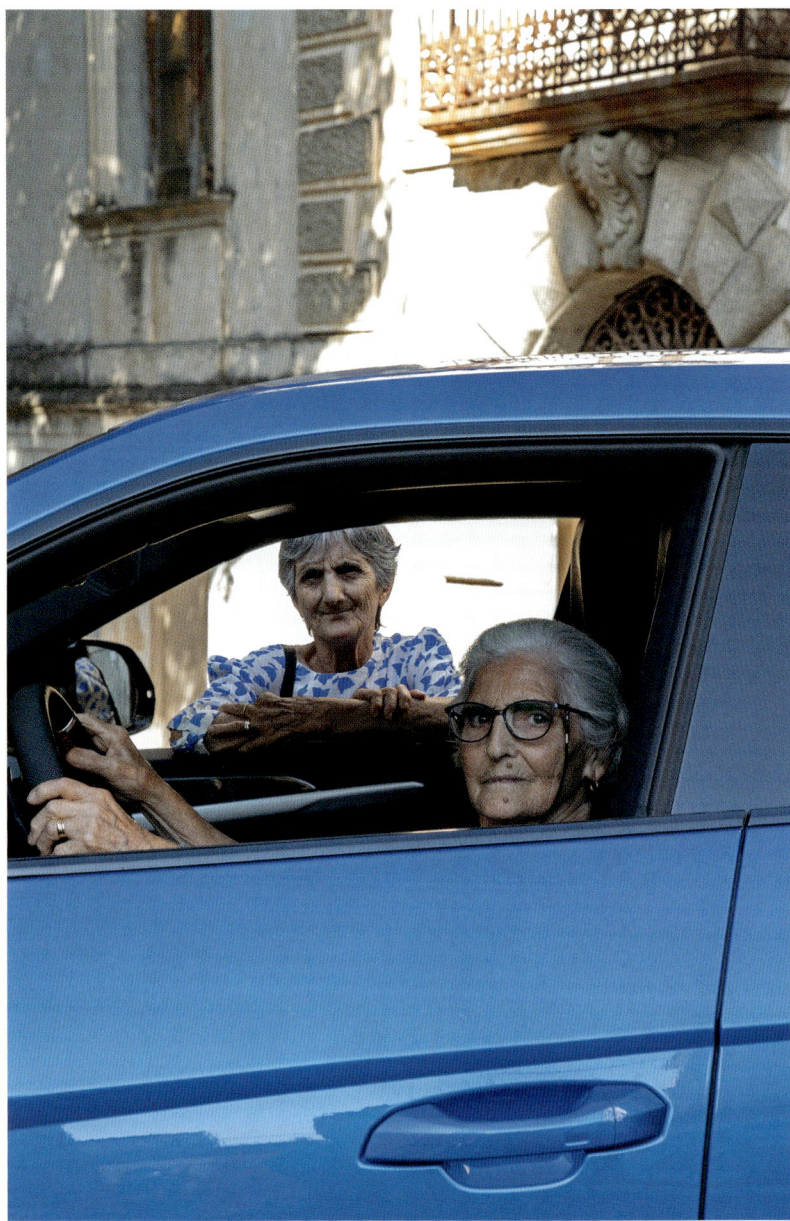

Carmela and Maria, Sant'Andrea Apostolo dello Ionio, Catanzaro

The Sersale Pacchiane in typical Calabrian costume at the Mother Church in Sersale, Catanzaro

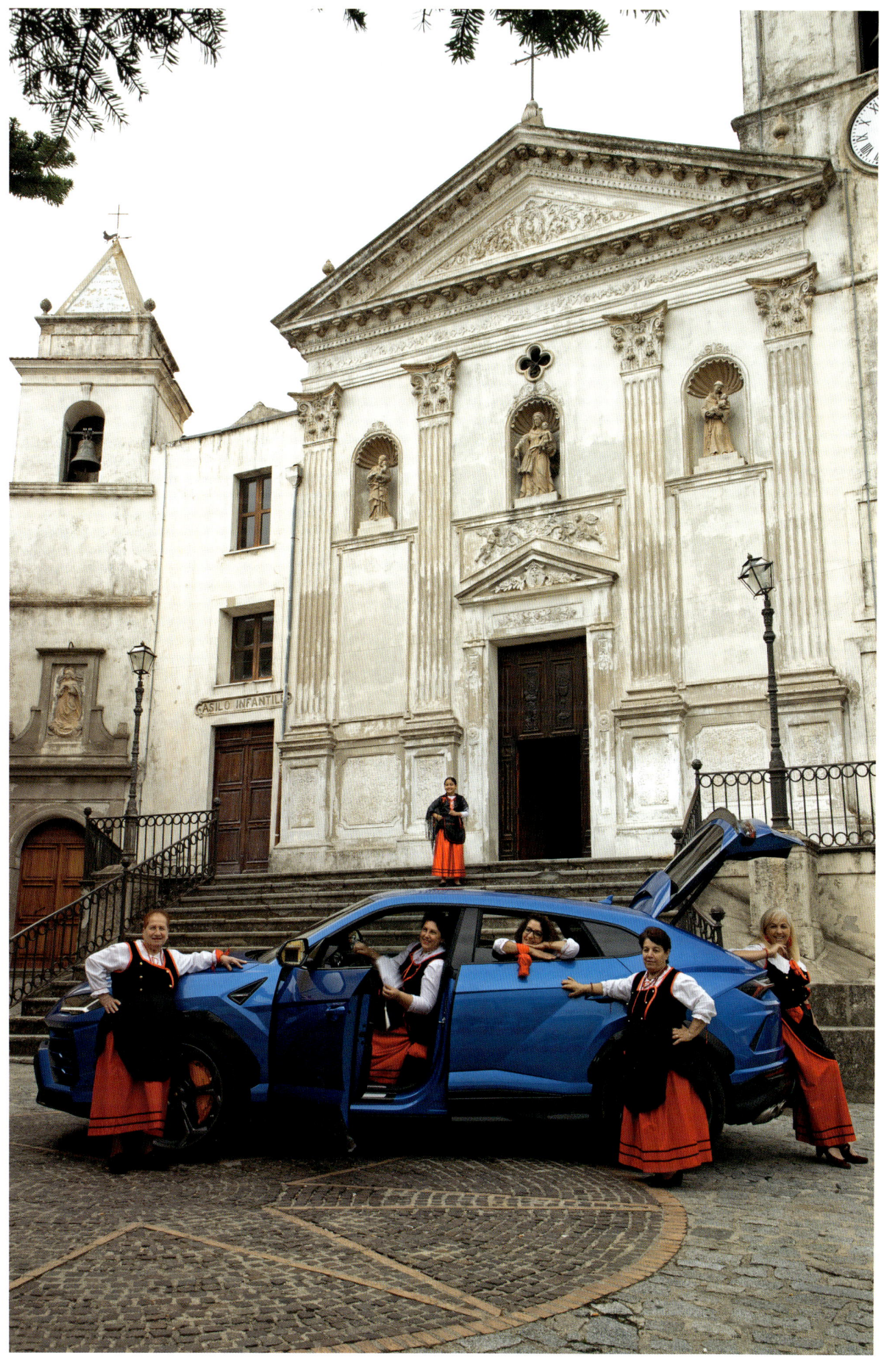

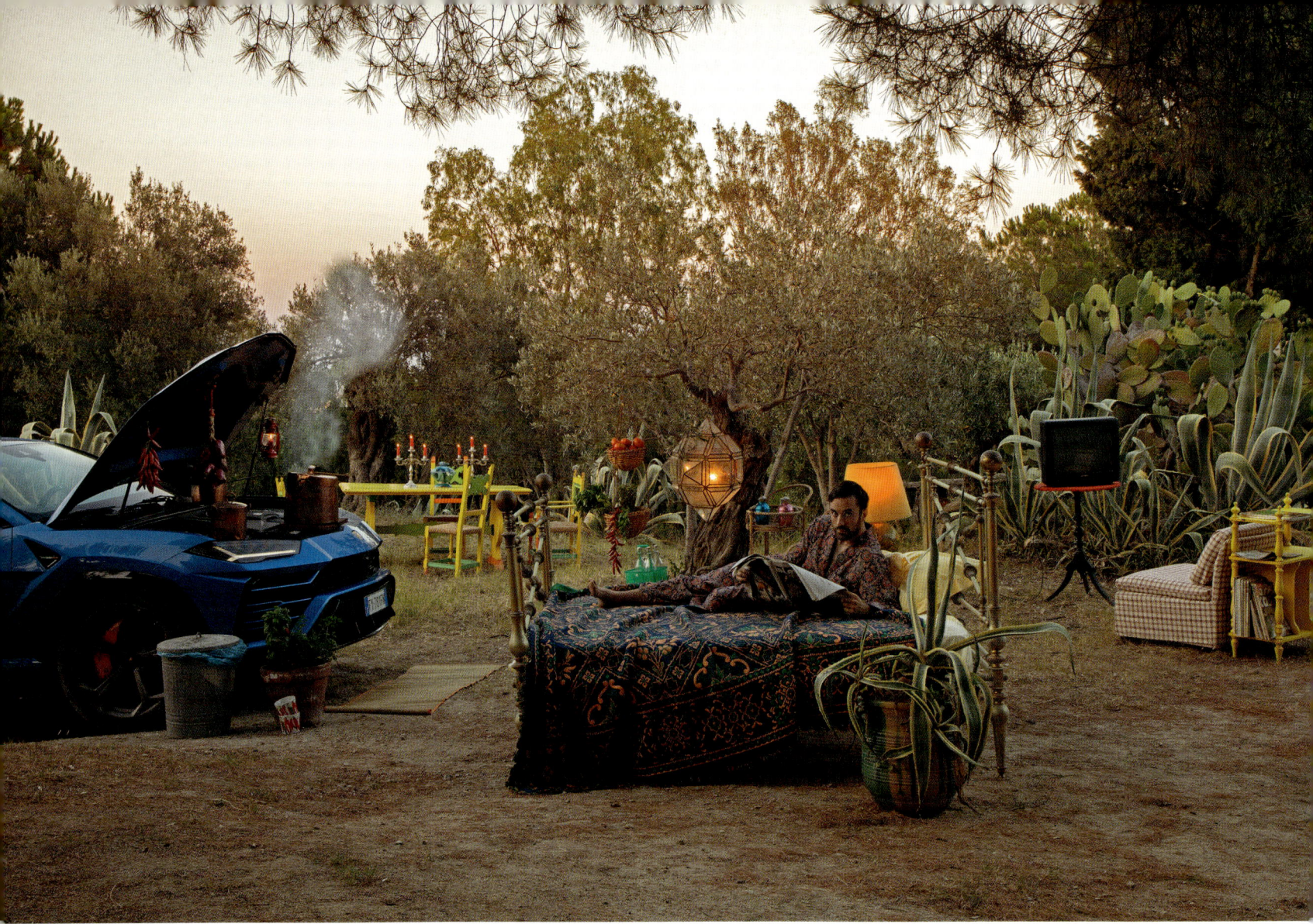

Self-portrait, Sant'Andrea Apostolo dello Ionio, Catanzaro

"The Lamborghini Urus, the very same blue as the Calabrian sea, first hides in nature, only slowly to reveal itself, becoming part of the daily life of the locals and their traditions, infused with the irony that is part and parcel of my photographic spirit"

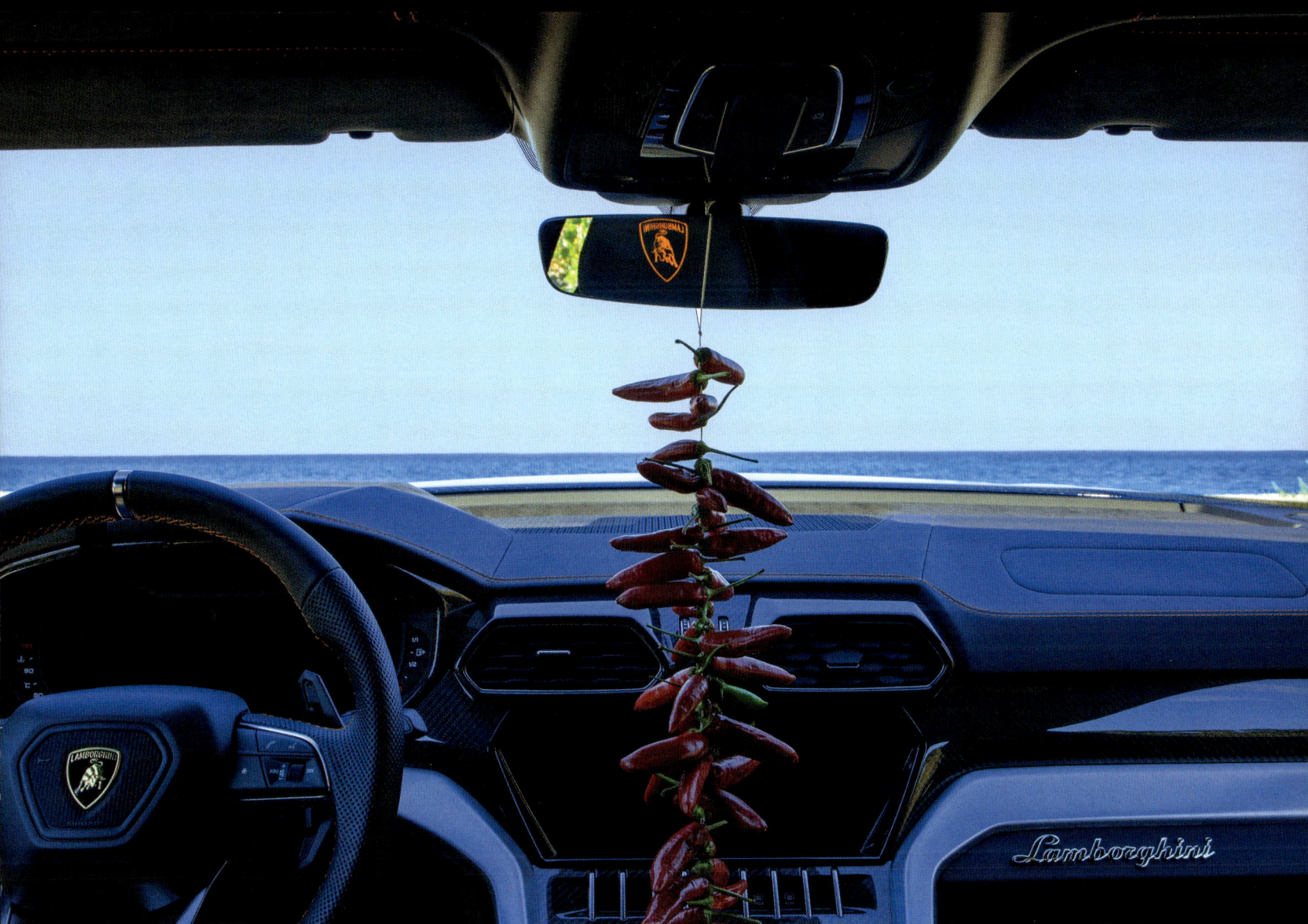
Propitiatory chili peppers, Ionian Coast

GABRIELE MICALIZZI
PUGLIA

Apulia has so many different faces because it is an anarchic, contentious, heterogeneous place, characterized by a multitude of backdrops and landscapes. My reportage is a fresco of genuine emotions in authentic places; it is a narrative of the passion, joy and wonder aroused by the passage of a unique and inimitable car like the Lamborghini Aventador SVJ Roadster. My photographic research, set against carefree, sometimes bizarre backdrops, is all about observing ordinary people's reactions at the sight of an extraordinary element: a car that qualifies as a legend.

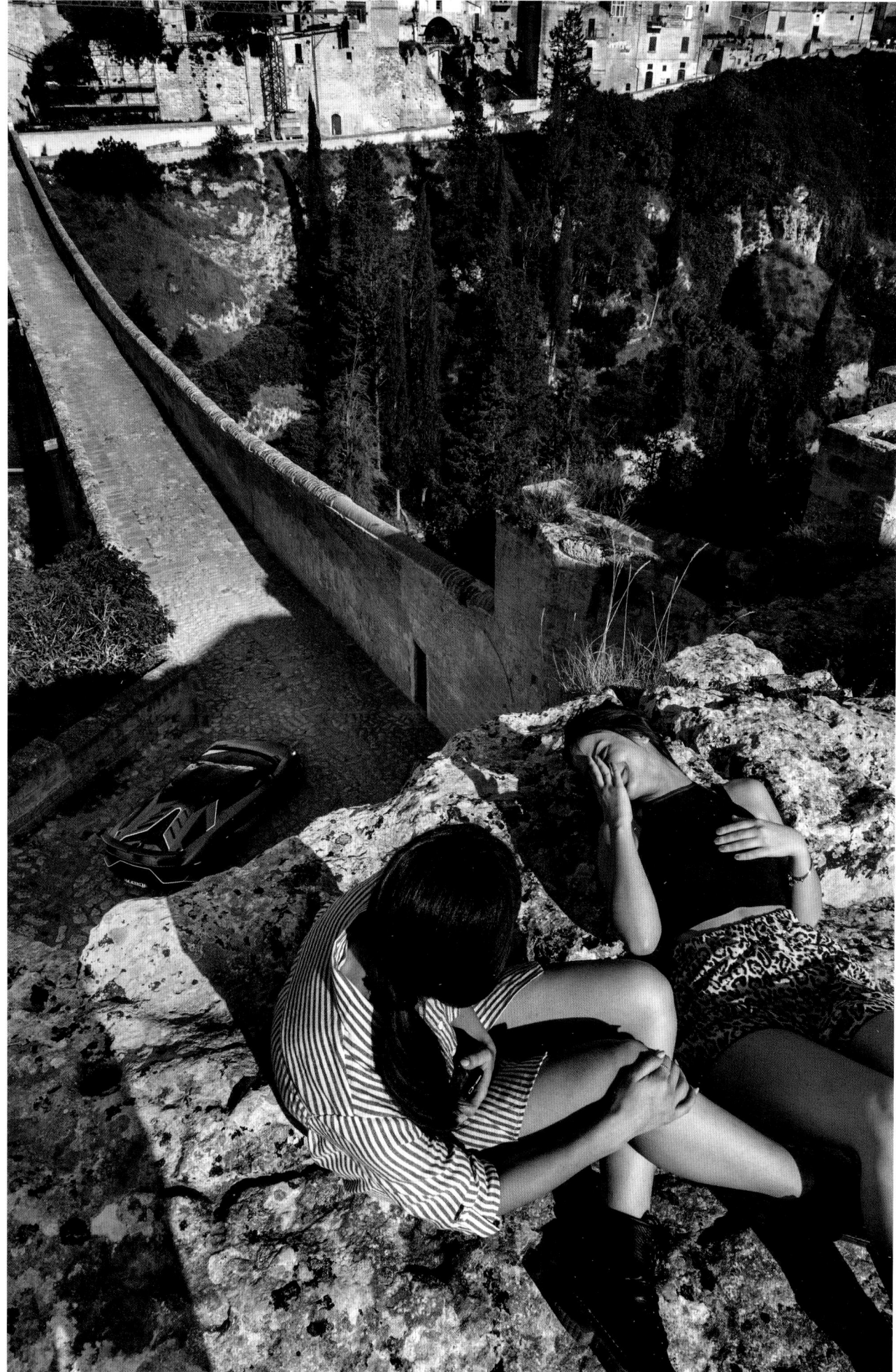

Gravina in Puglia, Bari

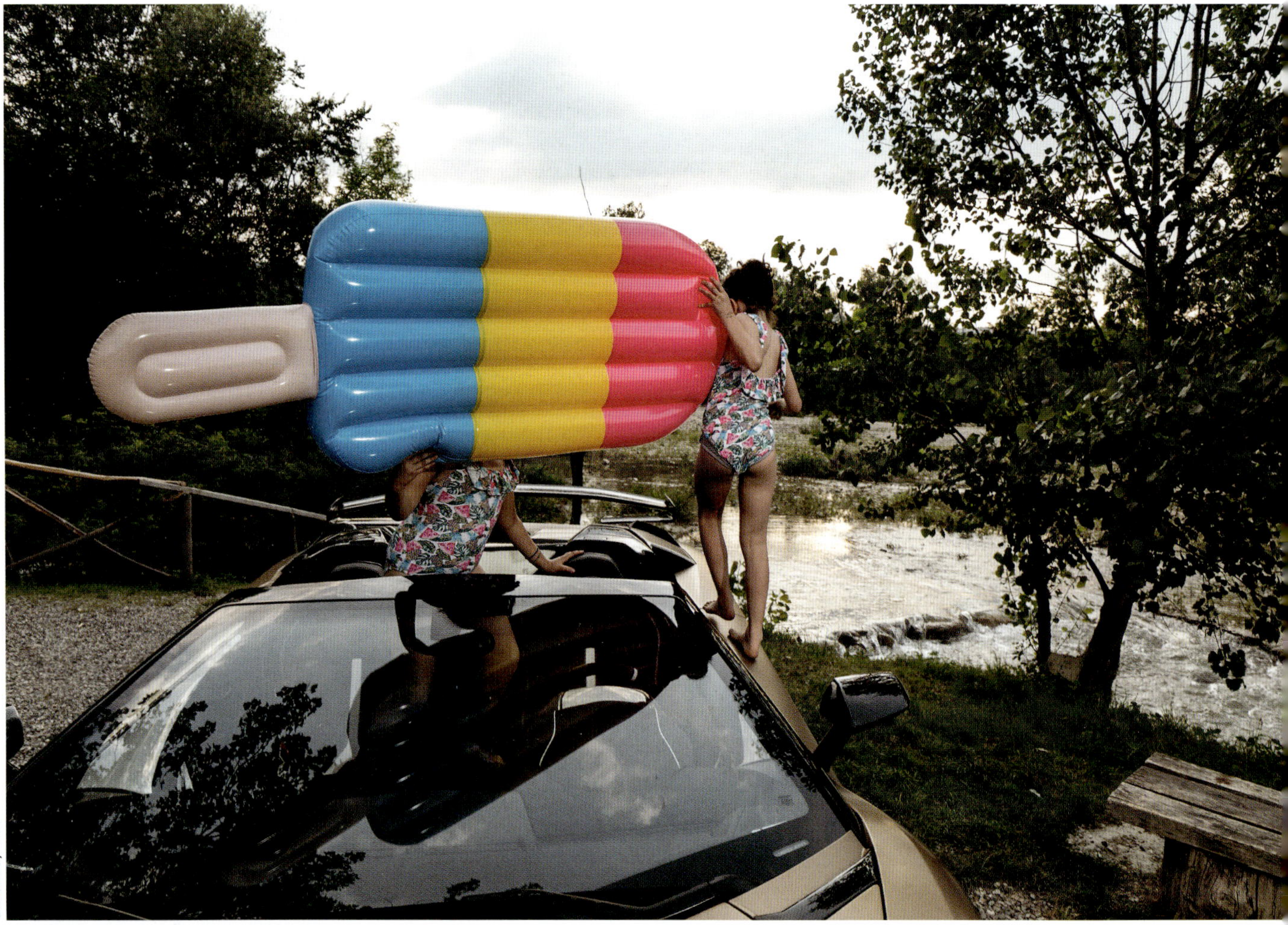

Bitonto, Bari

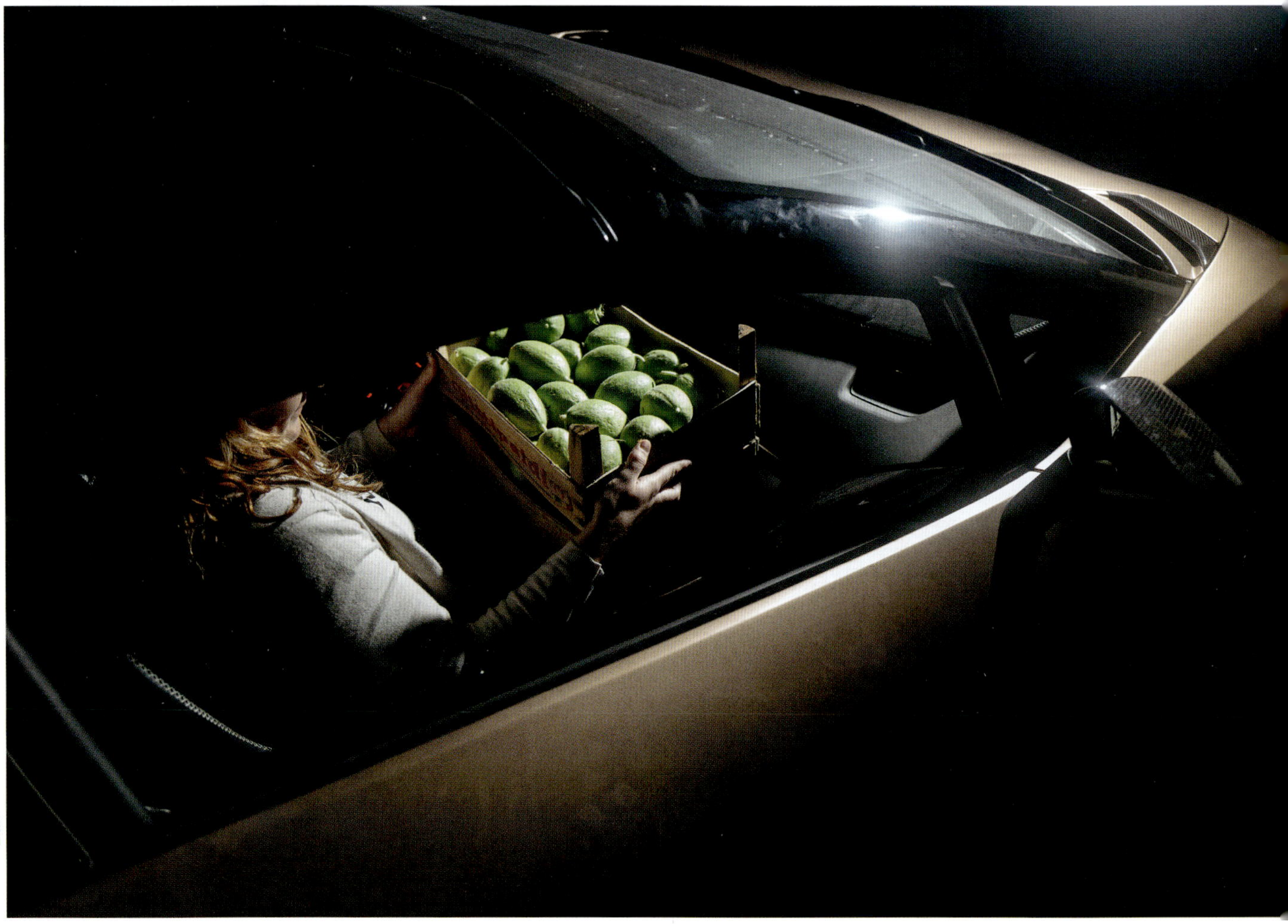

Bari

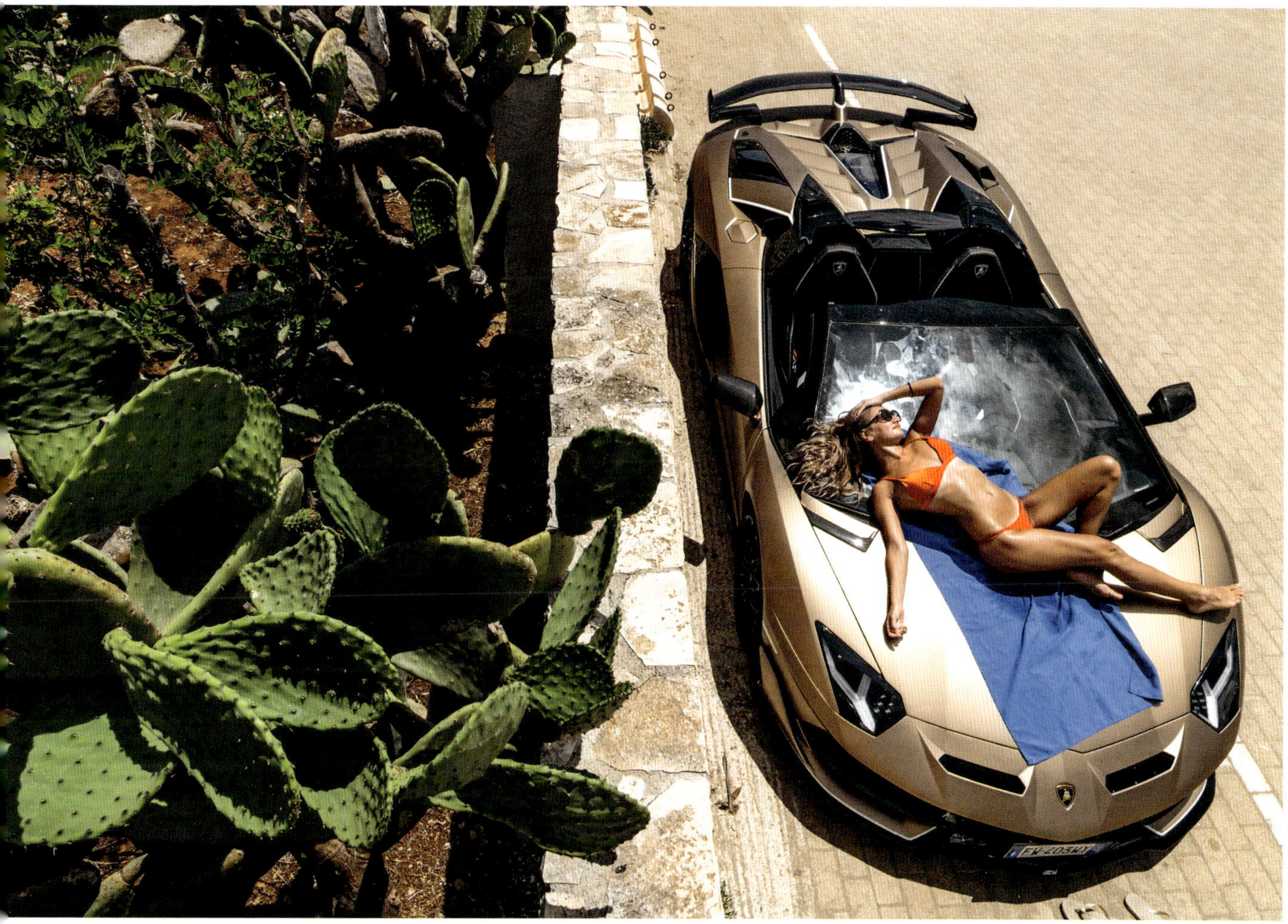

Marina di San Gregorio, Lecce

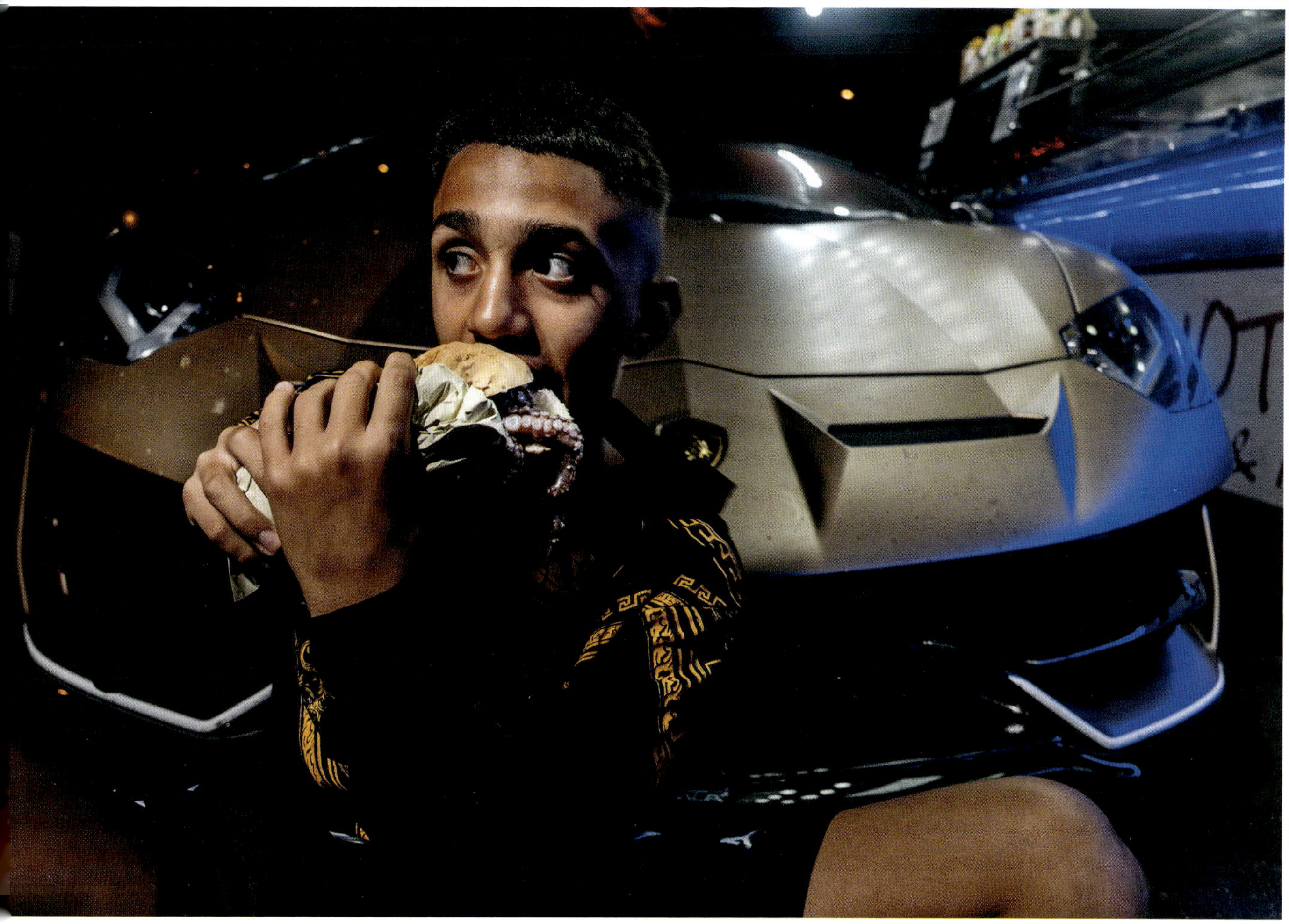

Bari

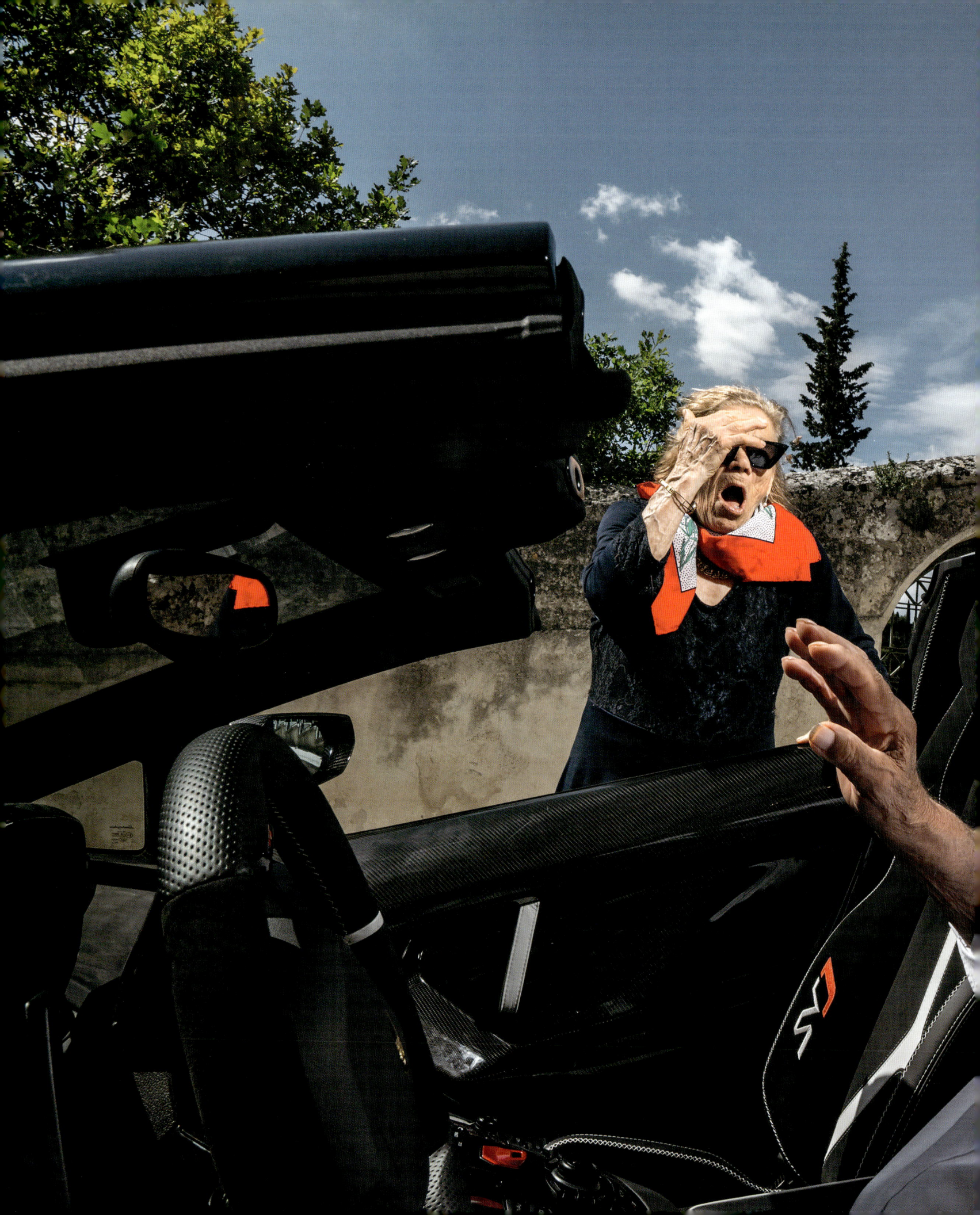

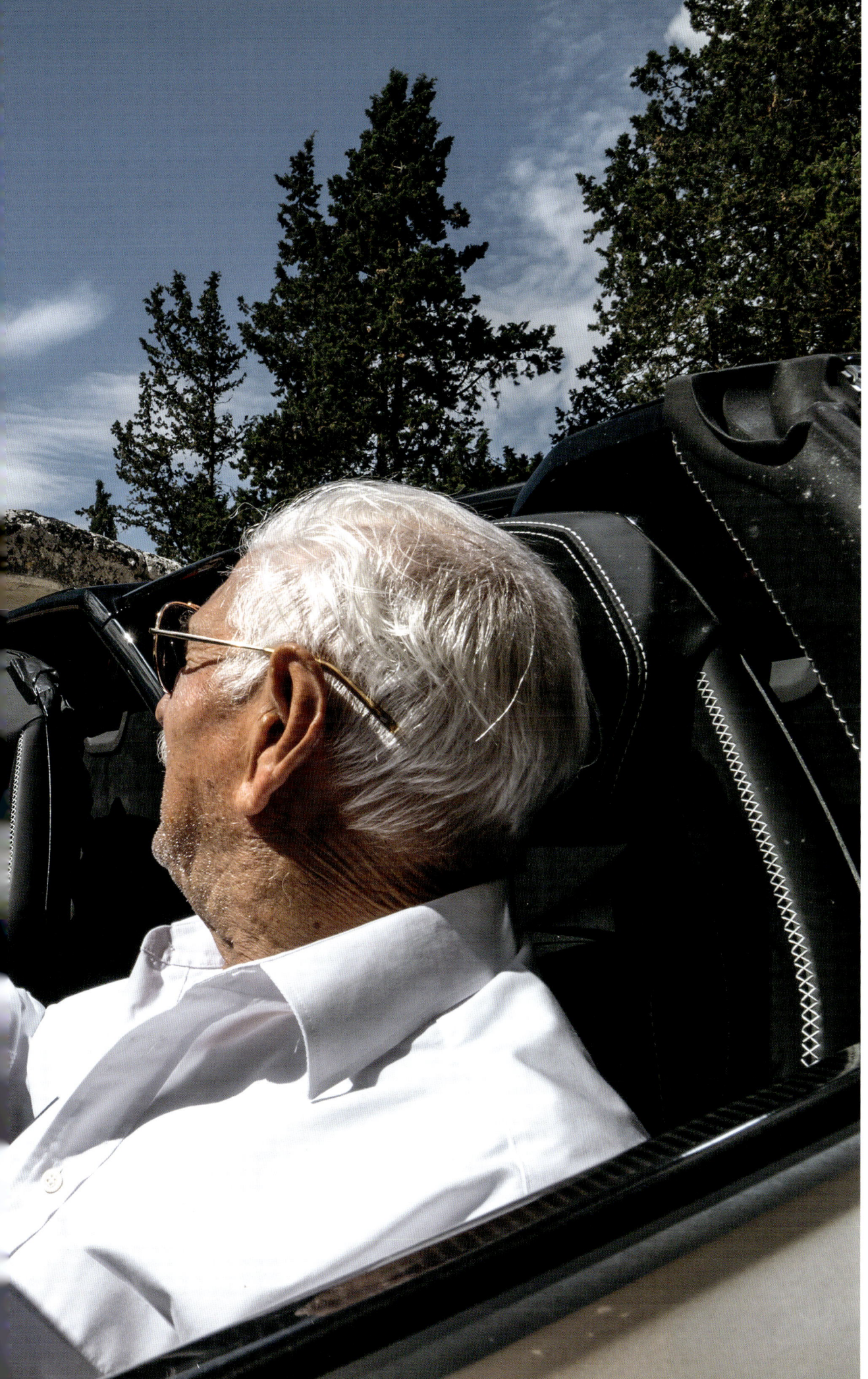

Gravina in Puglia, Bari

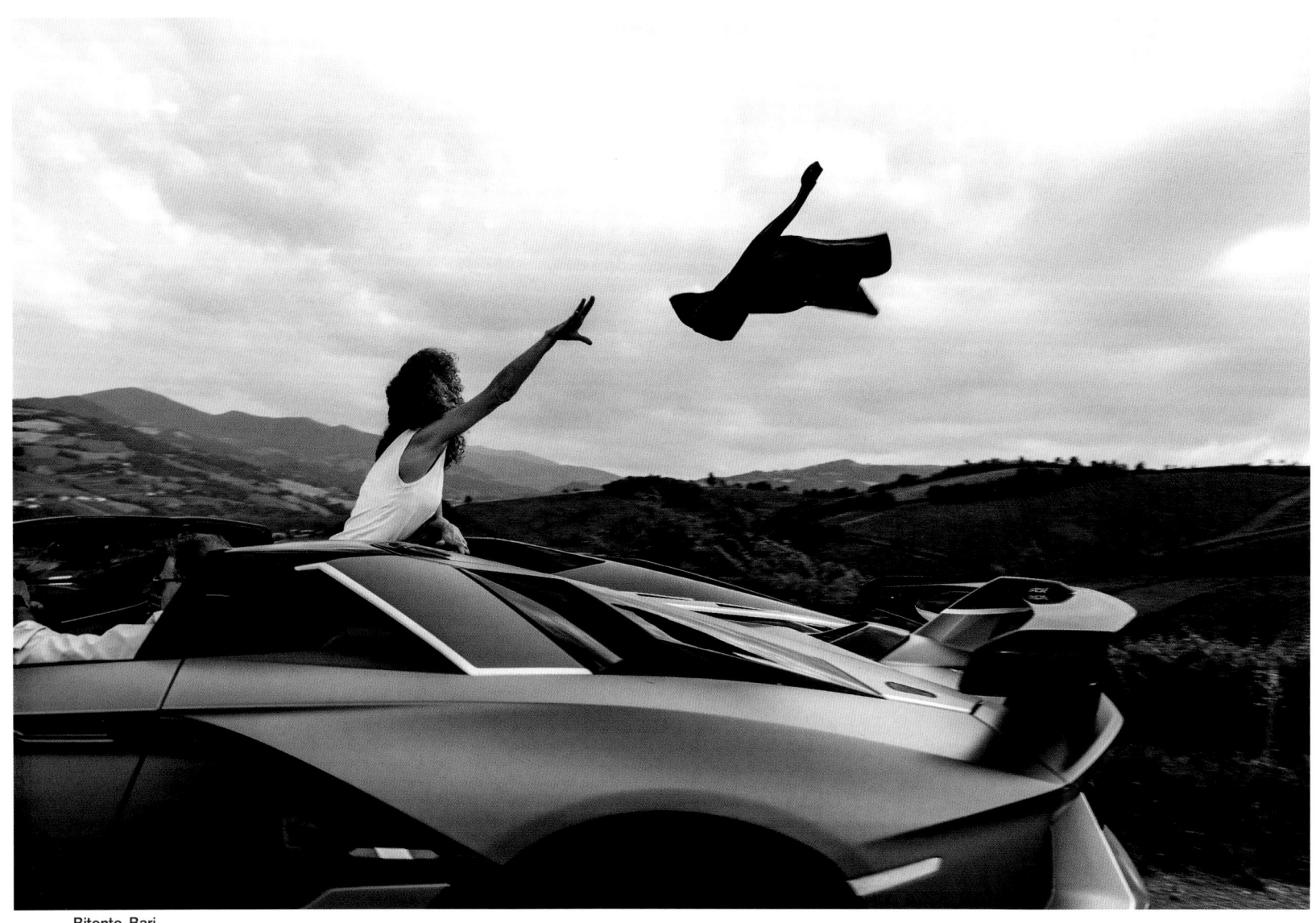

Bitonto, Bari

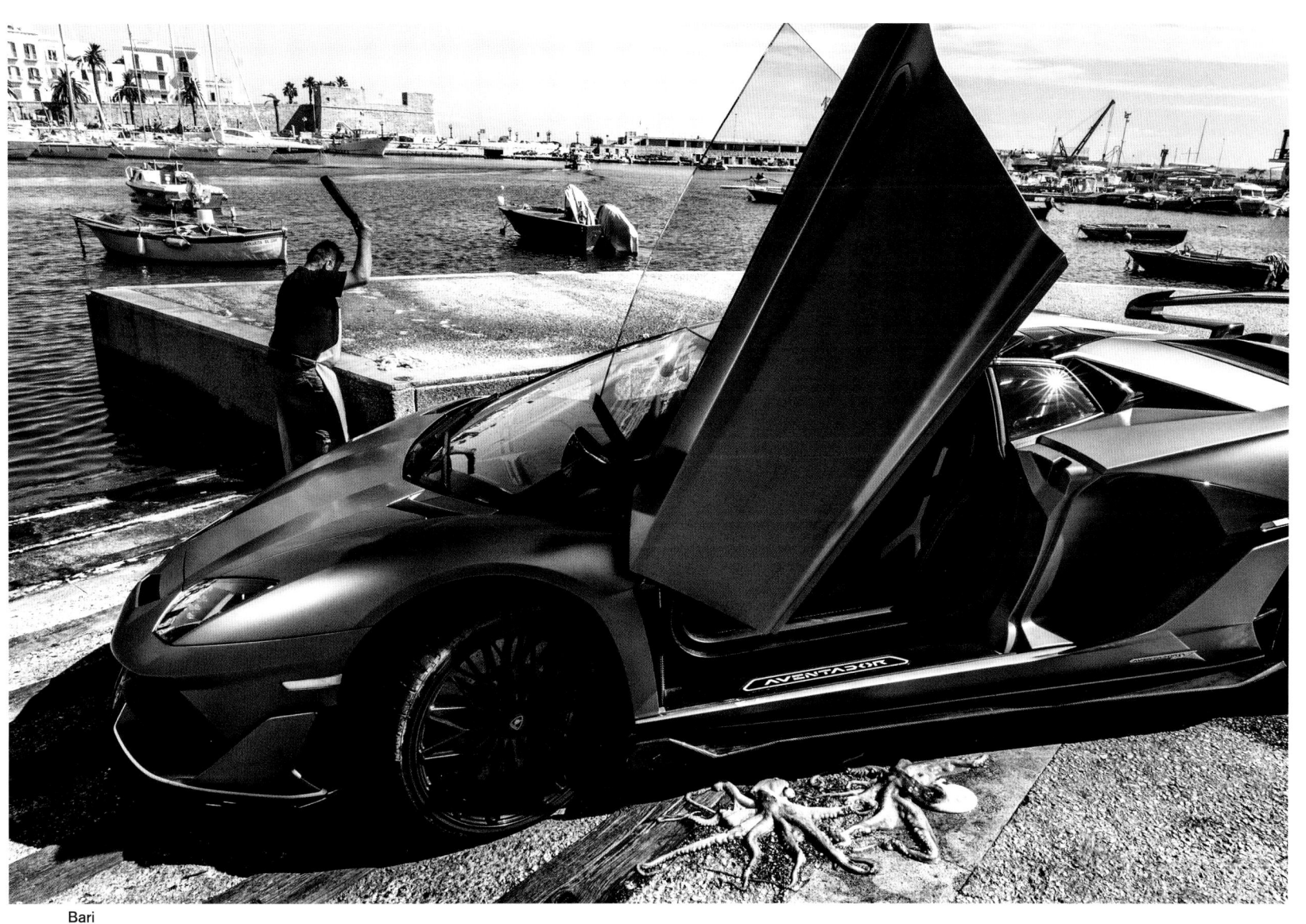
Bari

"My reportage is a fresco of genuine emotions in authentic places; it is a narrative of the passion, joy and wonder aroused by the passage of a unique and inimitable car"

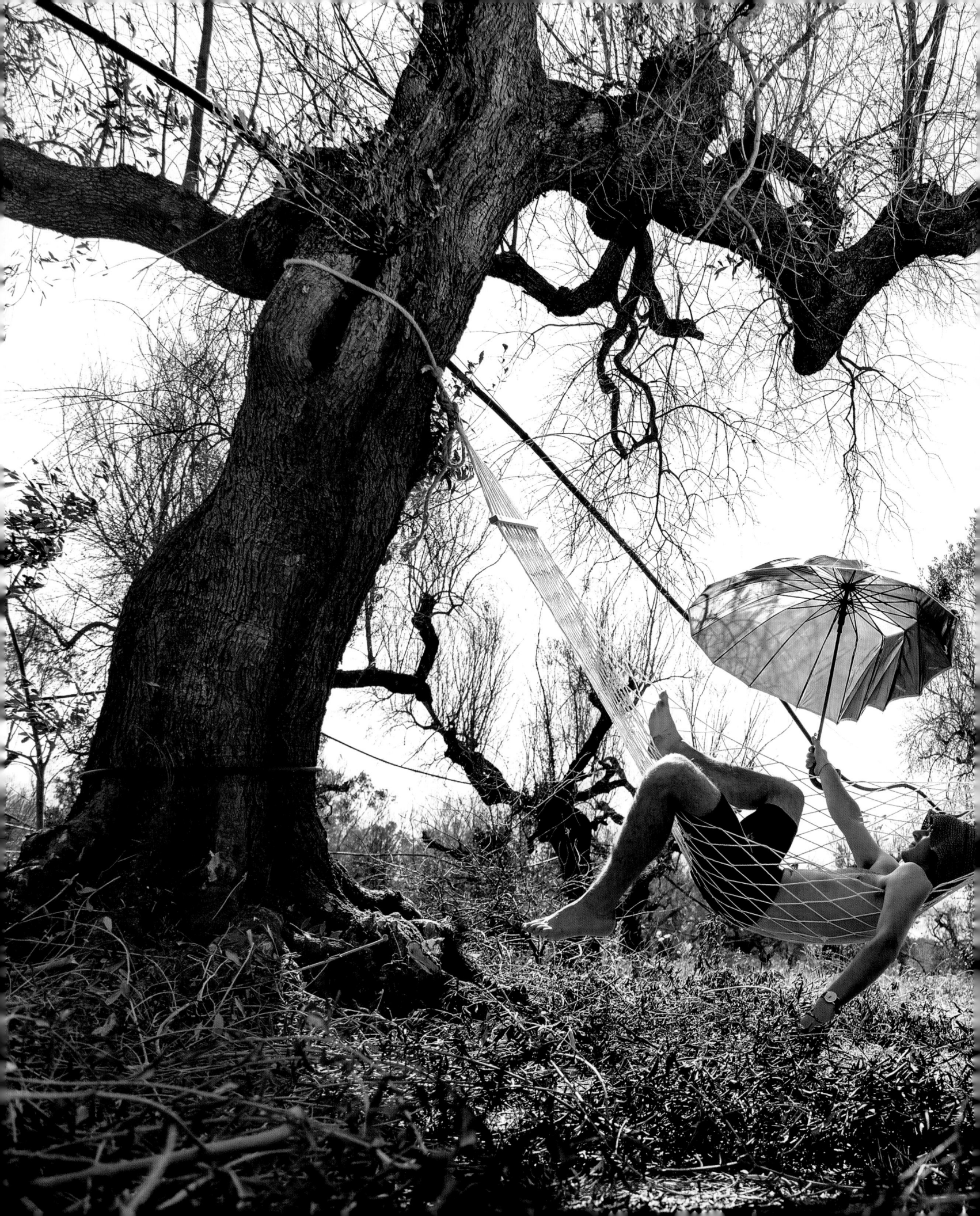

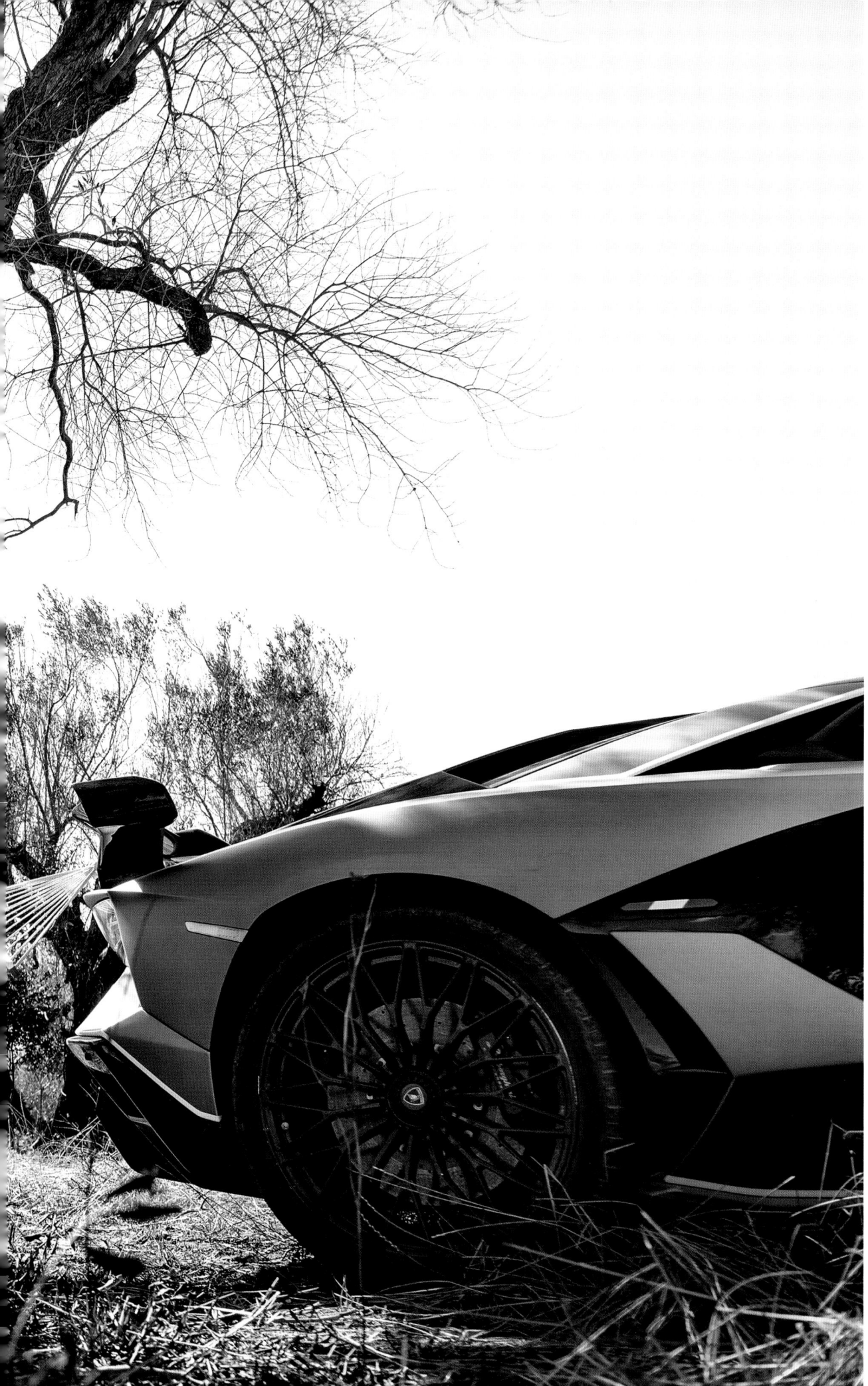

Ugento, Lecce

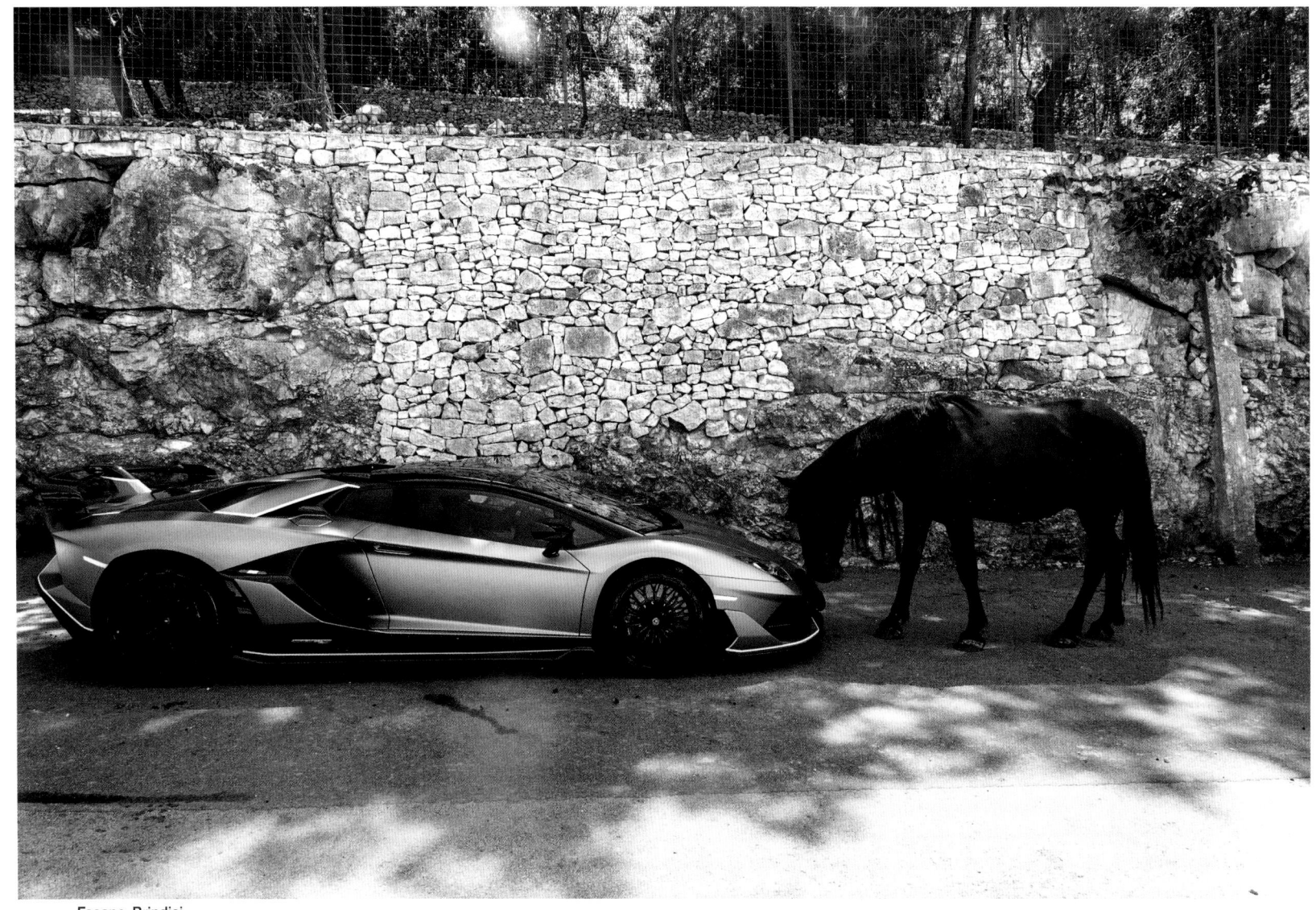

Fasano, Brindisi

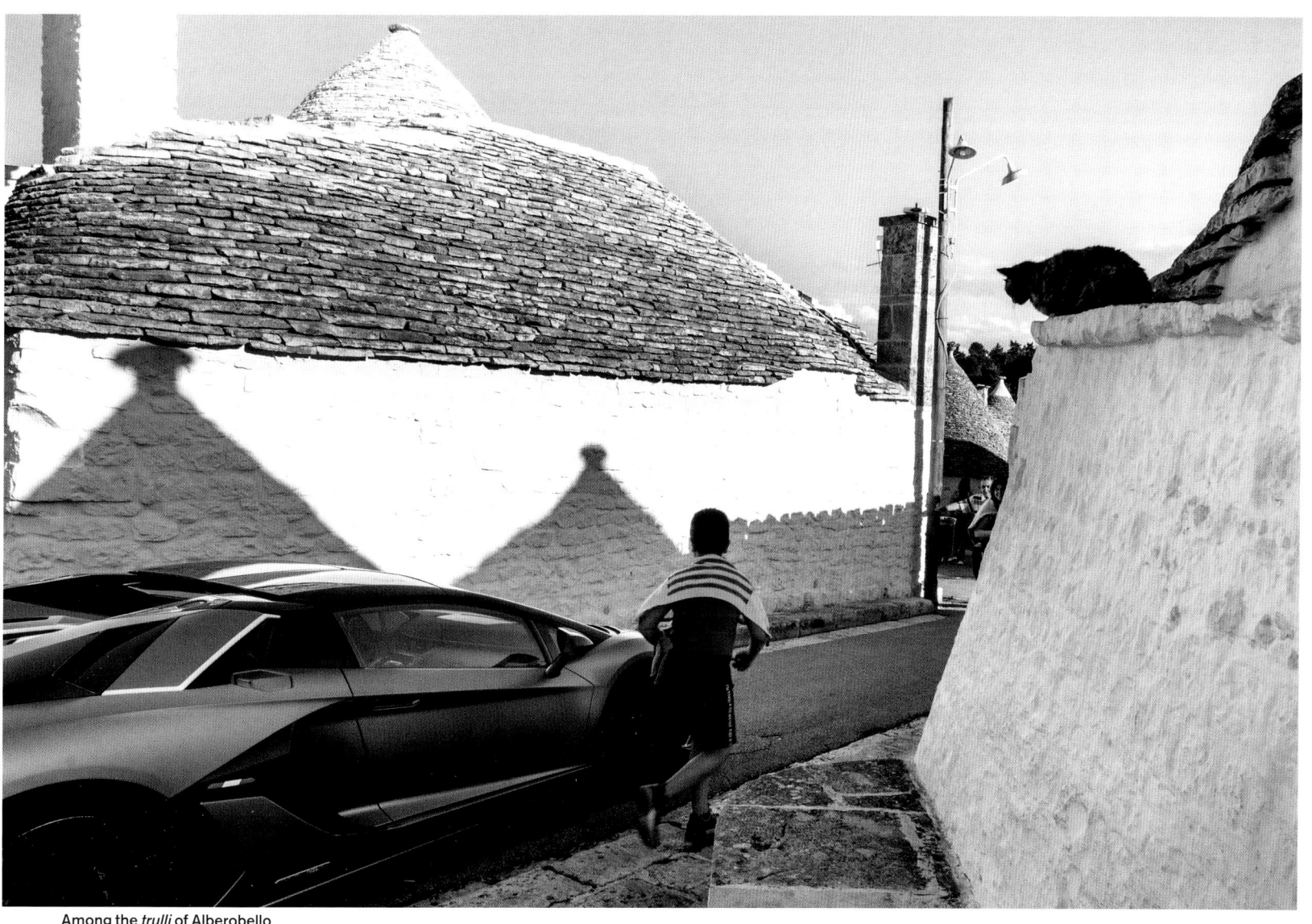
Among the *trulli* of Alberobello

CAMILLA FERRARI
BASILICATA

Basilicata is a land that drives itself like a wedge between earth and sky. It is the repository of an ancient past and traditions, but at the same time a leading power in highly-innovative projects. The images I wanted to create were intended to draw subtle, poetic parallels between the Lucanian soul and the essence of the Lamborghini Aventador S which, like Basilicata, is a combination of craftsmanship and cutting-edge technology. The narrative develops in a crescendo: from evocative details of the car, harmoniously part of the landscape, it becomes more and more decisive until, bearing witness to a hypothetical timeline that runs from the past to the future, the Aventador S takes its full place in the limelight.

Matera

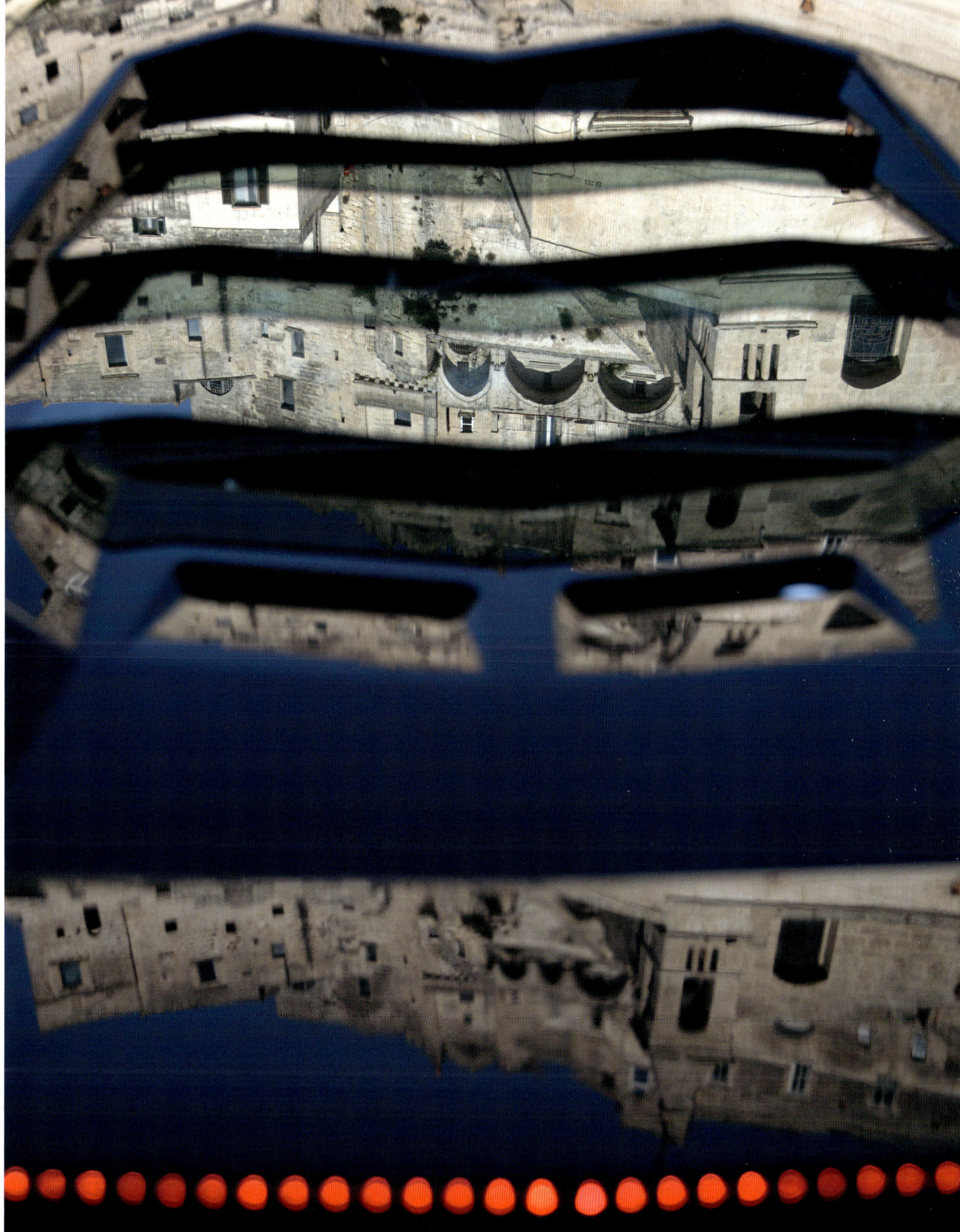

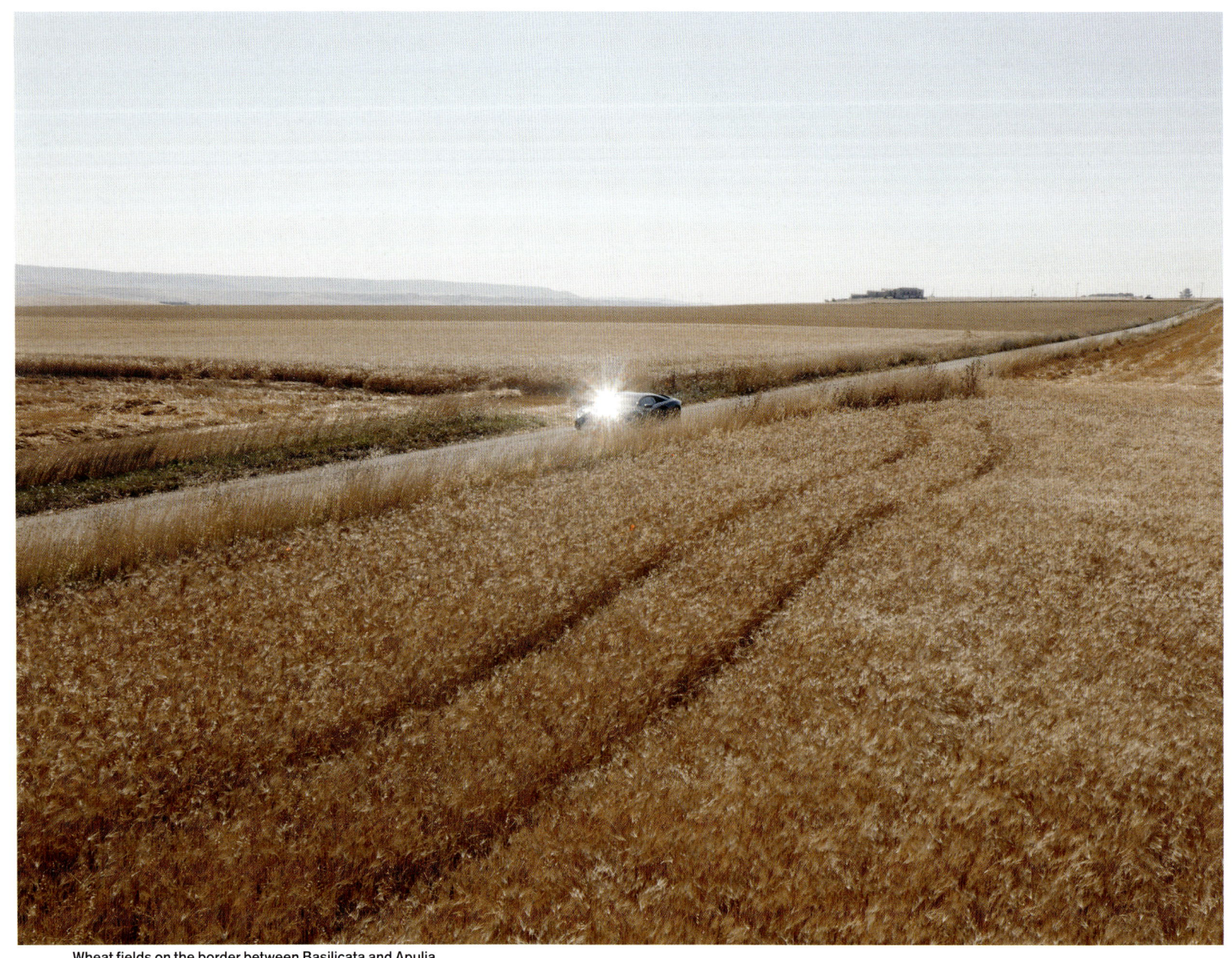
Wheat fields on the border between Basilicata and Apulia

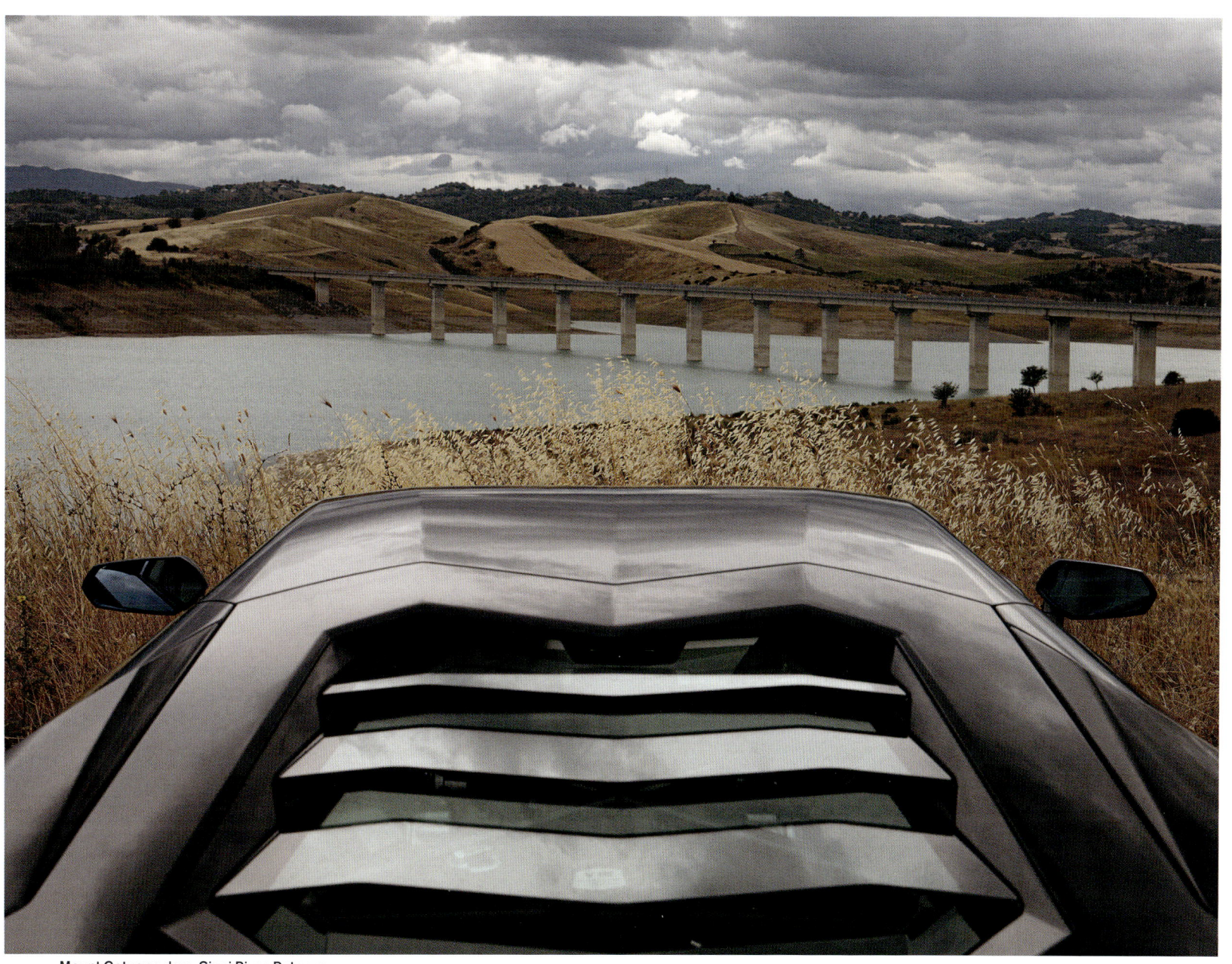

Mount Cotugno dam, Sinni River, Potenza

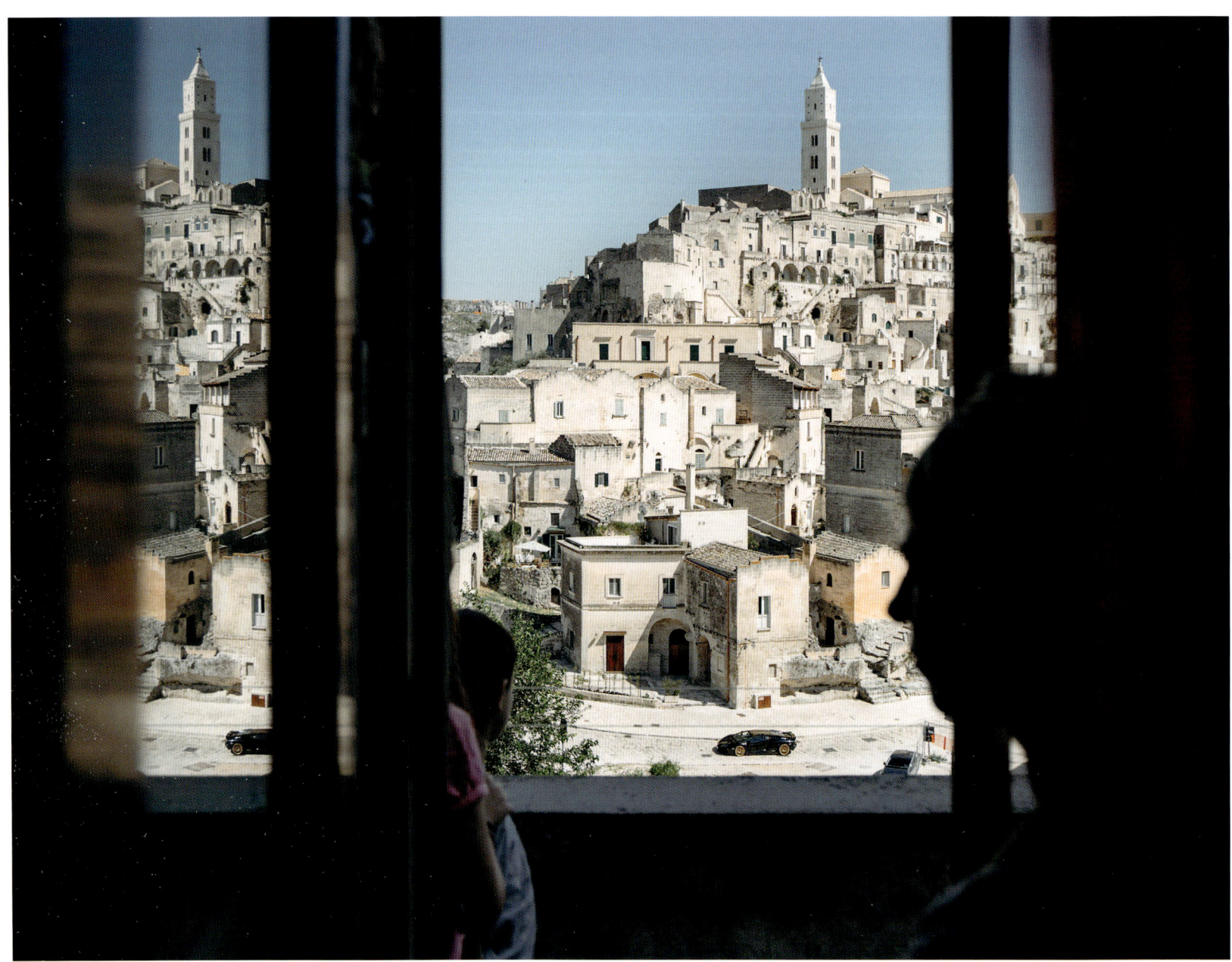

Matera

Aliano, Matera

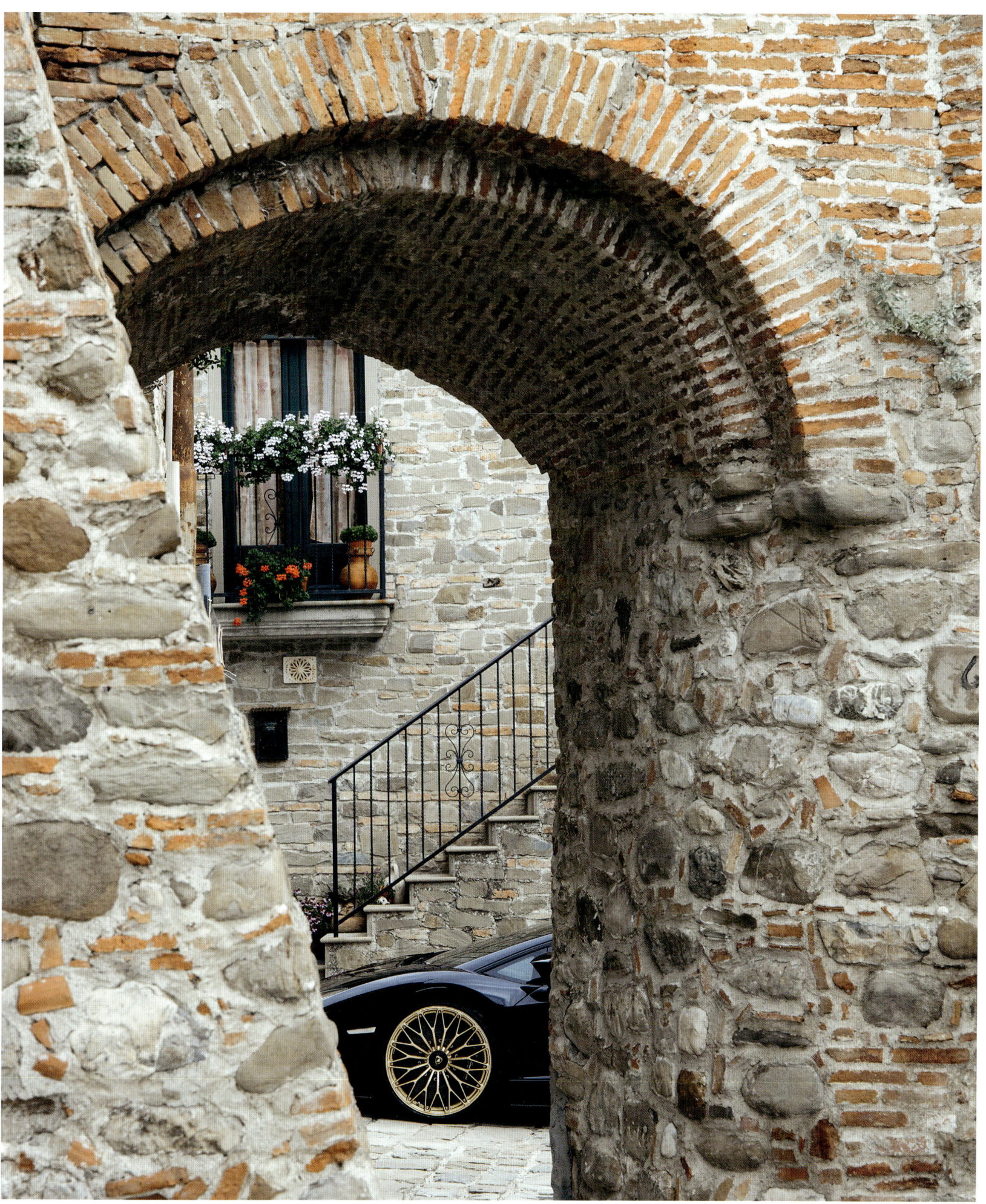

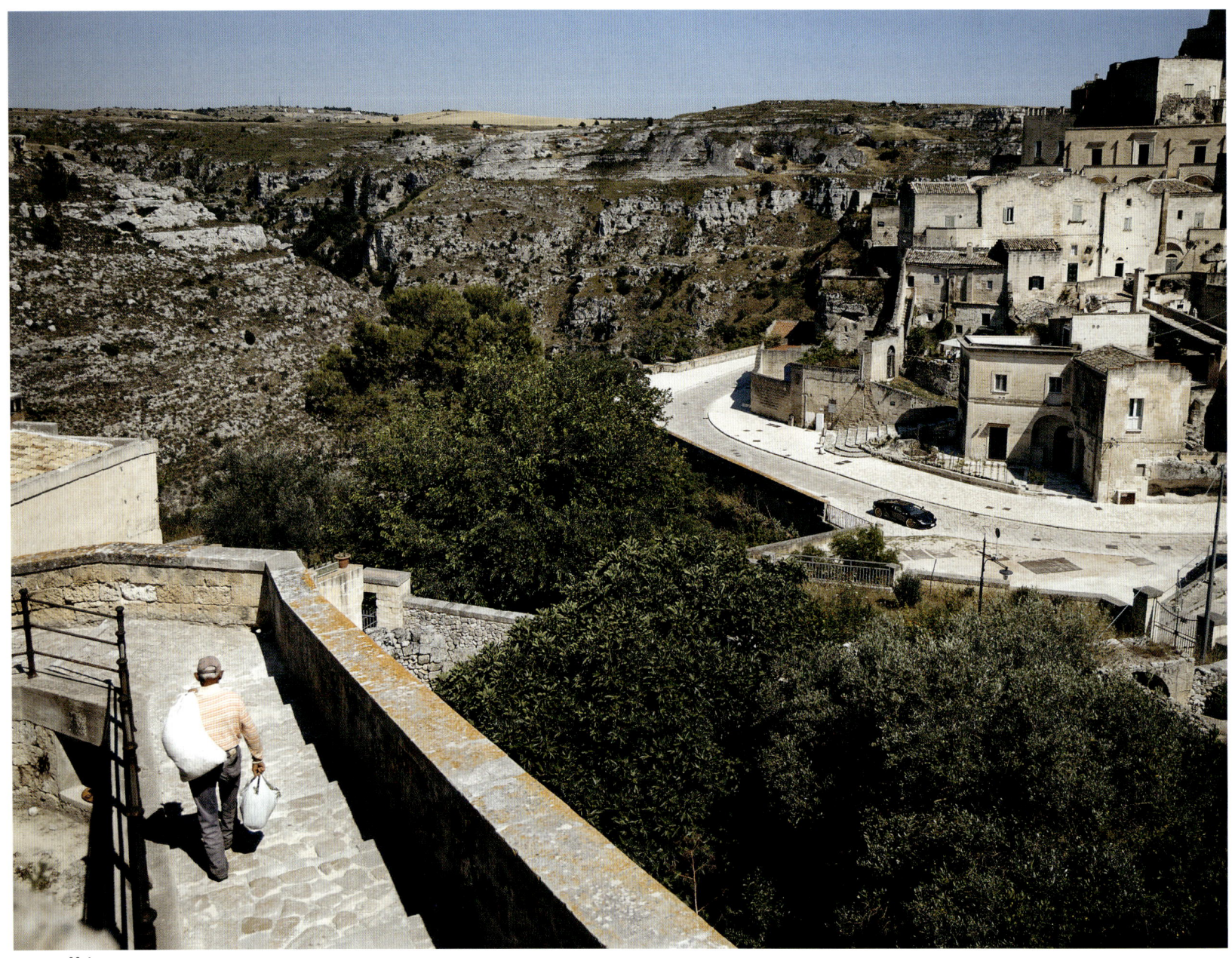

Matera

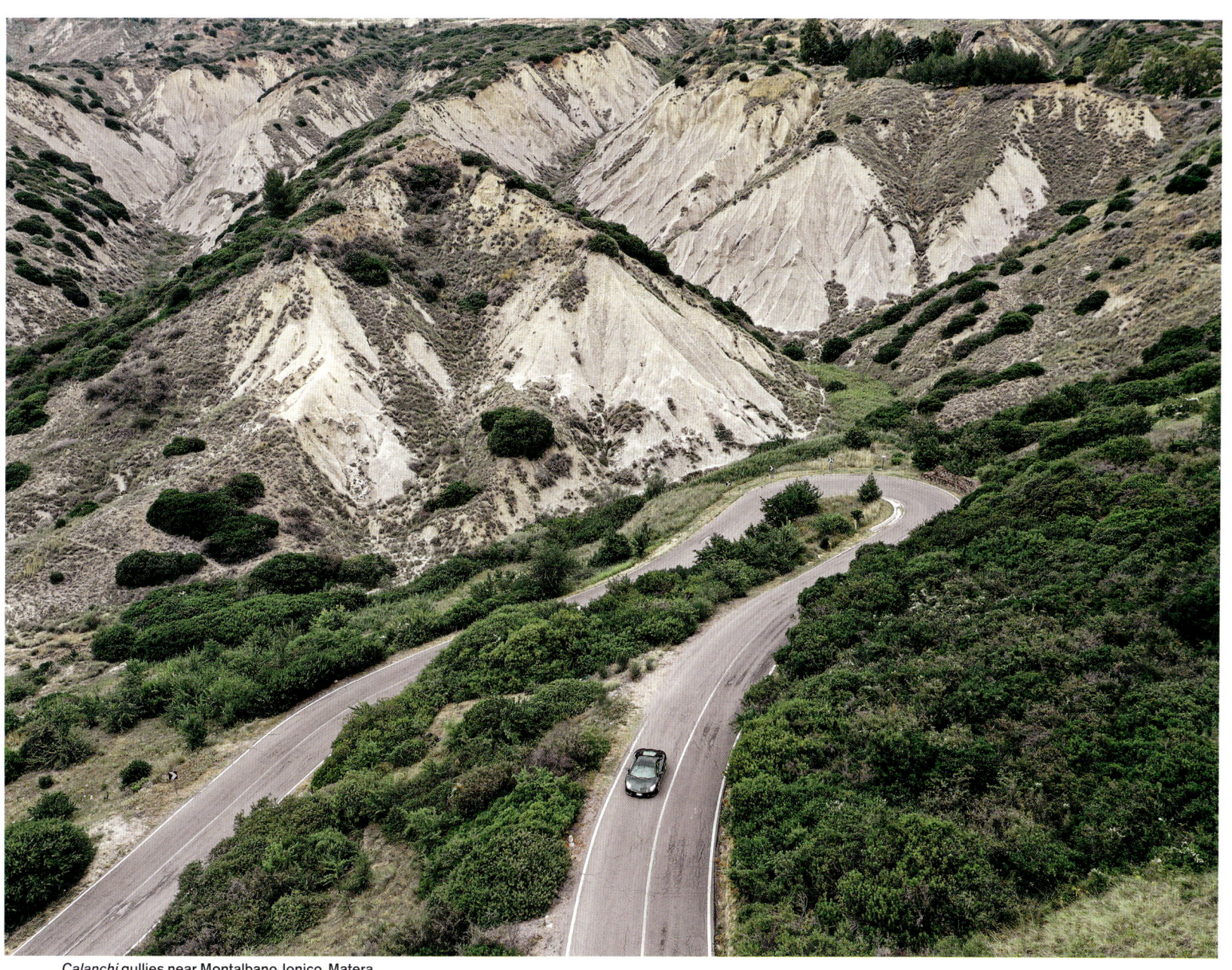
Calanchi gullies near Montalbano Jonico, Matera

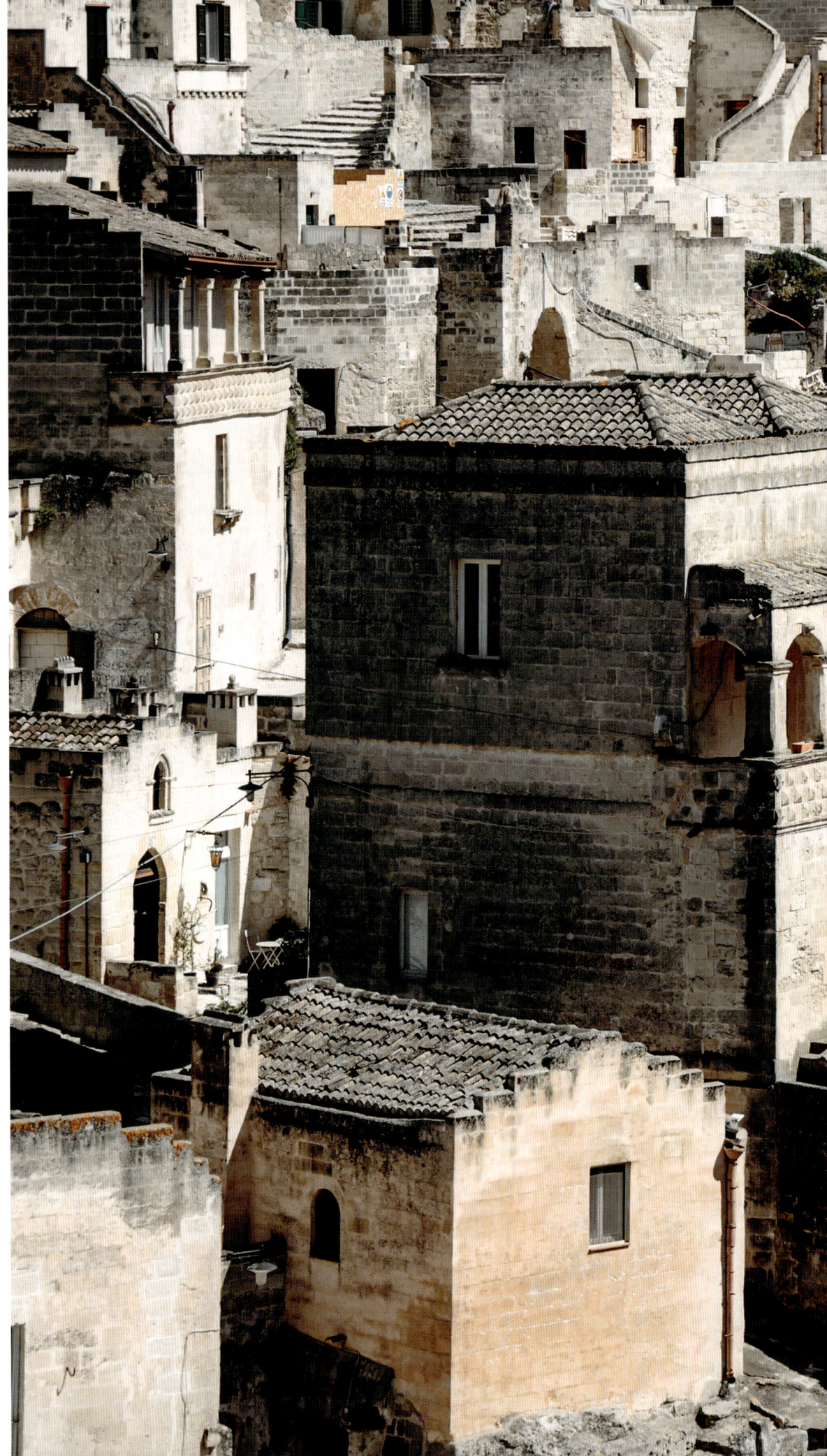

Matera

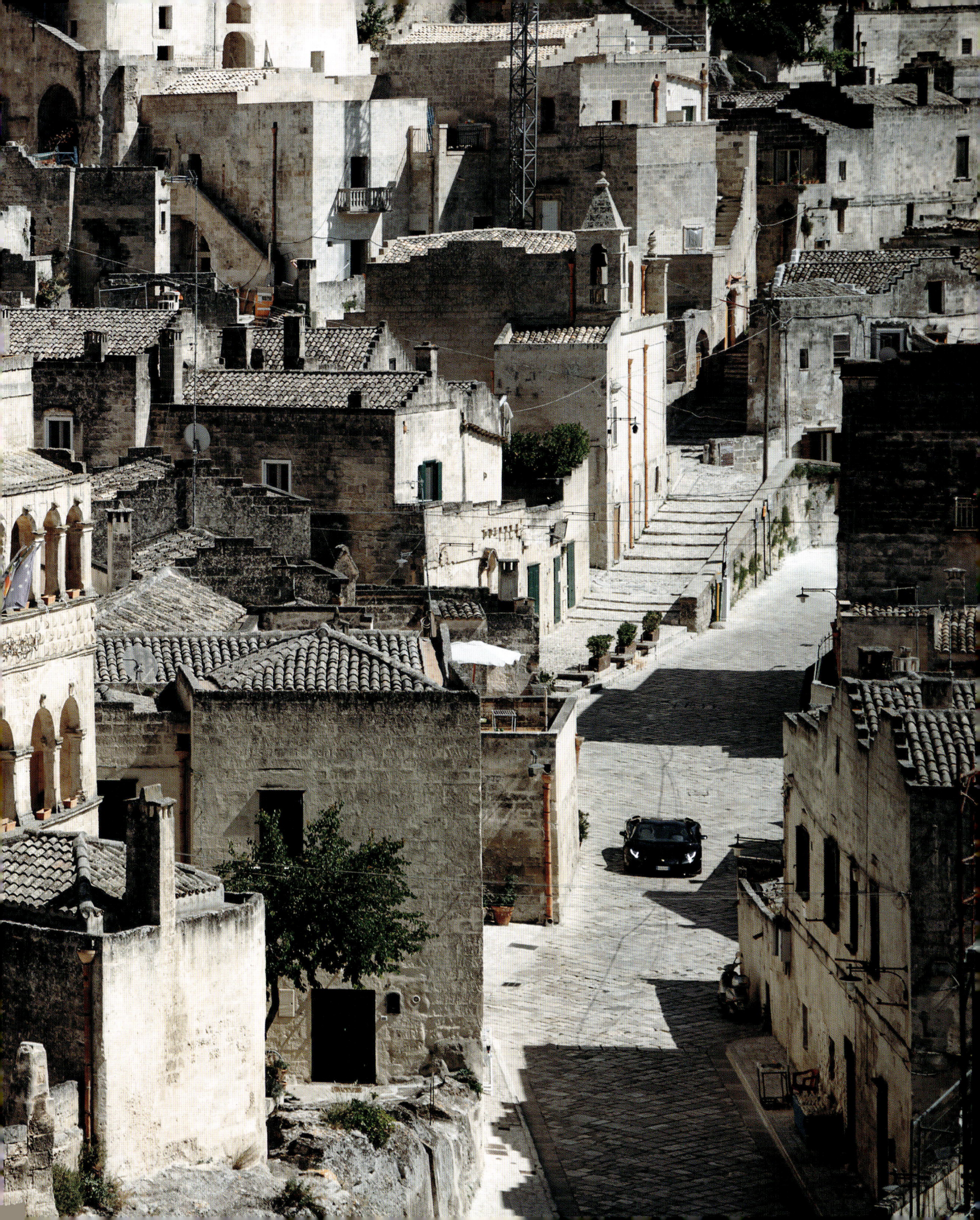

MARCO CASINO
CAMPANIA

My journey with the Lamborghini Huracán EVO Spyder is the story of a region, my land, a compendium of unique beauty and powerful contrasts. In typical documentary photography style, I was keen to portray the heterogeneity of the landscapes, all of the urban and suburban architectural stratifications. Through a narrative playing on nuance and detail, I wanted to bring out not just the extraordinary appeal of this land, but its authentic essence, liveliness and great temperament.

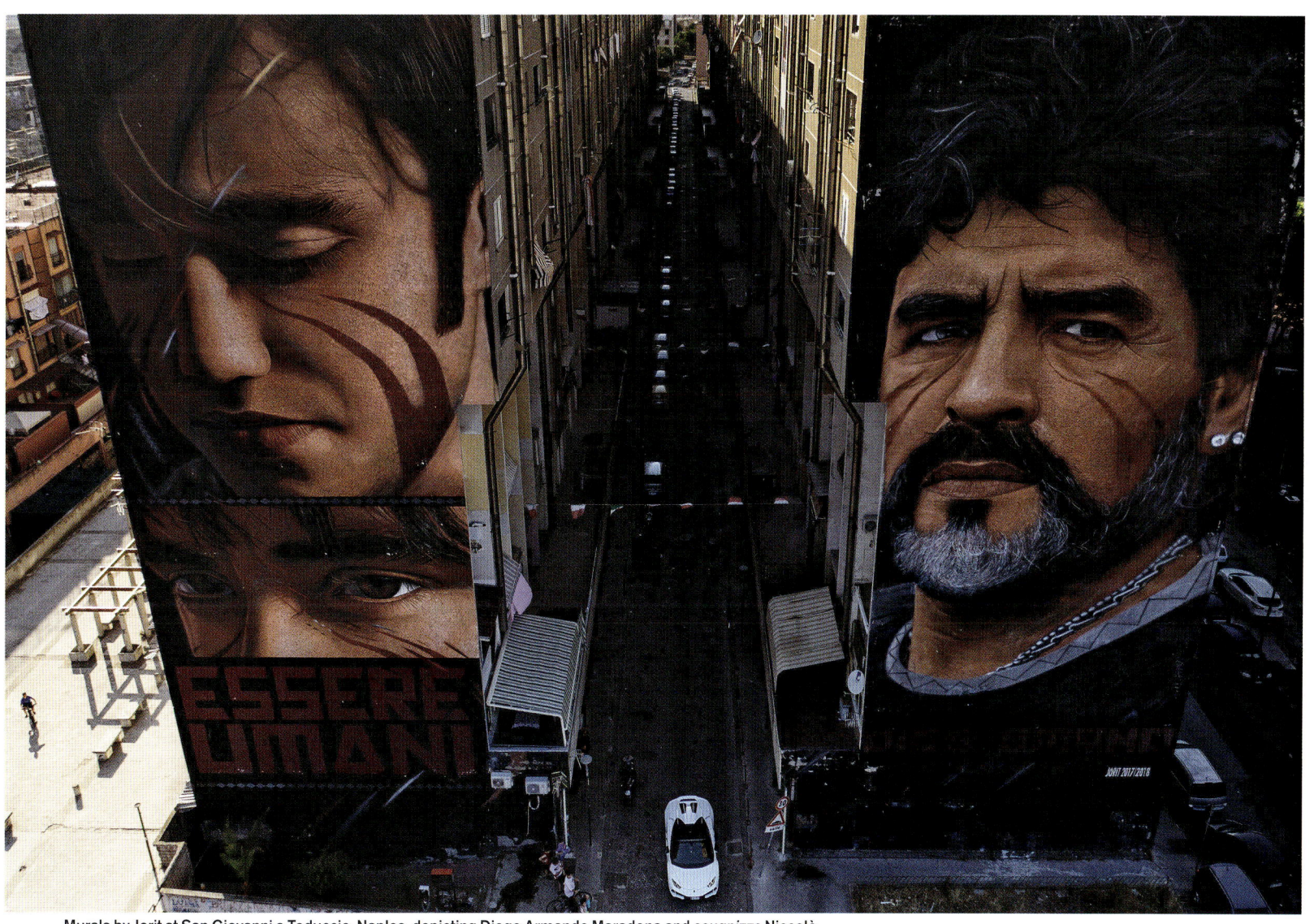

Murals by Jorit at San Giovanni a Teduccio, Naples, depicting Diego Armando Maradona and *scugnizzo* Niccolò

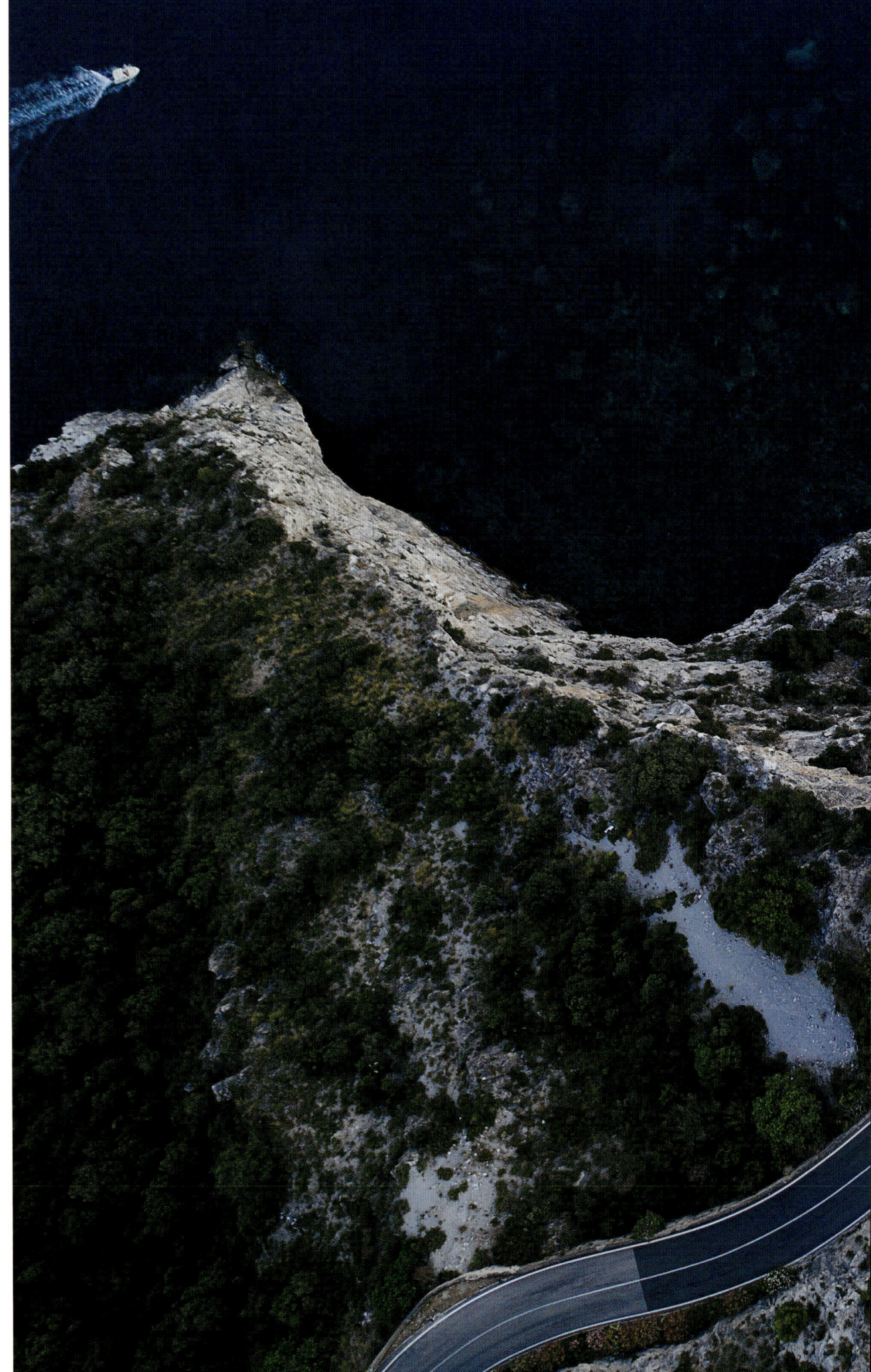

The western fringes of the Amalfi Coast

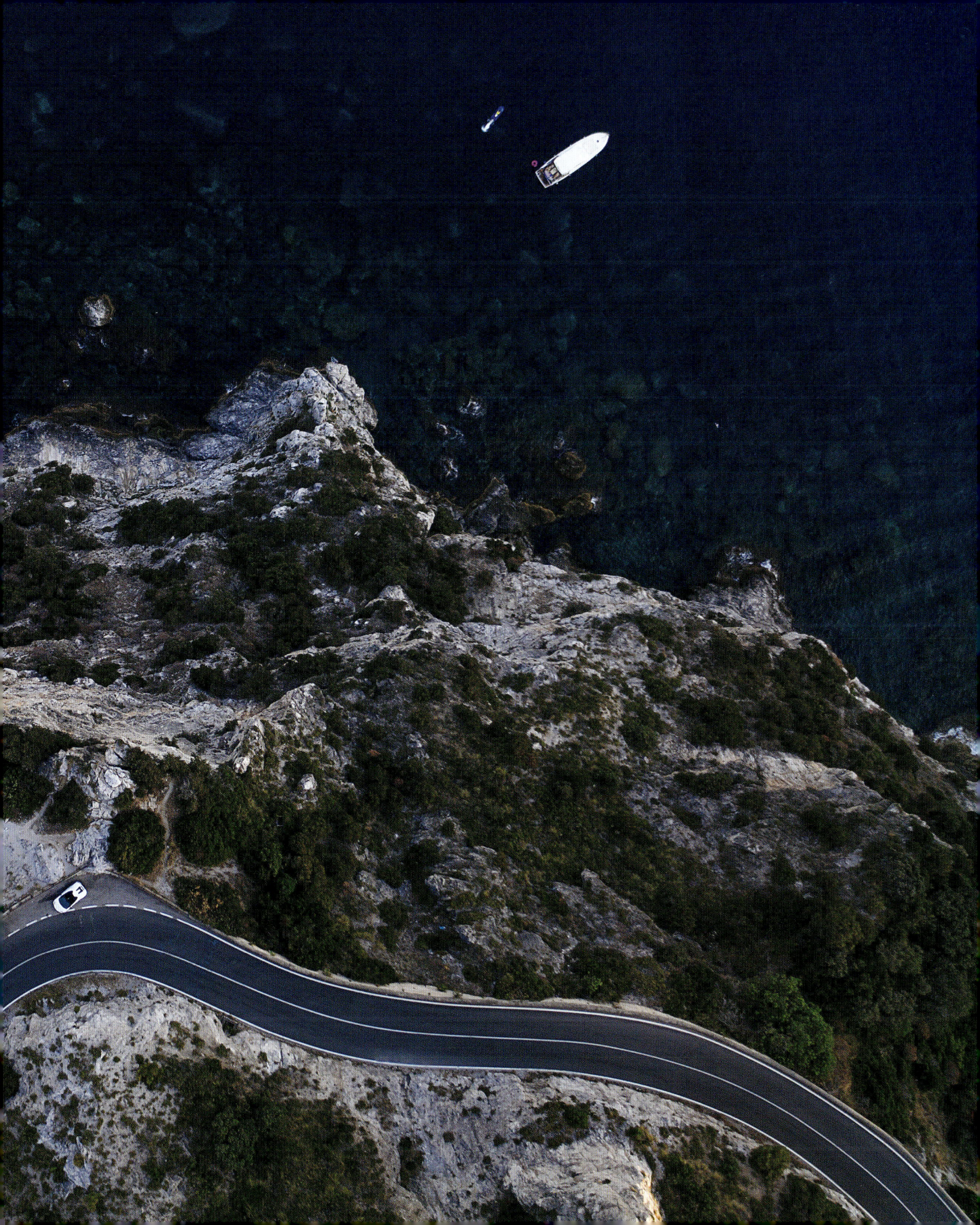

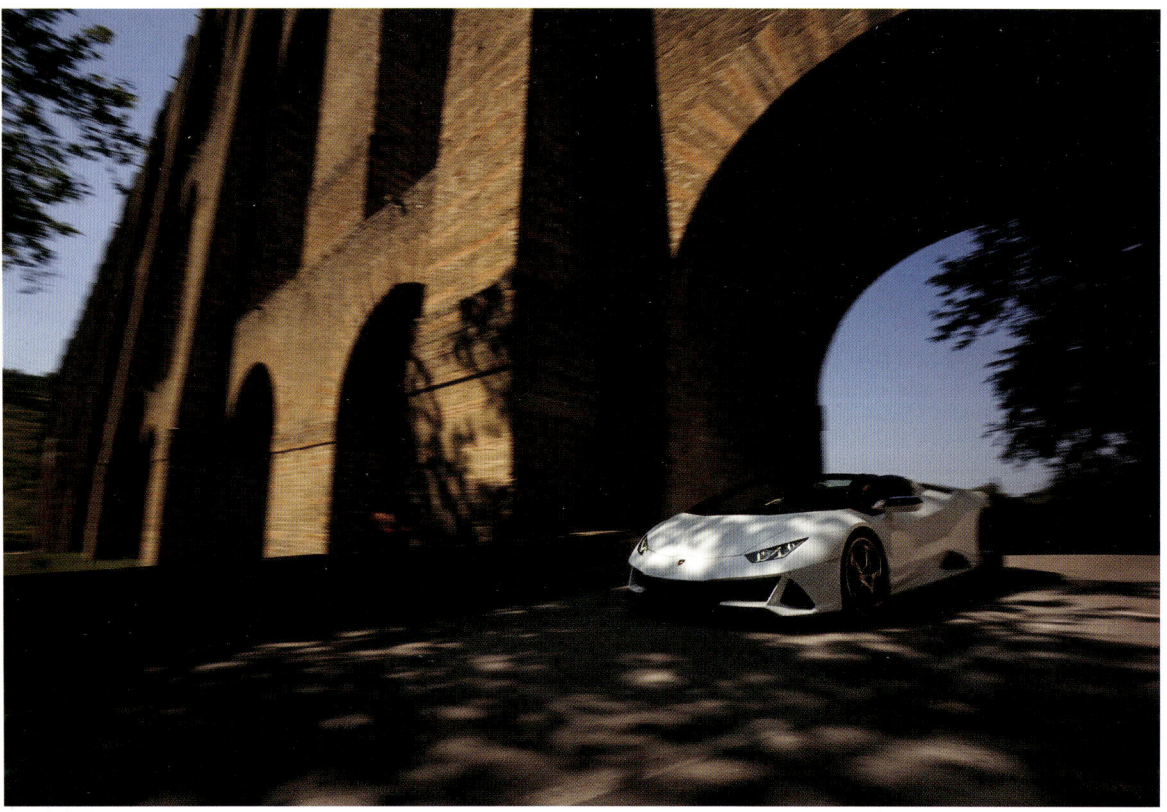

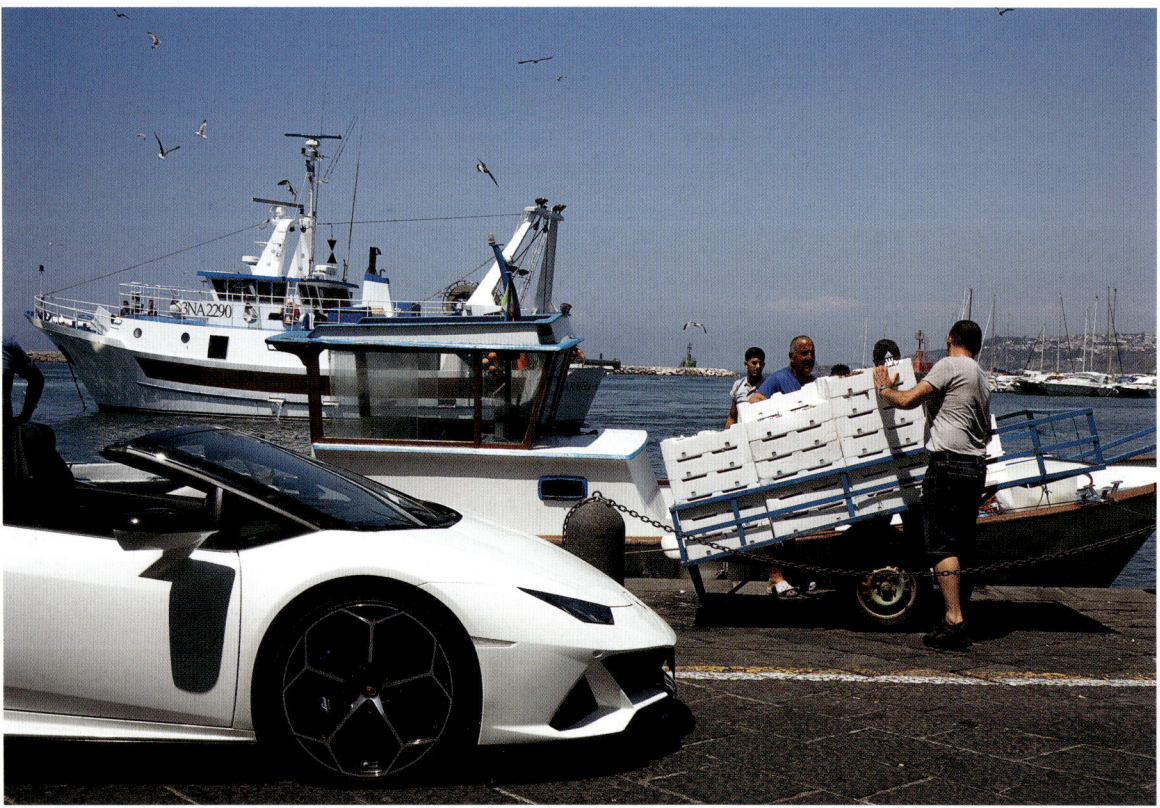

Carolingian aqueduct, Bucciano, Benevento
Procida Port, Naples

La Casina Vanvitelliana, Bacoli, Naples

Pages 90 and 91 Naples seafront, taralli-sellers. Amalfi Port

Pages 92–93 View of Naples and municipalities around Vesuvius

"In typical documentary photography style, I was keen to portray the heterogeneity of the landscapes, all of the urban and suburban architectural stratifications"

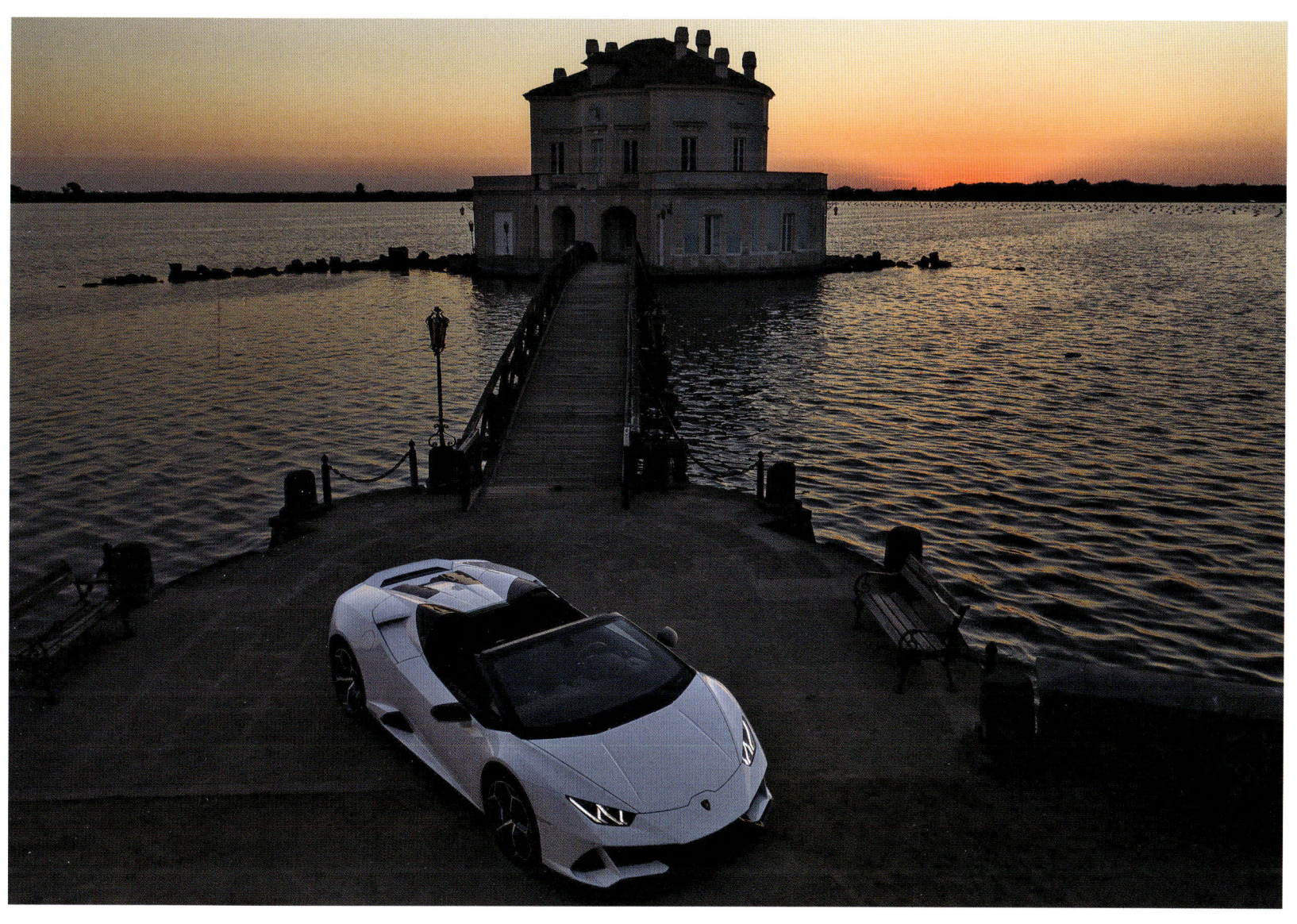

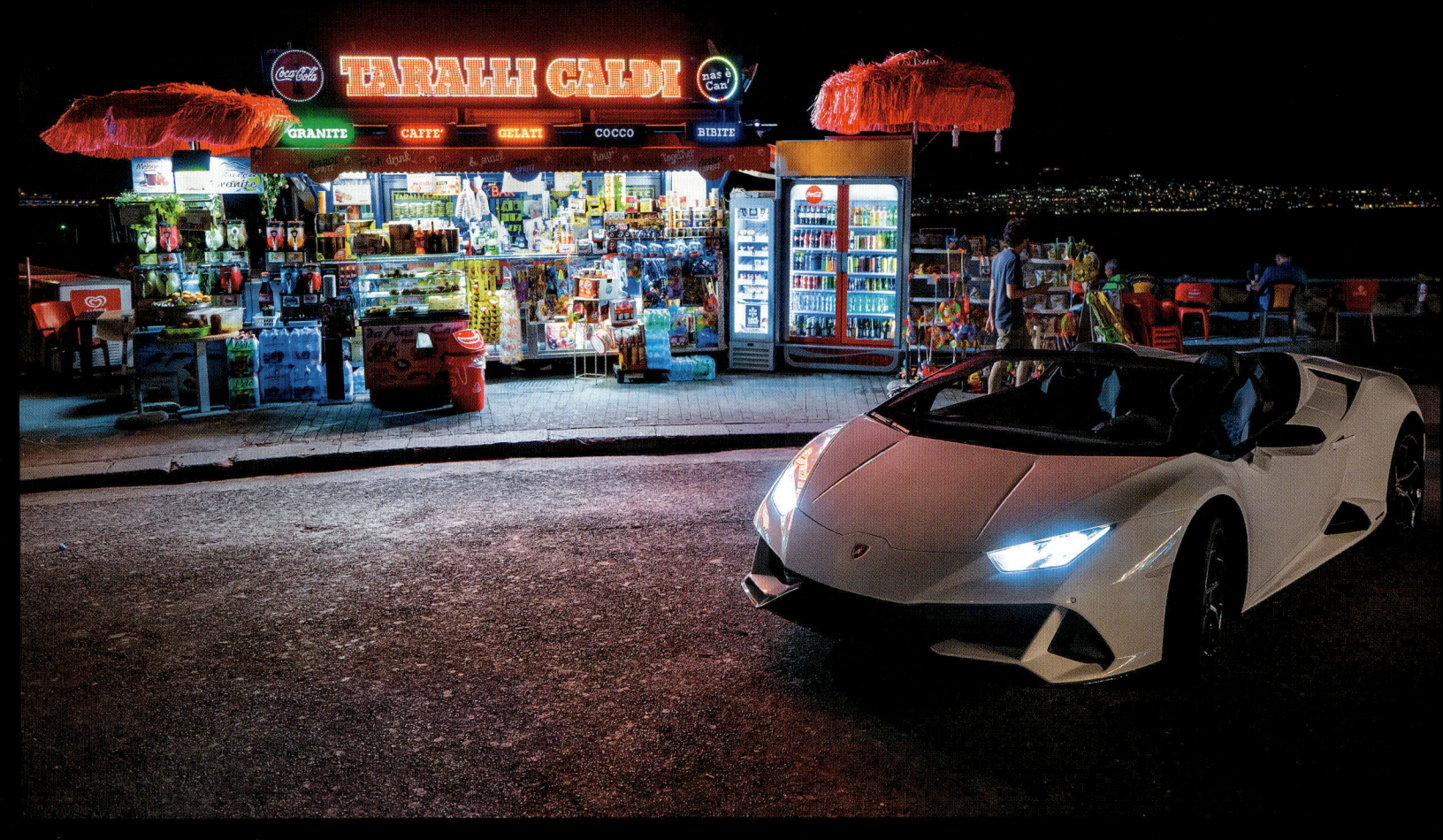

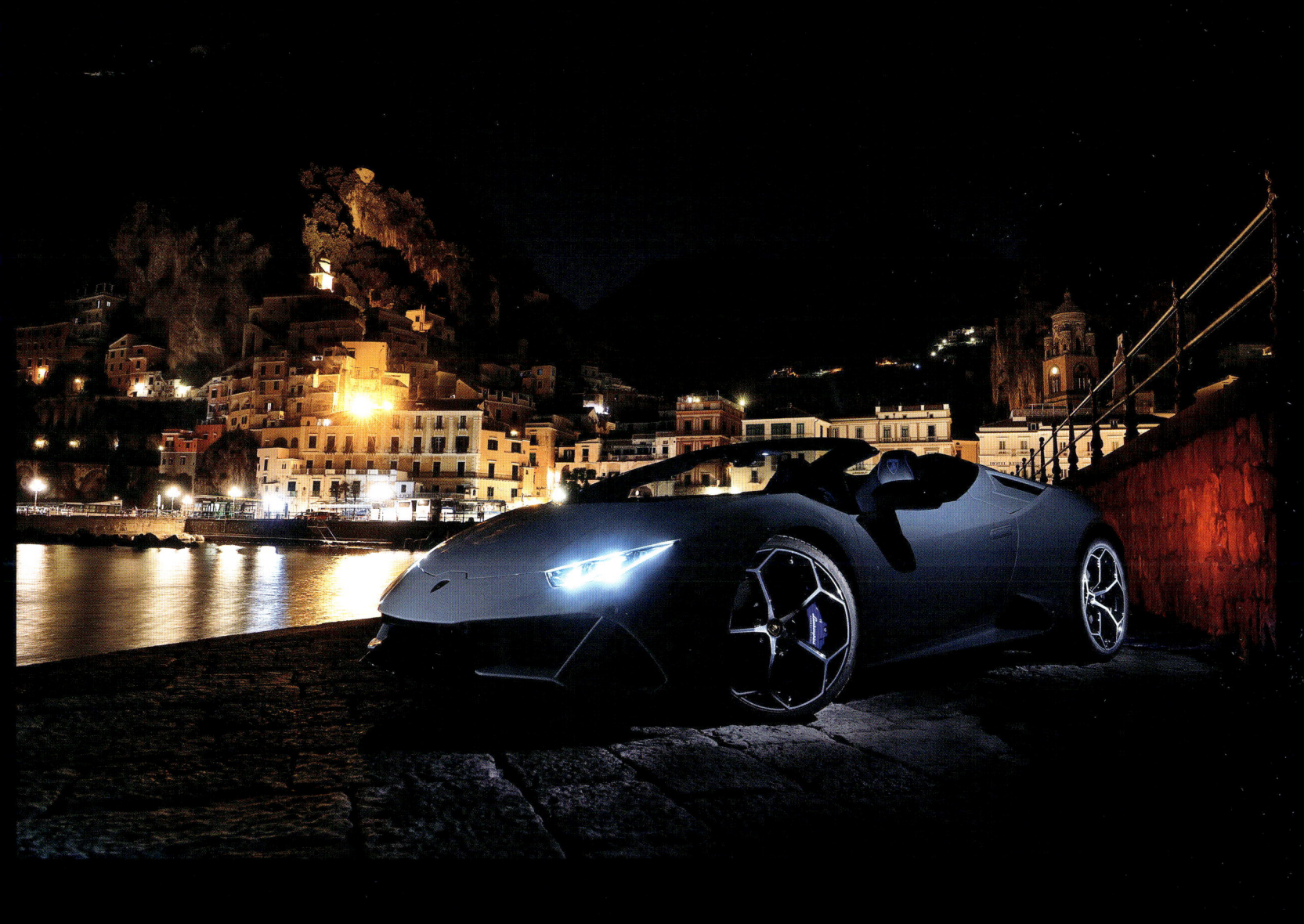

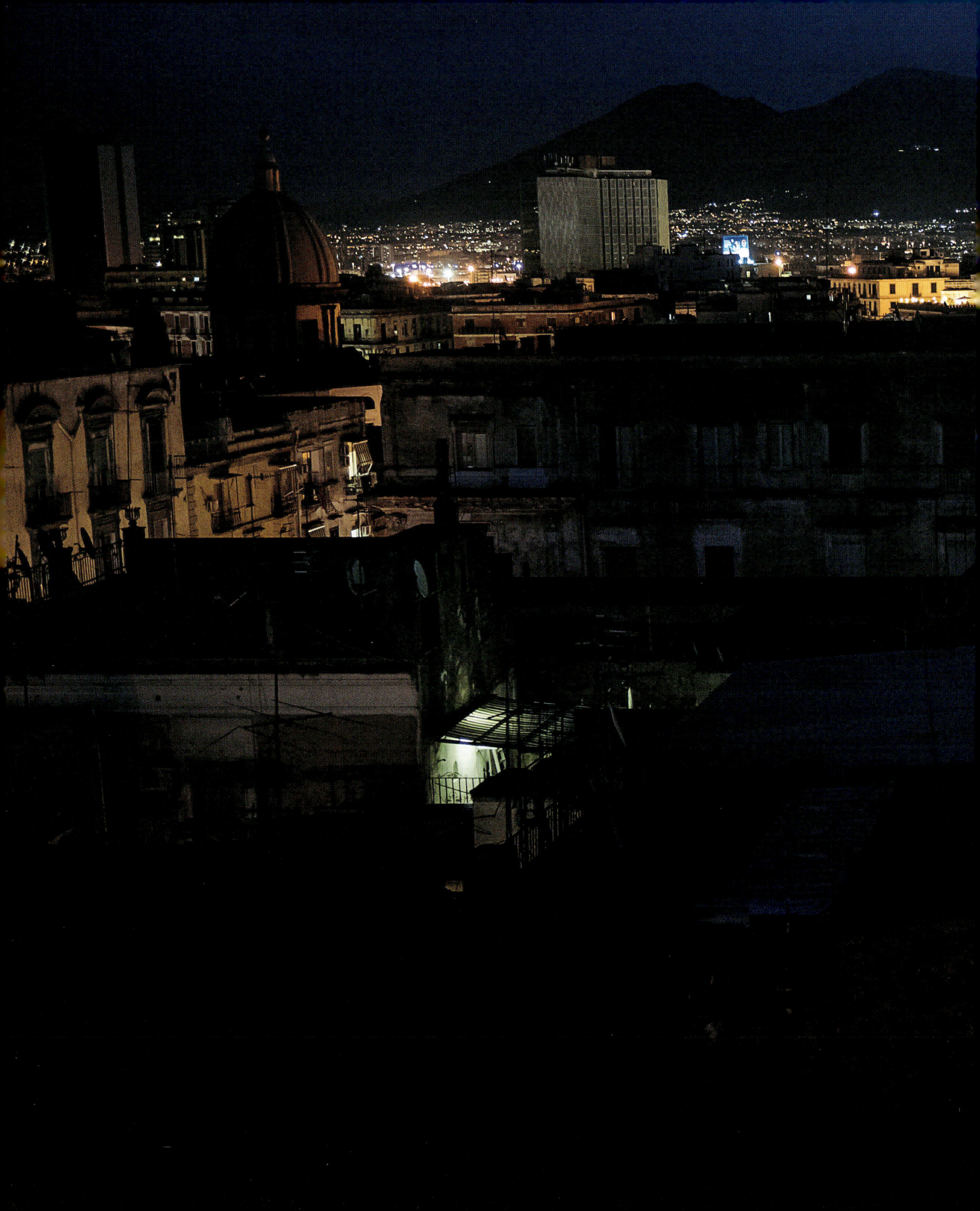

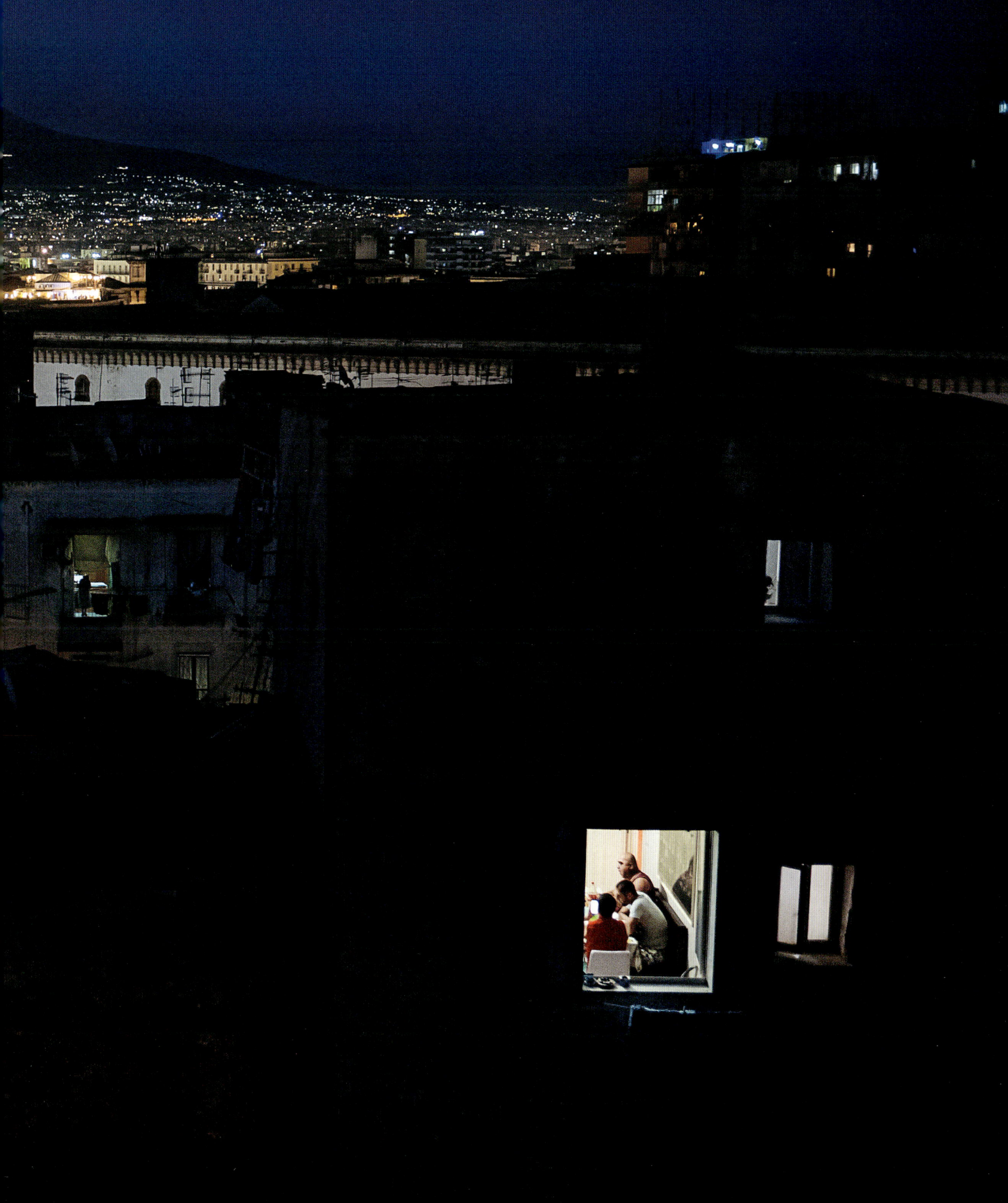

ROSELENA RAMISTELLA
MOLISE

I wanted to tell the story of Molise and the vast variety of its landscape. I drove the Lamborghini Urus through a multitude of different backdrops, from the woods to the sea, from wheat- and sunflower-filled valleys to archaeological sites, embarking on a social exploration of a region with a low demographic presence. I chose to portray the younger members of the population, teenagers, who represent the vital energy of a part of Italy that is gradually losing its population. "I want to stay here" is the phrase I hear most often from these boys and girls, their declaration of love and hope for a land that, like the Lamborghini, is built around a powerful engine.

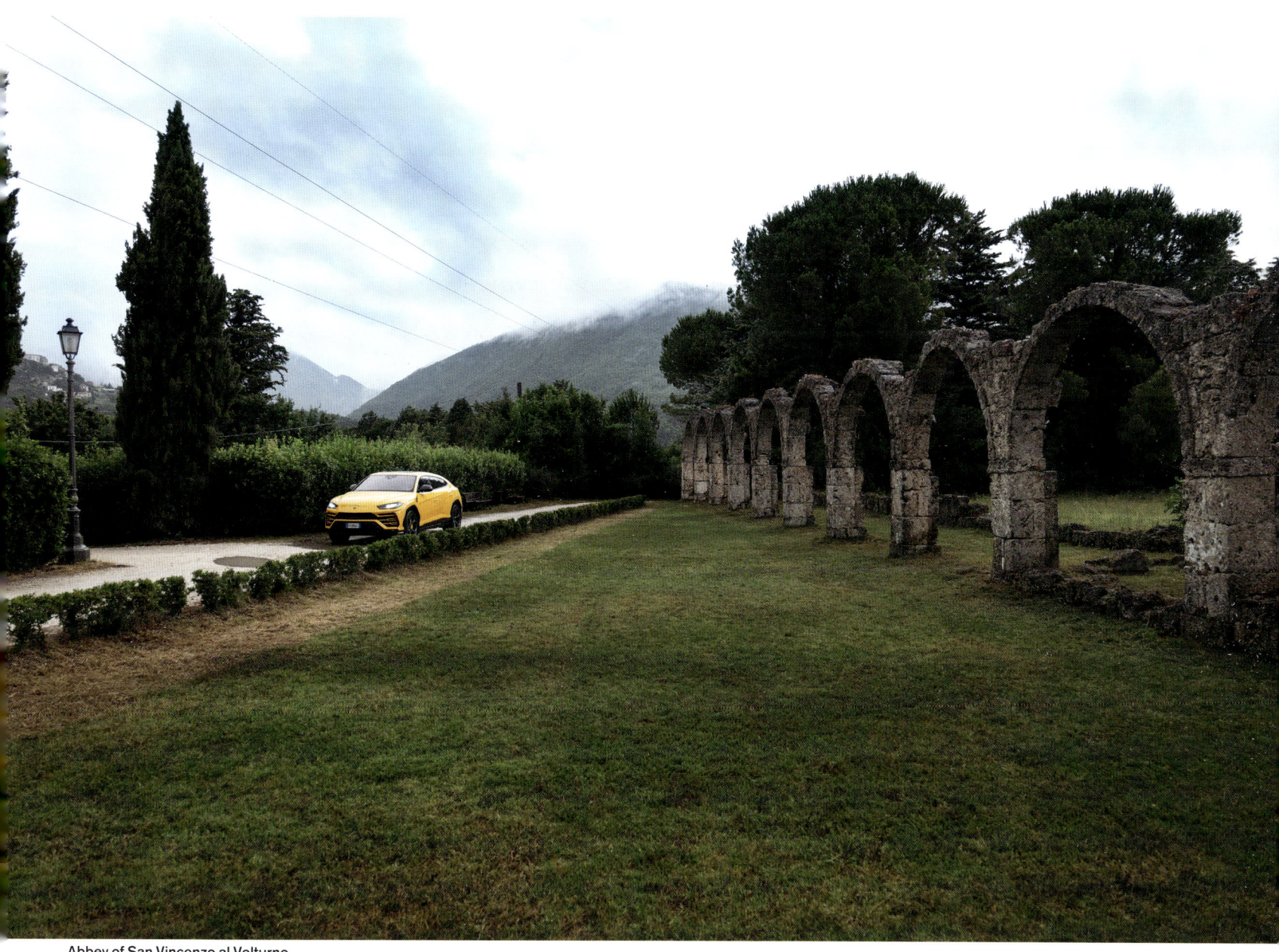
Abbey of San Vincenzo al Volturno

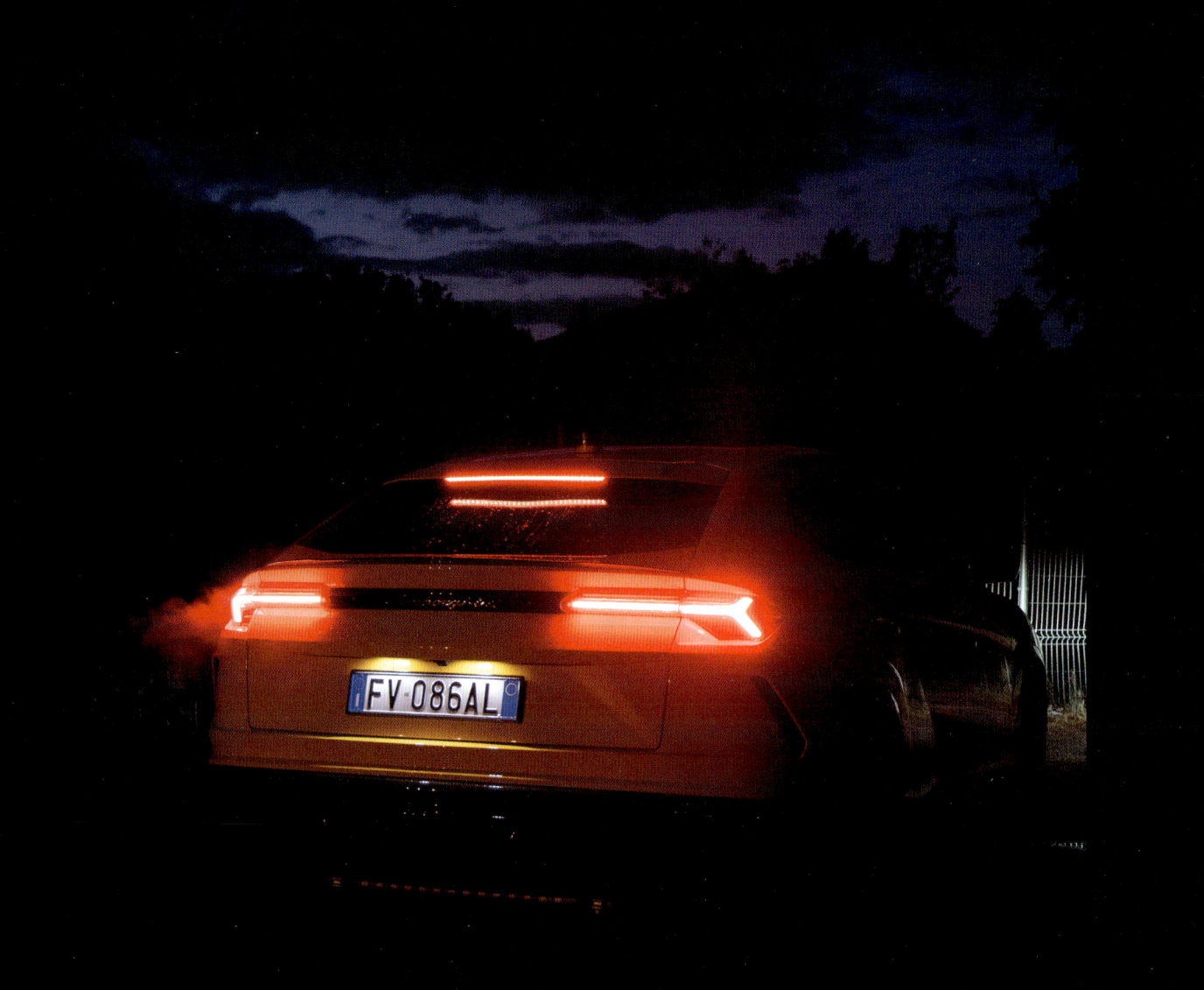

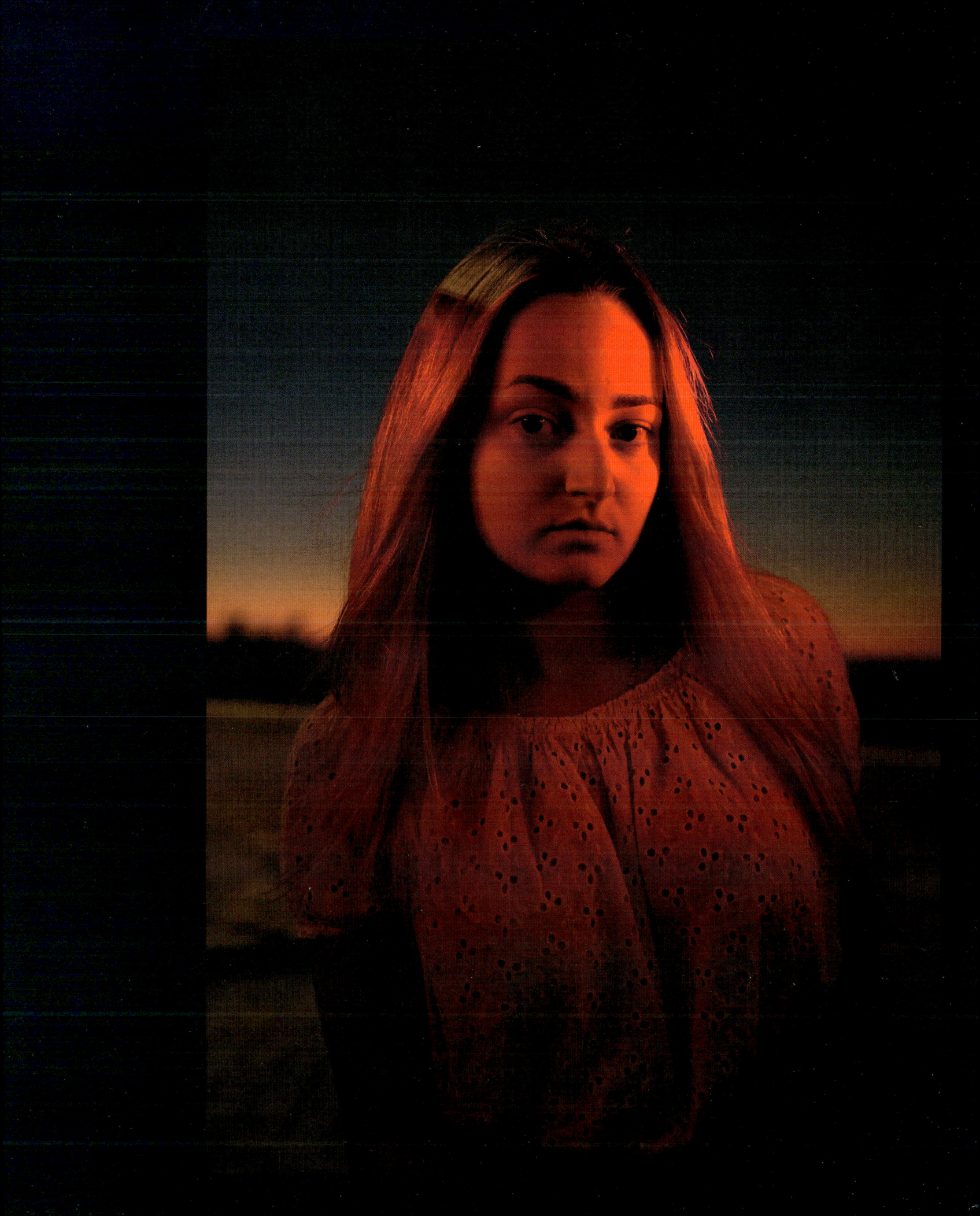

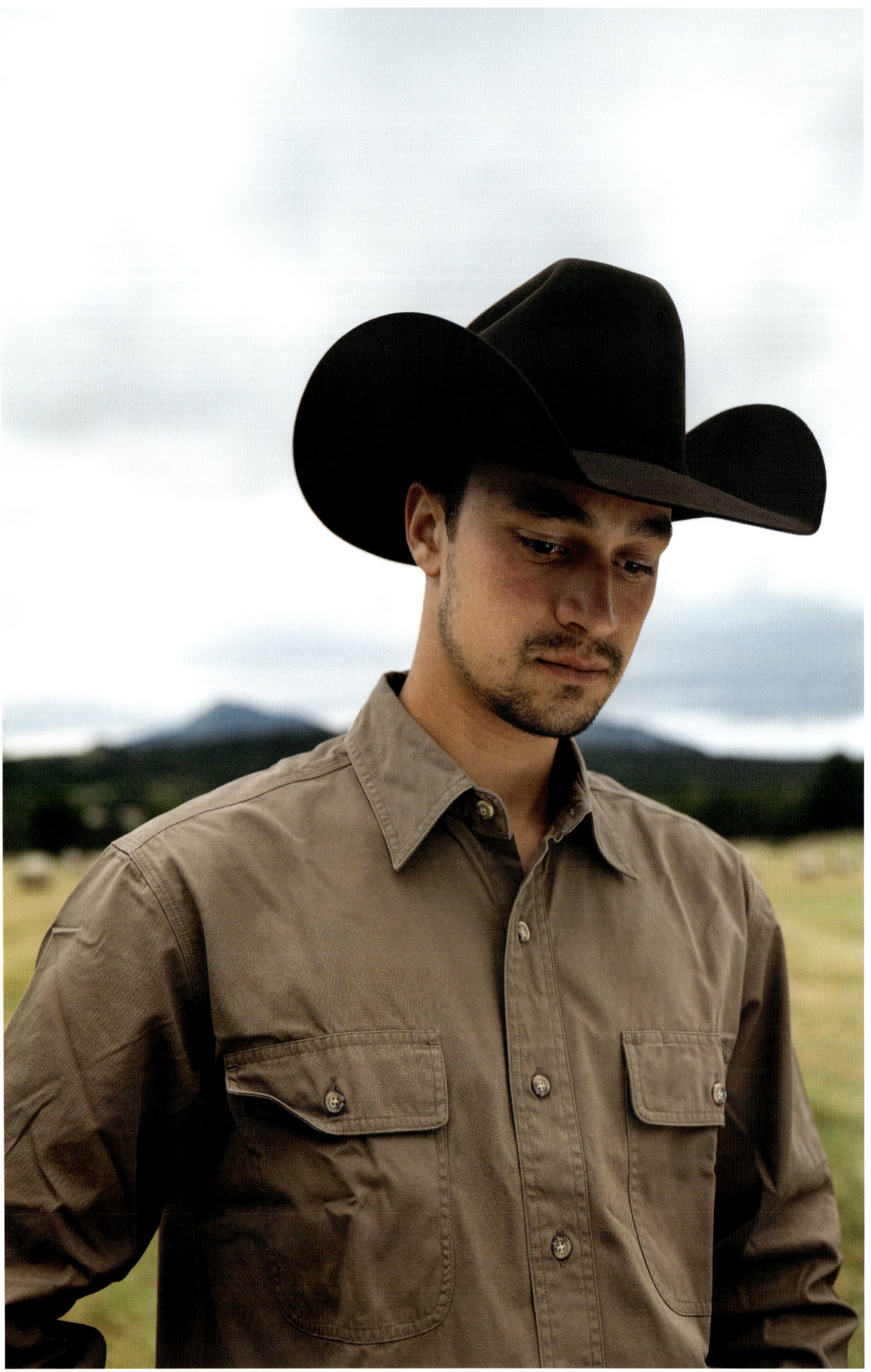

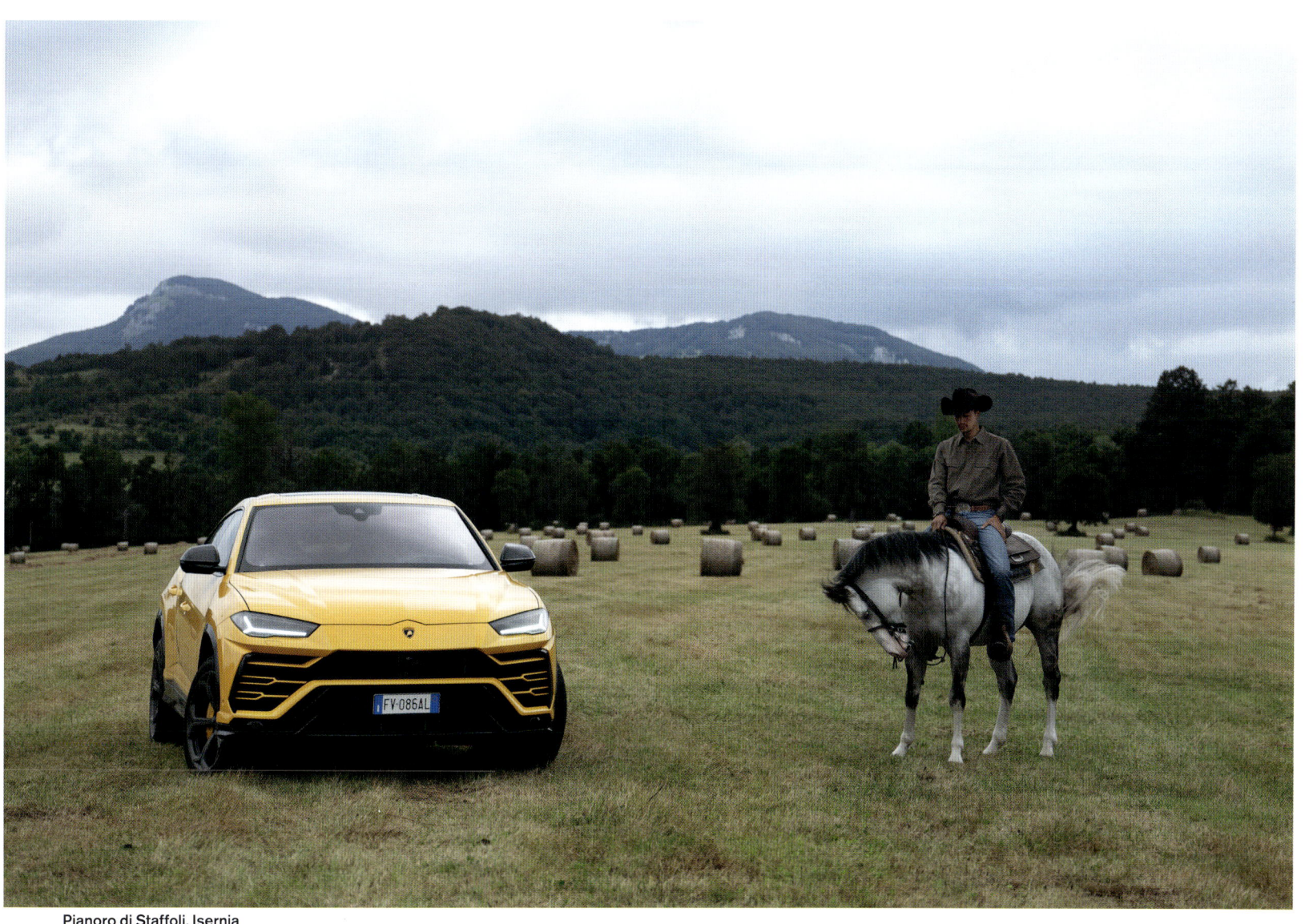
Pianoro di Staffoli, Isernia

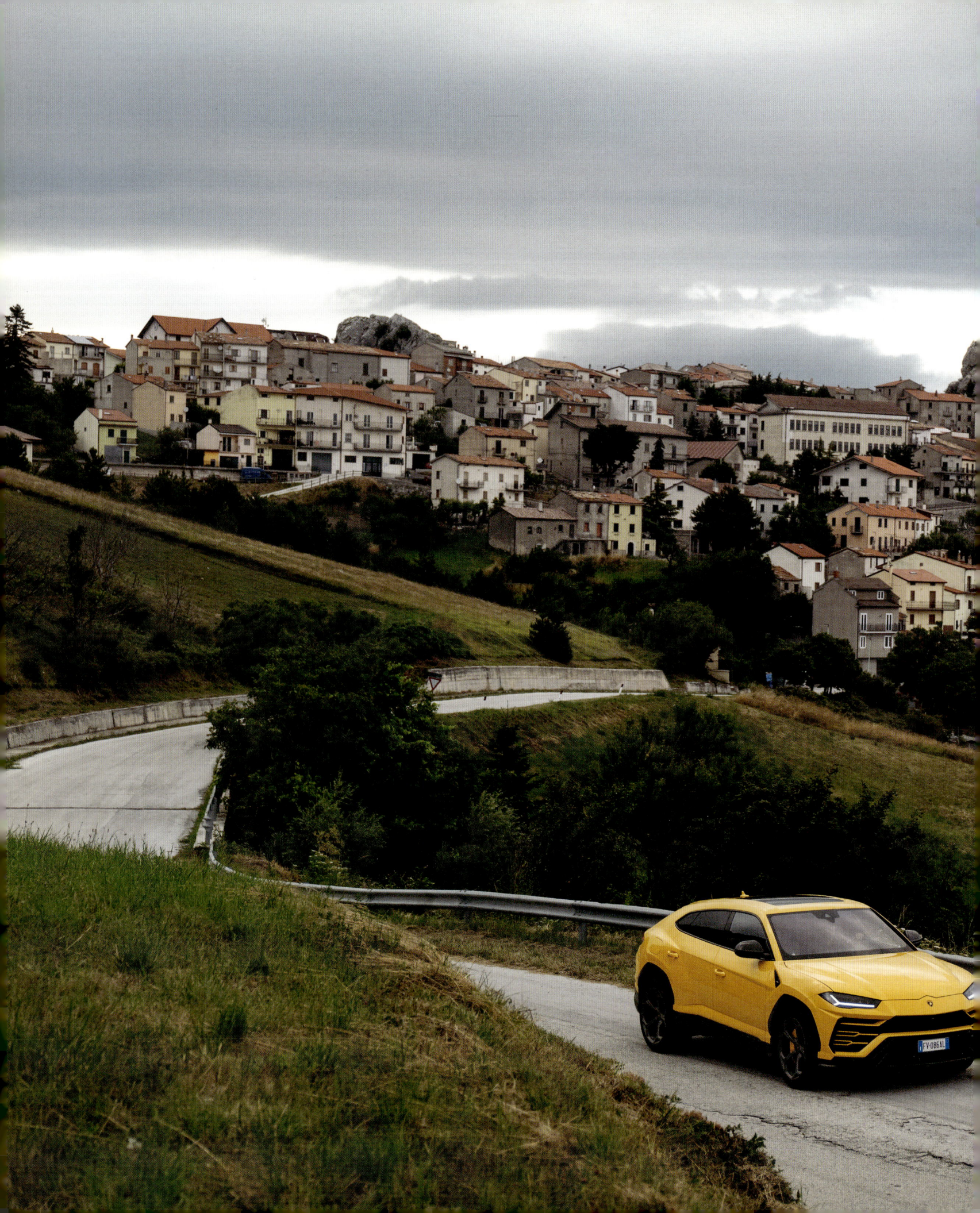

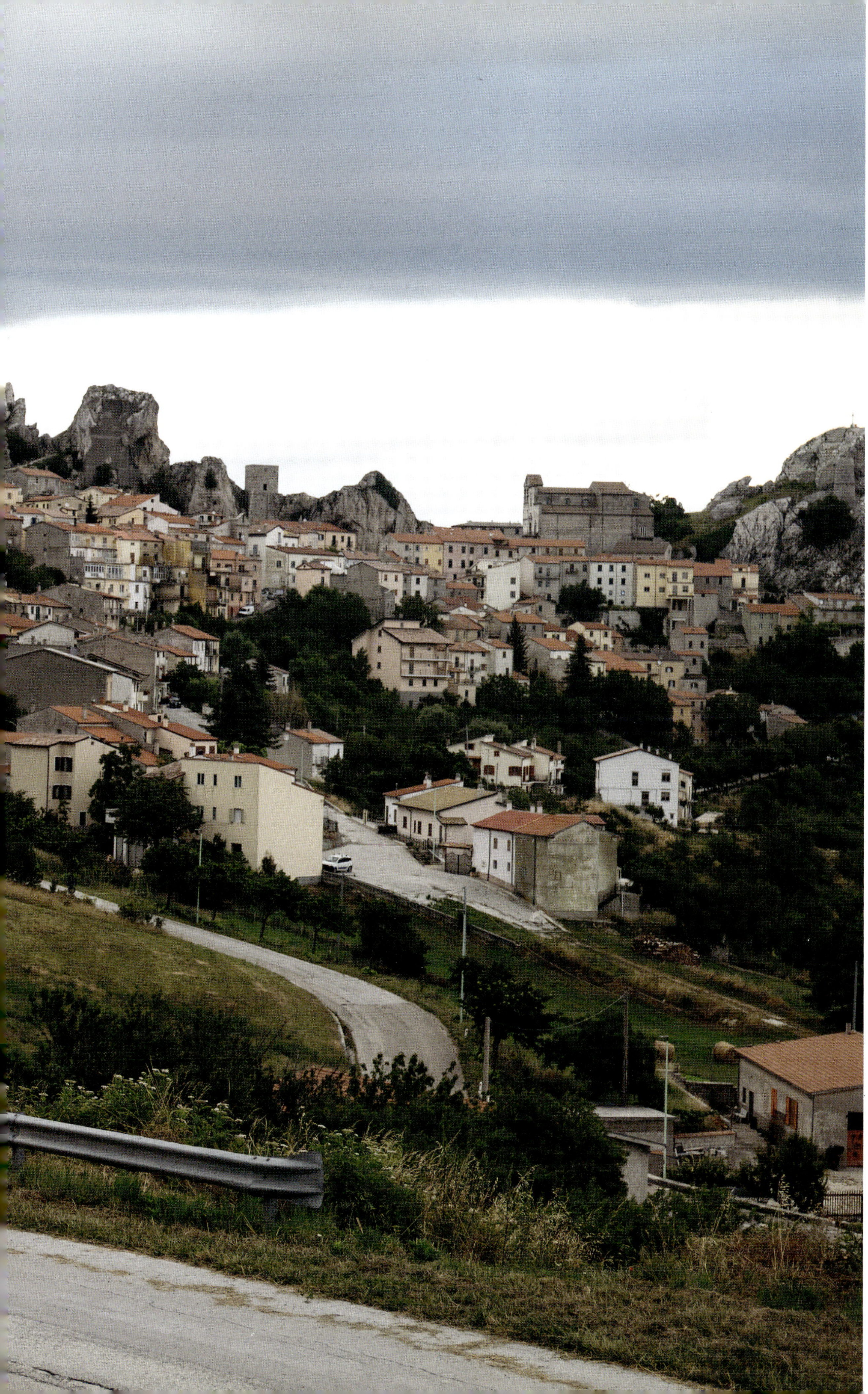

Pietrabbondante, Isernia

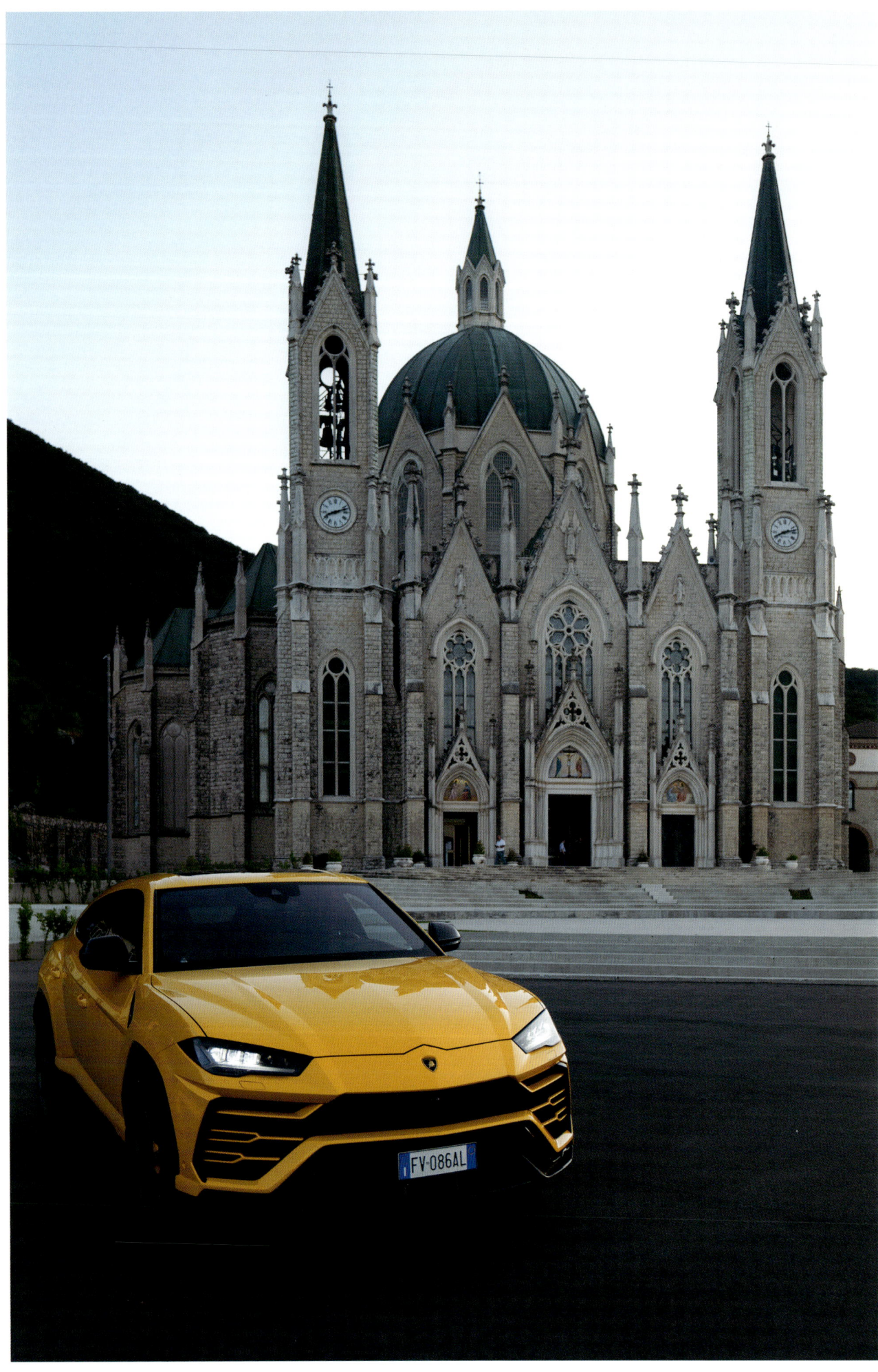

Sanctuary of Our Lady of Sorrows, Isernia

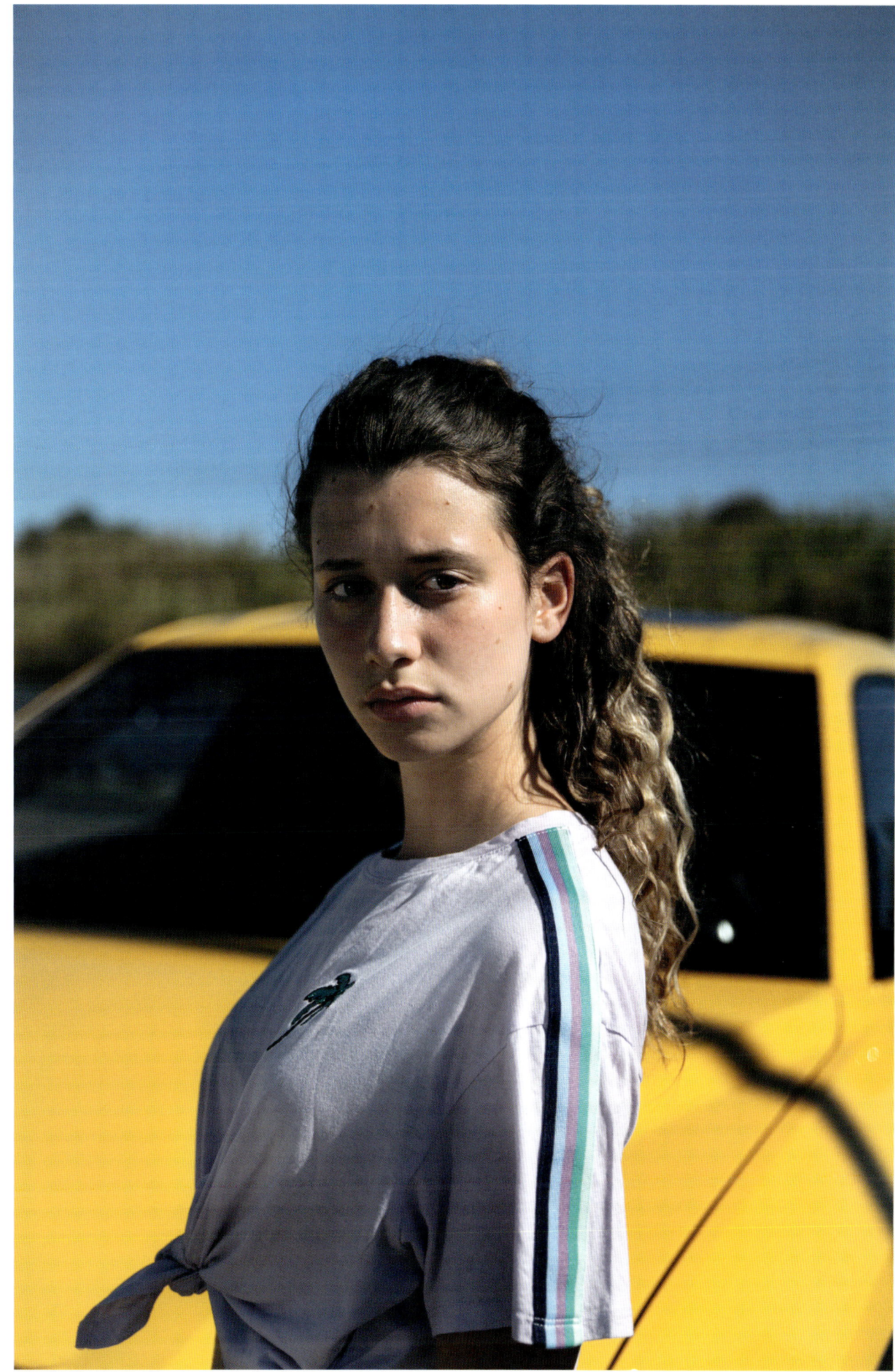

Colombo seafront,
Termoli

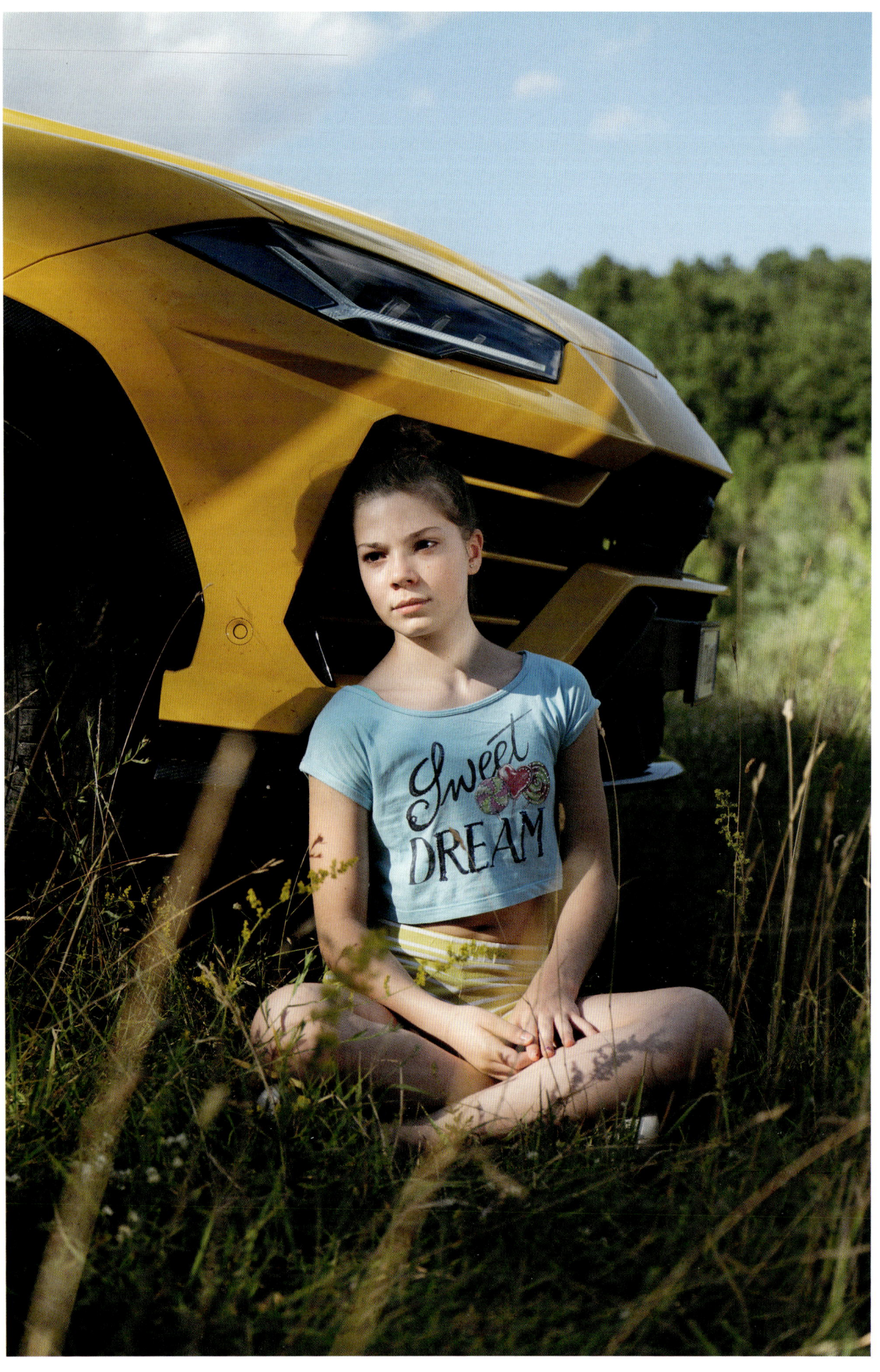

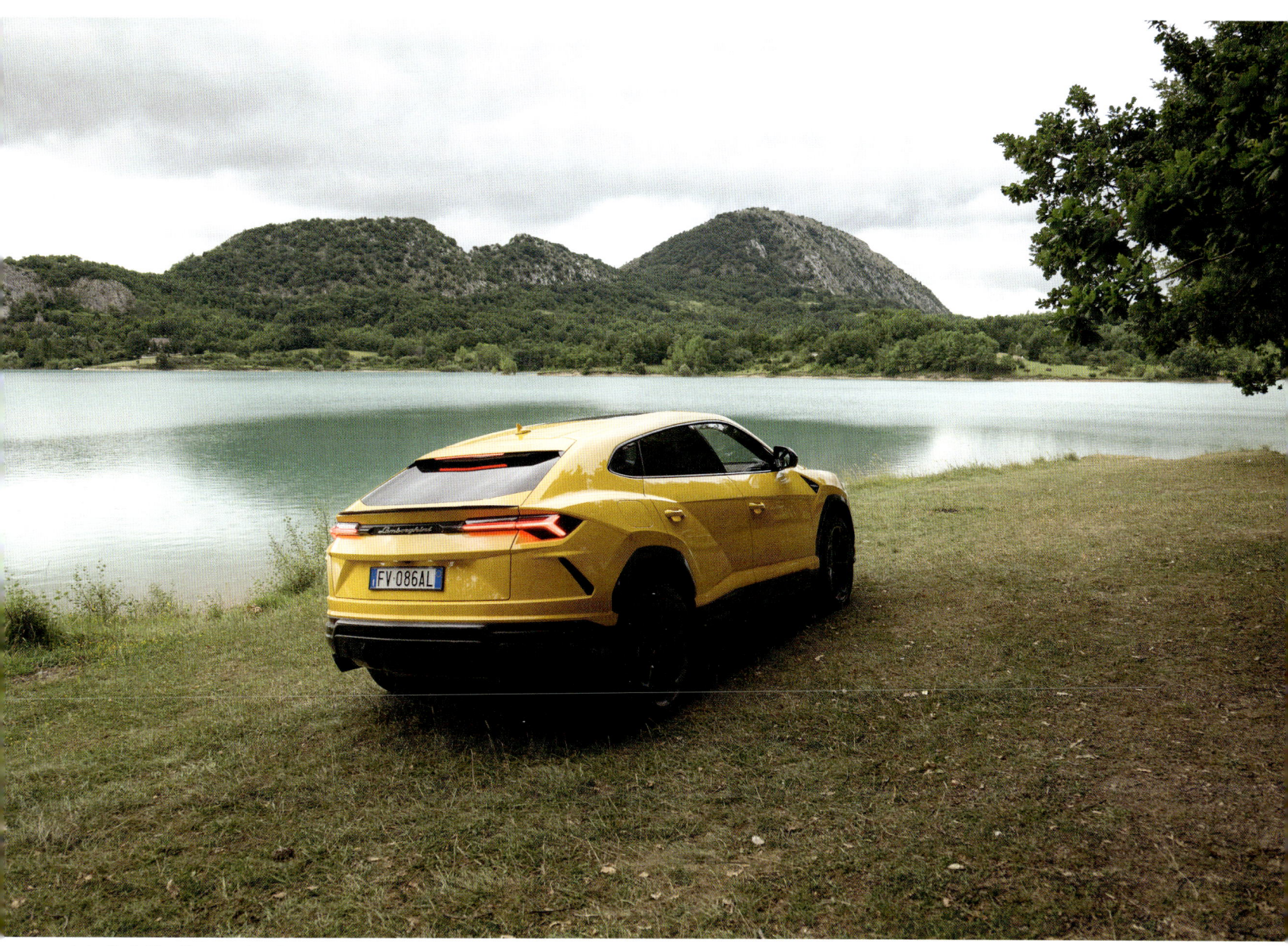

Lake Castel San Vincenzo, Isernia

"'I want to stay here' is the phrase
I hear most often from these boys and girls,
their declaration of love and hope
for a land that, like the Lamborghini,
is built around a powerful engine"

VALENTINA SOMMARIVA
ABRUZZO

The beauty of Abruzzo lies partly in its isolation. It is a place still largely to be discovered, with powerful nature and landscapes so vast you cannot make out where they end, mapping out in the collective imagination the concept of some new world. This futuristic vision of Abruzzo is what inspired my narrative, starring two aliens on board an unusual spacecraft, a Huracán EVO Spyder that accidentally wound up on planet Earth. The unique appeal of the Campo Imperatore plateau, the Fucino Space Centre and areas of the Abruzzo hinterland provide the backdrop for a space story in a never-before-seen land, among strange life forms and attempts at exploration and acclimatization, searching for a space-time portal to return home.

Campo Imperatore: on the mistery planet

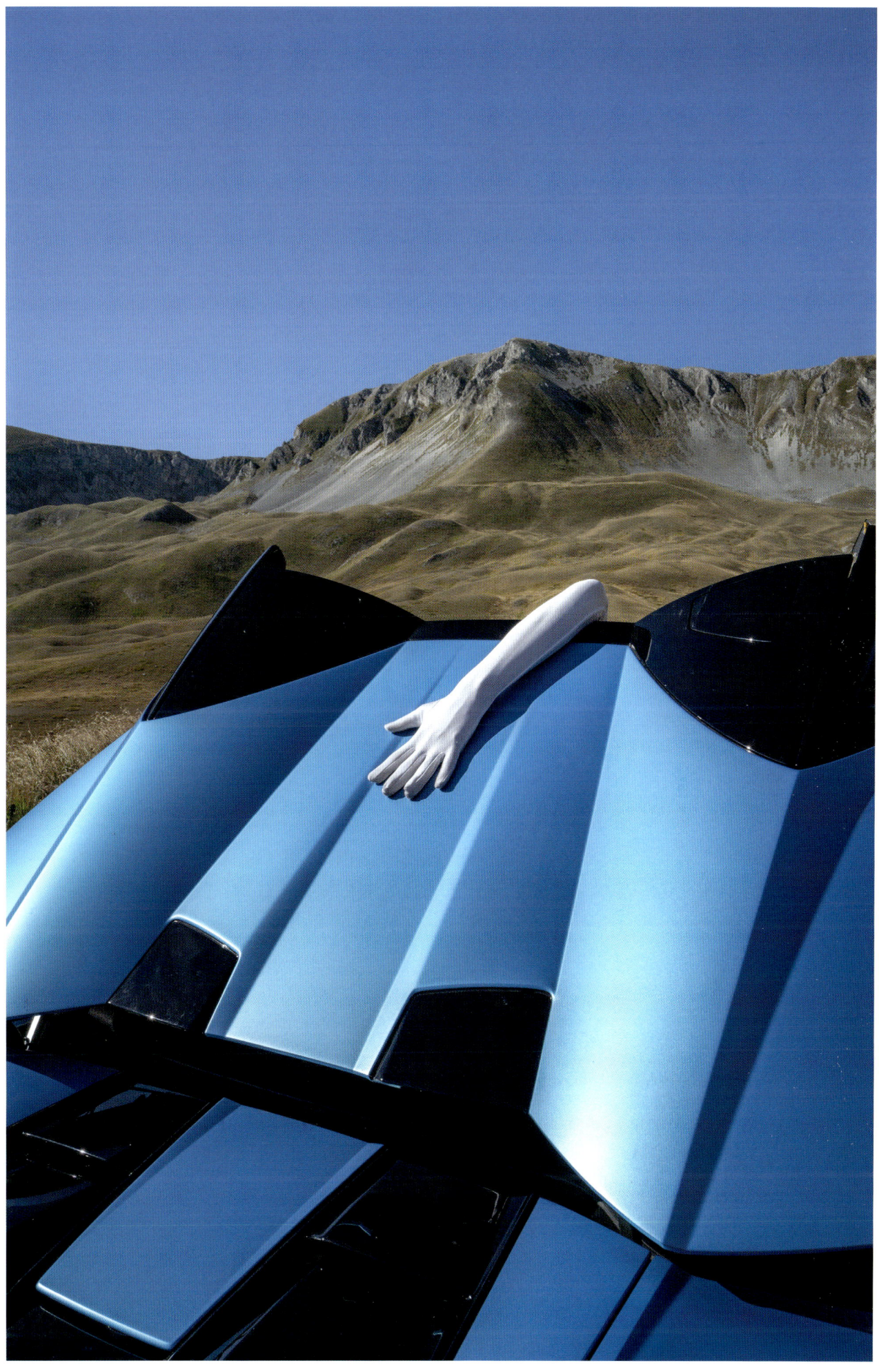

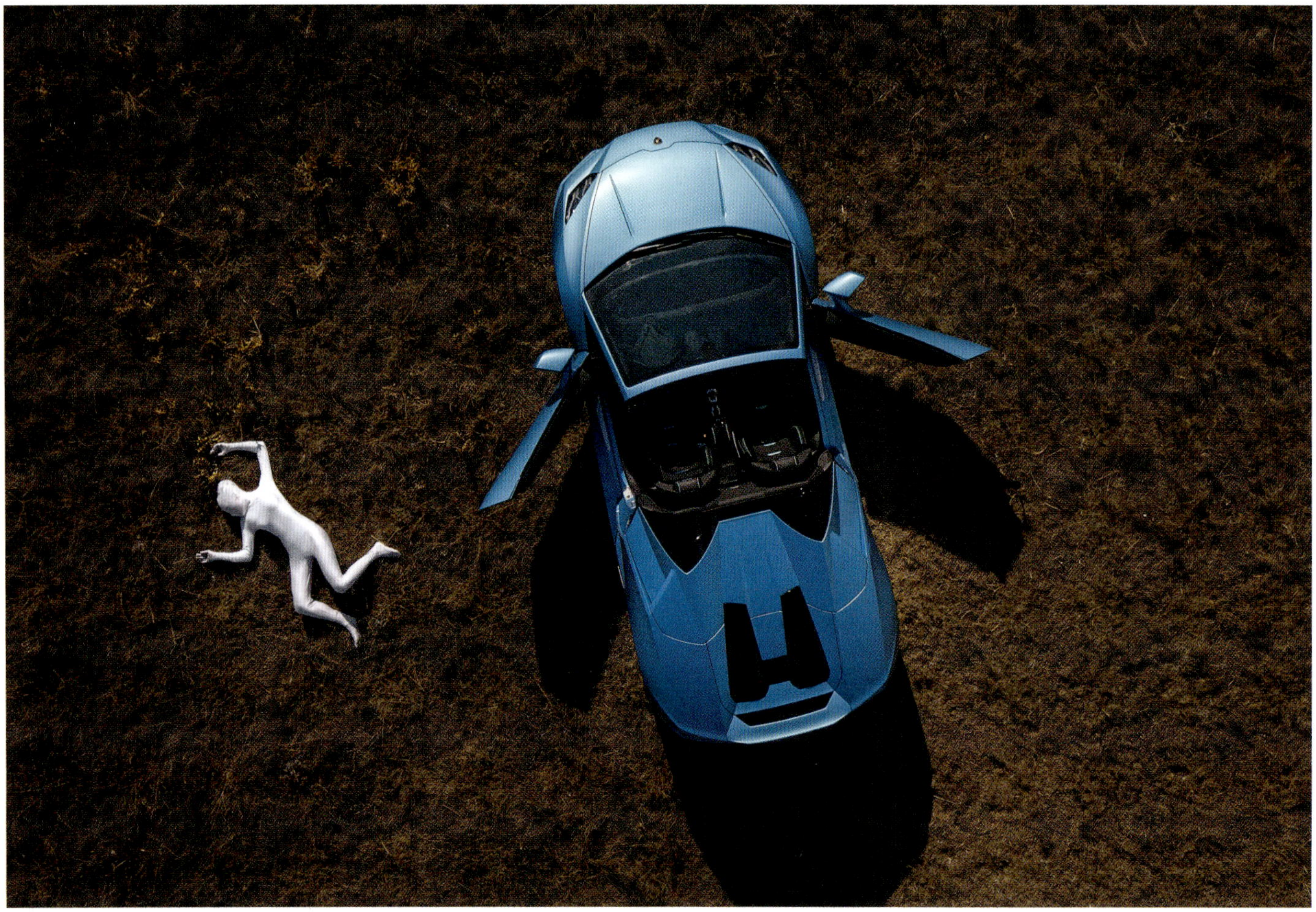

Campo Imperatore: impact and first attempt at imitation

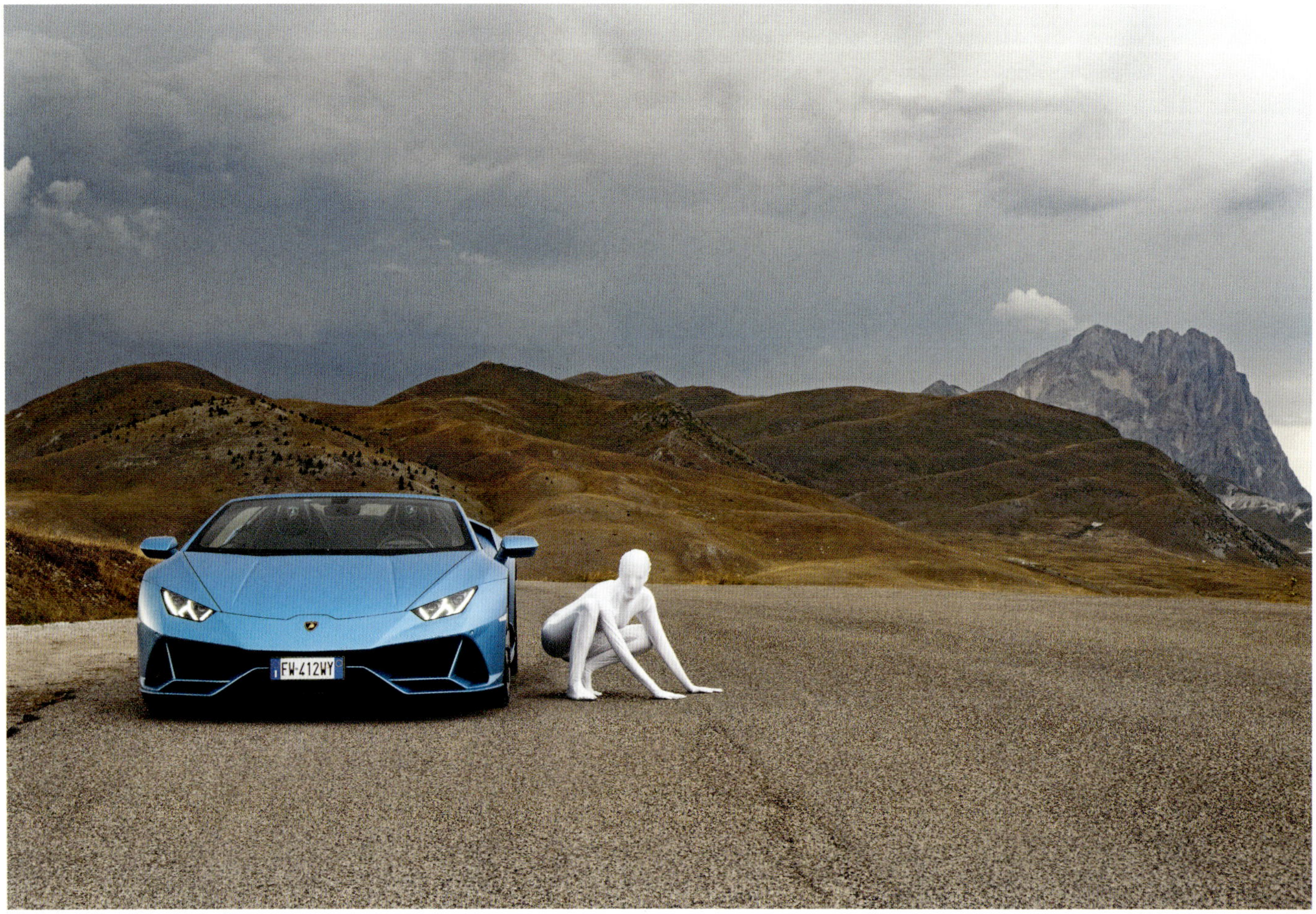

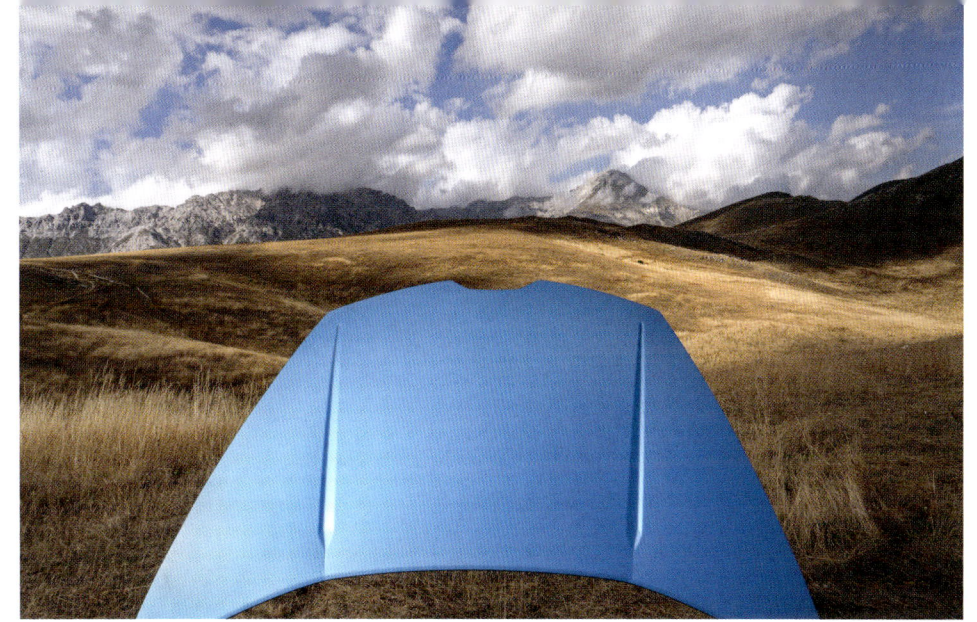

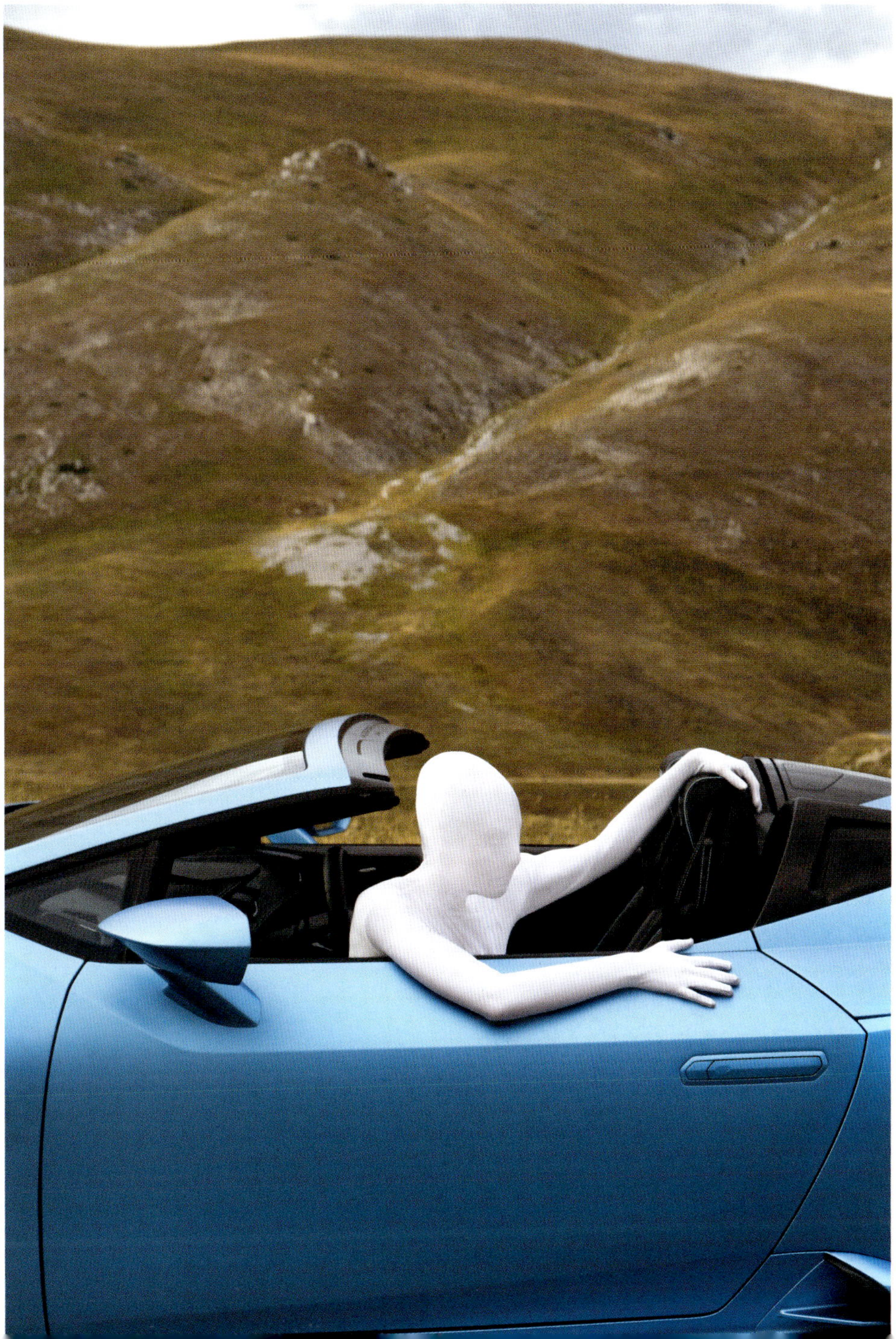

Campo Imperatore: flight test

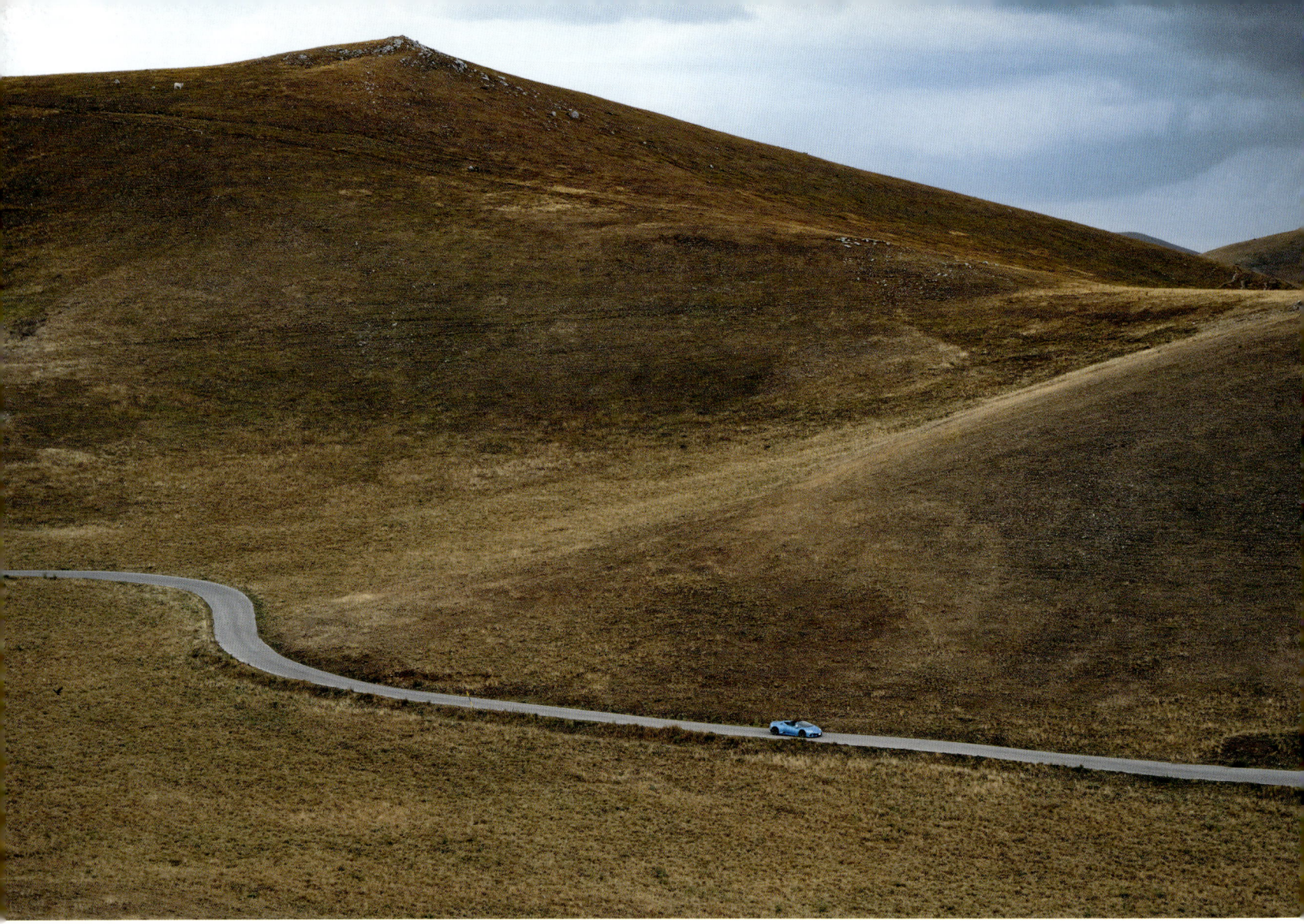
Campo Imperatore: cruising through space

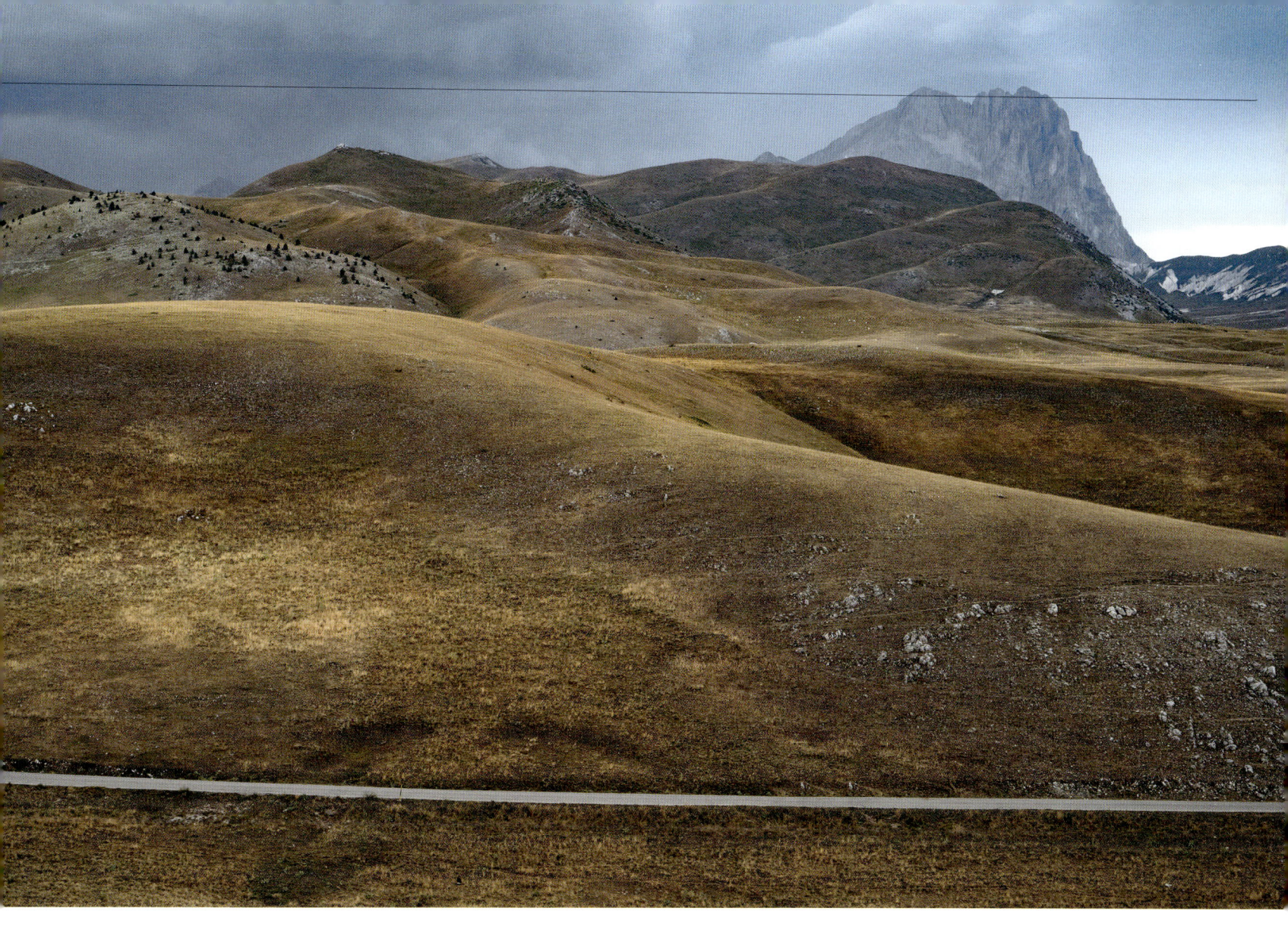

"The beauty of Abruzzo lies partly in its isolation. It is a place still largely to be discovered, with powerful nature and landscapes so vast you cannot make out where they end, mapping out in the collective imagination the concept of some new world"

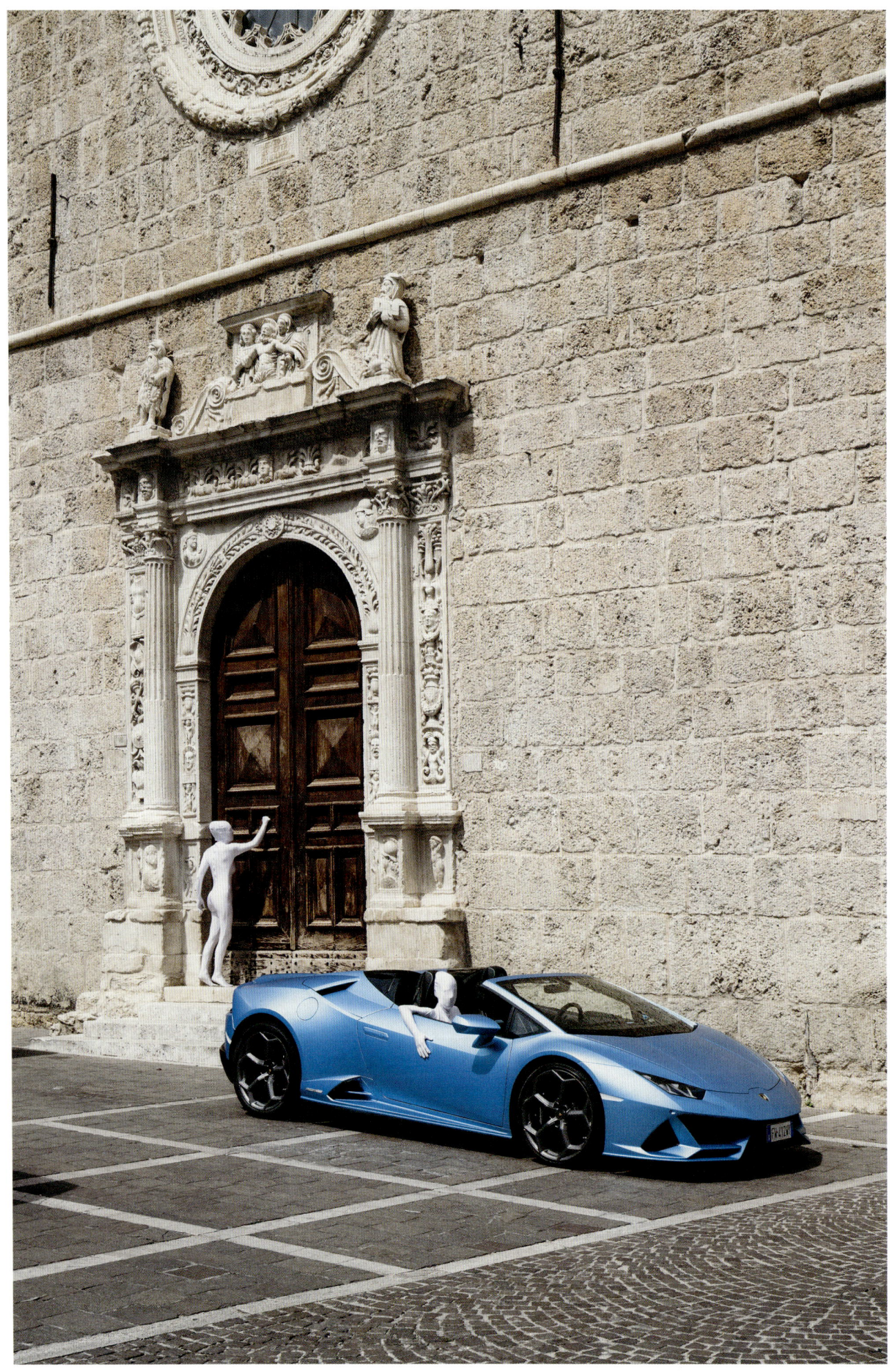

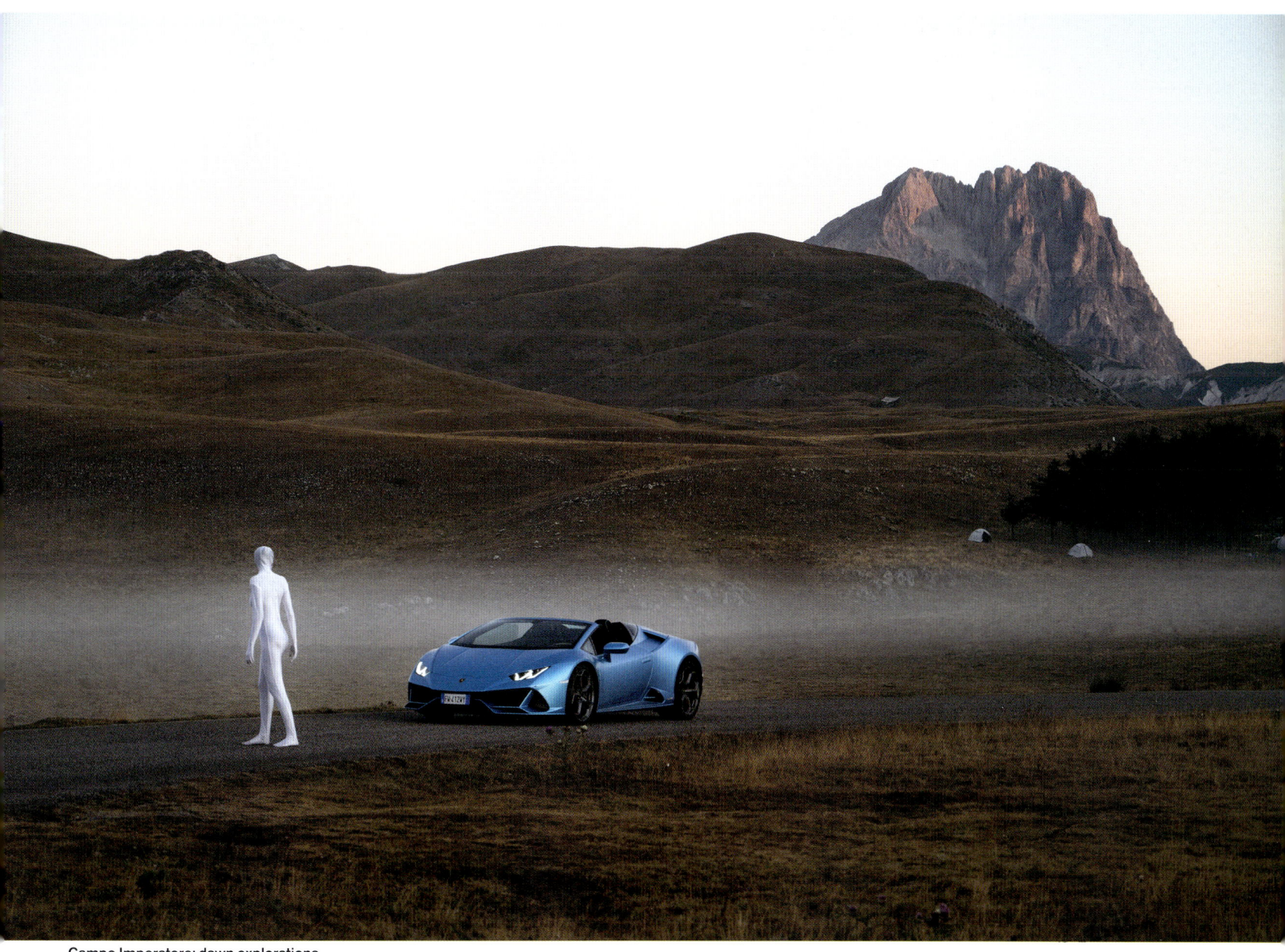

Campo Imperatore: dawn explorations

Facade of the Church of Santa Maria delle Grazie, Anversa degli Abruzzi: "Hello?"

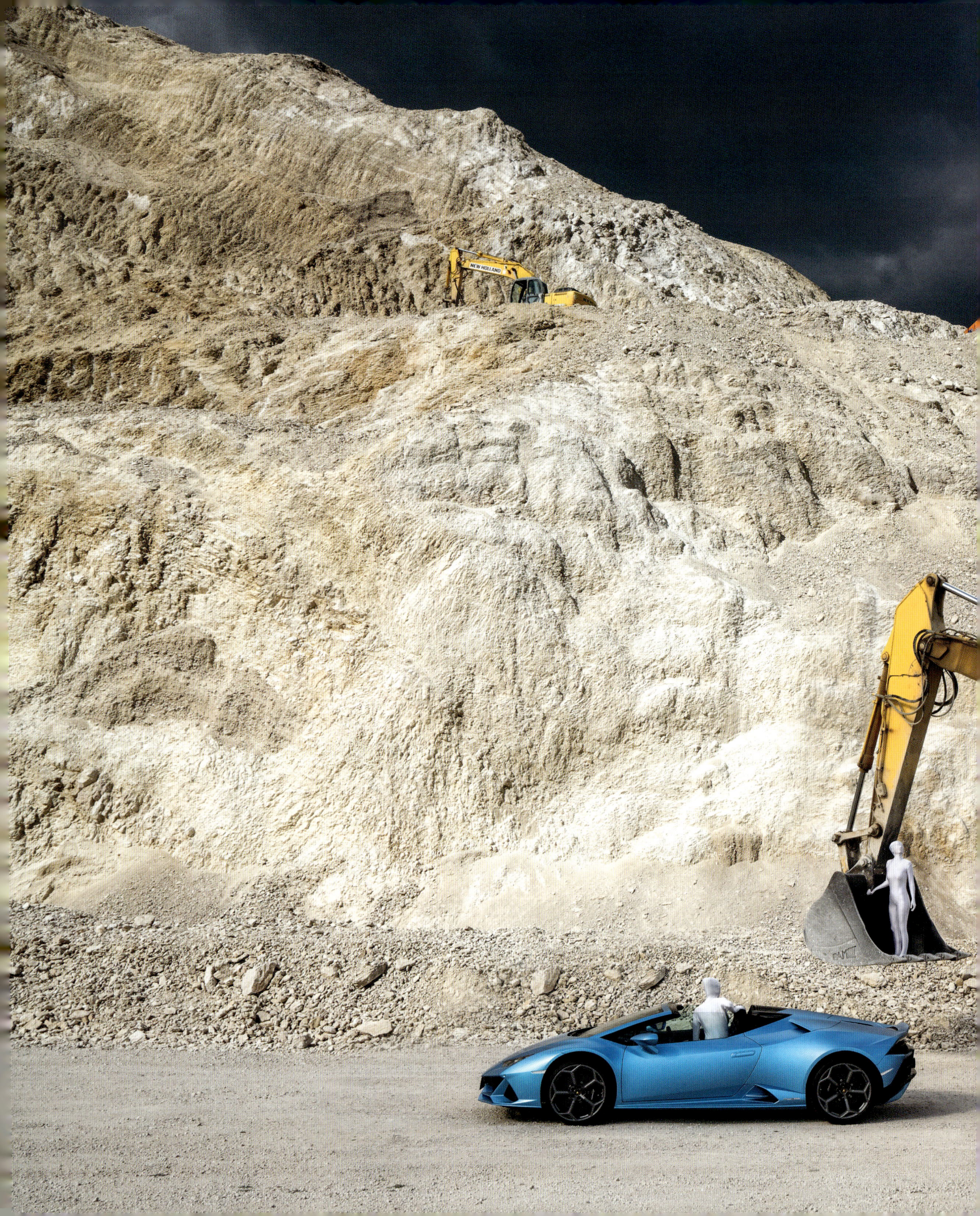

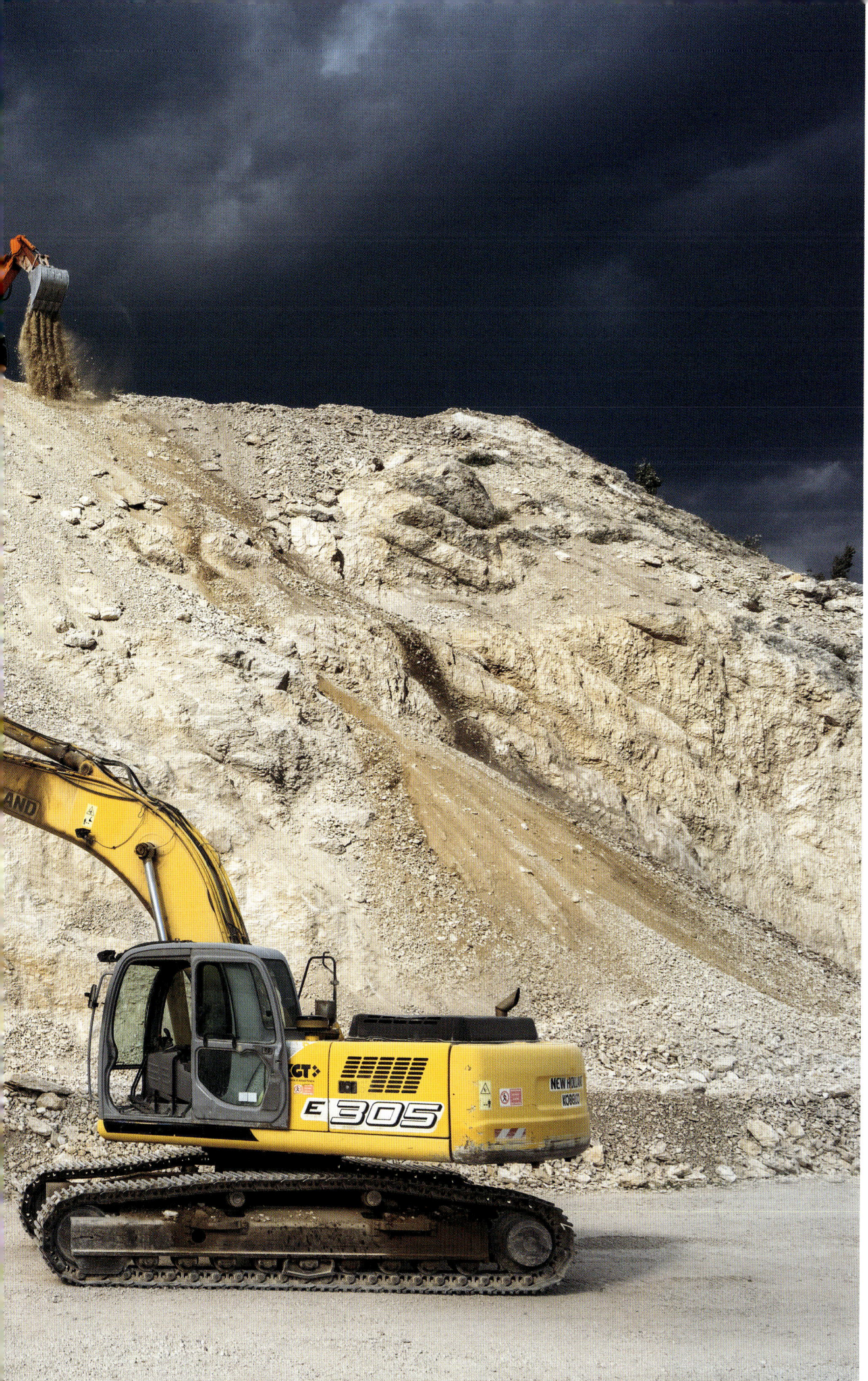

Quarry near Cocullo, L'Aquila: "What's down here?"

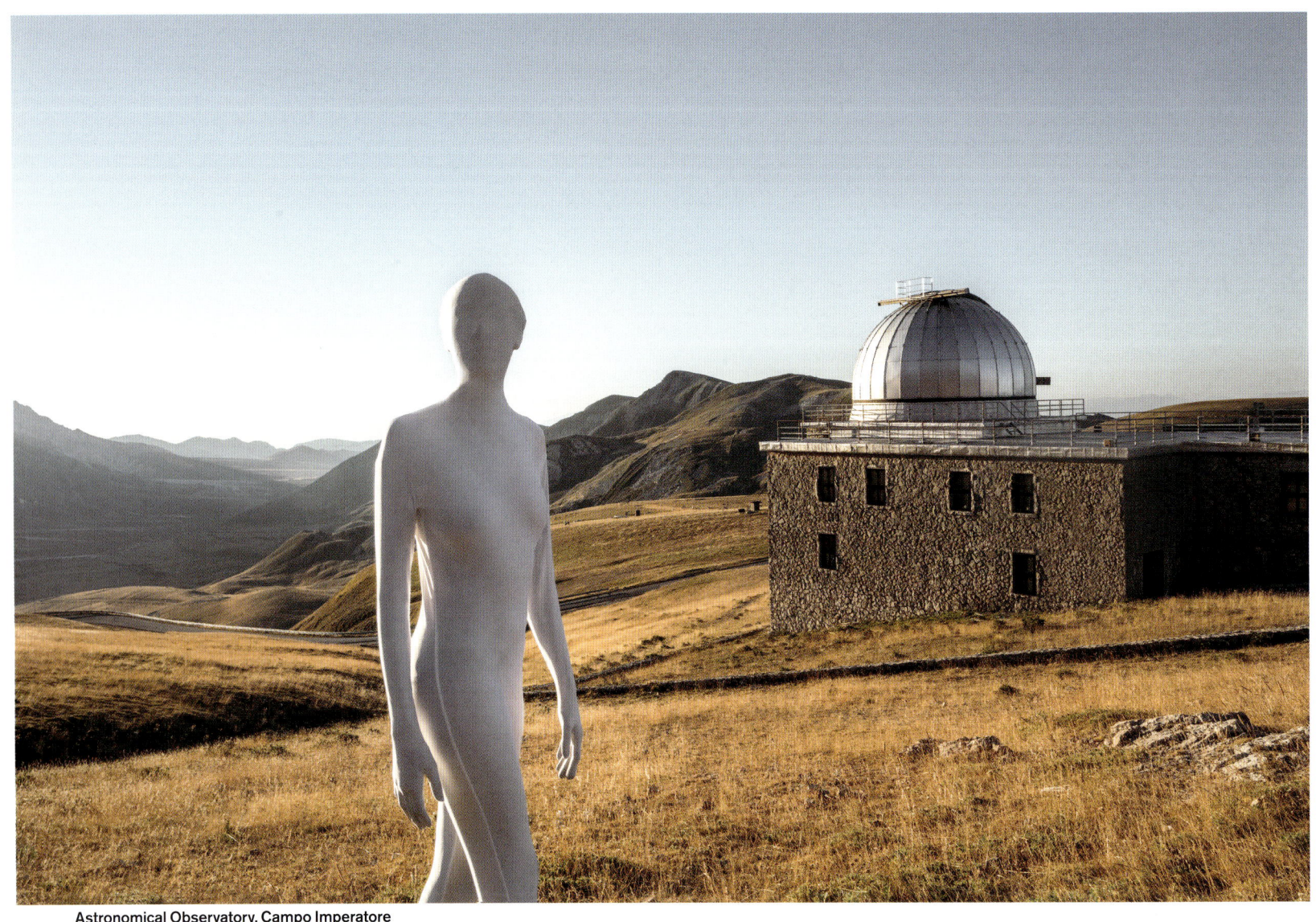
Astronomical Observatory, Campo Imperatore

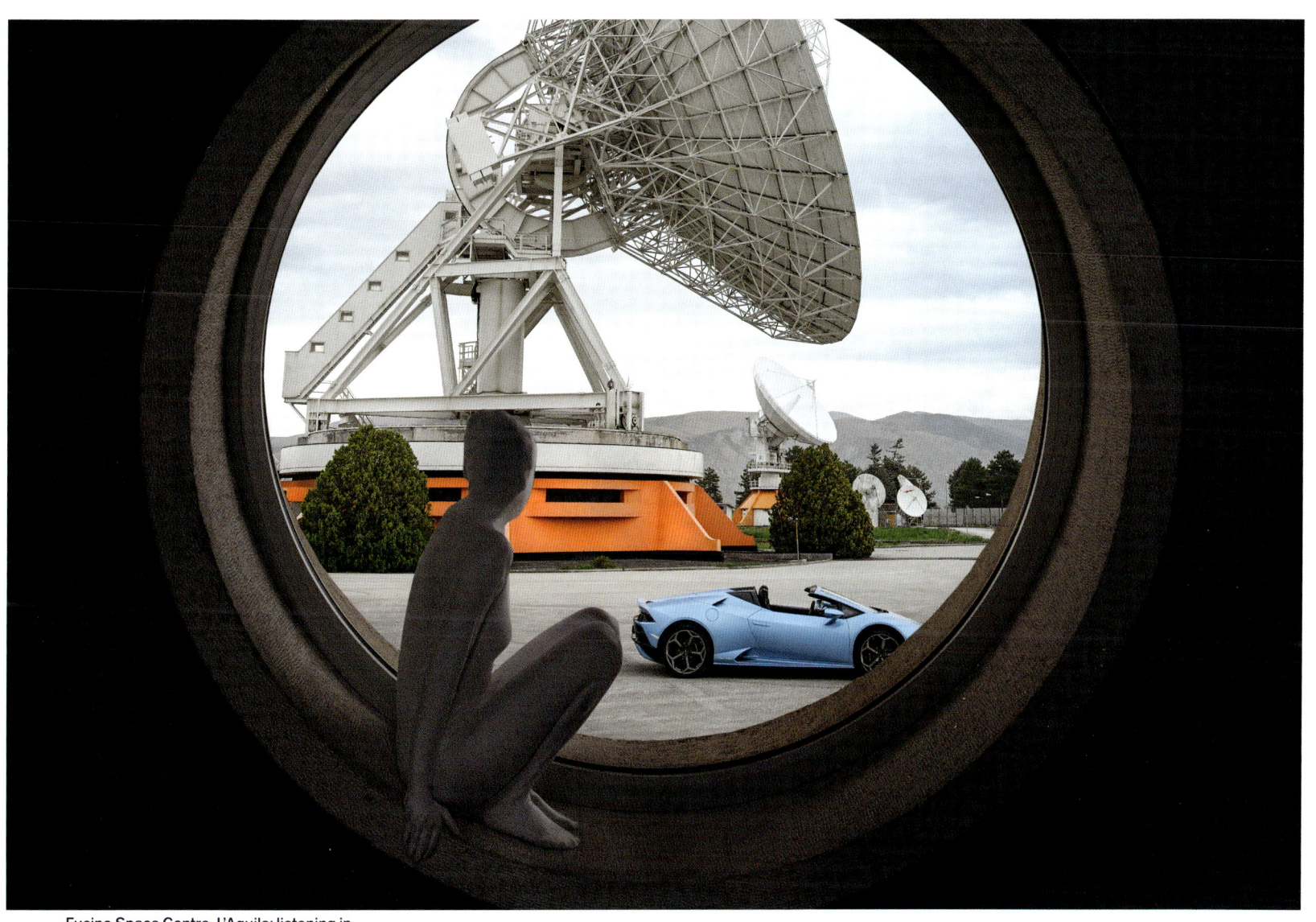

Fucino Space Centre, L'Aquila: listening in

"A man who has not been in Italy, is always conscious of an inferiority"

Samuel Johnson

ANNA DI PROSPERO
LAZIO

I wanted to take the Aventador S to the mythical origins of Italy, retracing Aeneas' heroic journey through Lazio, from Gaeta to Rome, in a modern key, and training its headlights on the art of the glorious Roman Empire that made the Eternal City the capital of the world. The Gulf of Gaeta, the Circeo promontory and the history of Rome provide the backdrop for a poetic dialogue between a woman and the Lamborghini's red profiles, studying the relationship between human beings and the environment. The route includes enriching visits to iconic Lazio heritage sites, such as Sperlonga and the Ninfa Gardens. This journey through Lazio ends up in front of the Palazzo della Civiltà Italiana in Rome, a celebration of Italian identity, brilliance and the arts. The excellence Lamborghini expresses is the *fil rouge*, the engine of a visual journey conceived to showcase our history and redesign the future, as ever with humankind front and centre.

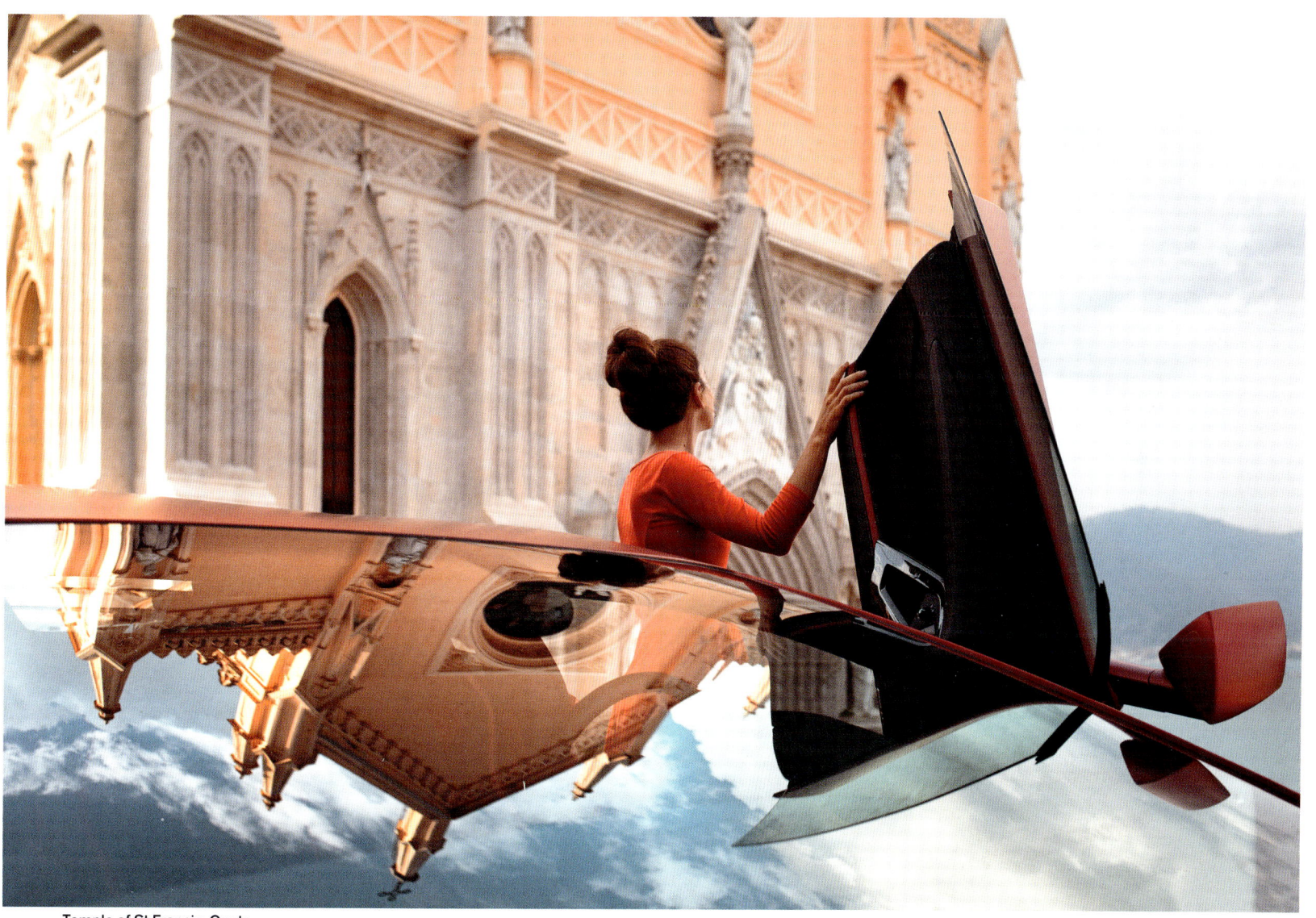
Temple of St Francis, Gaeta

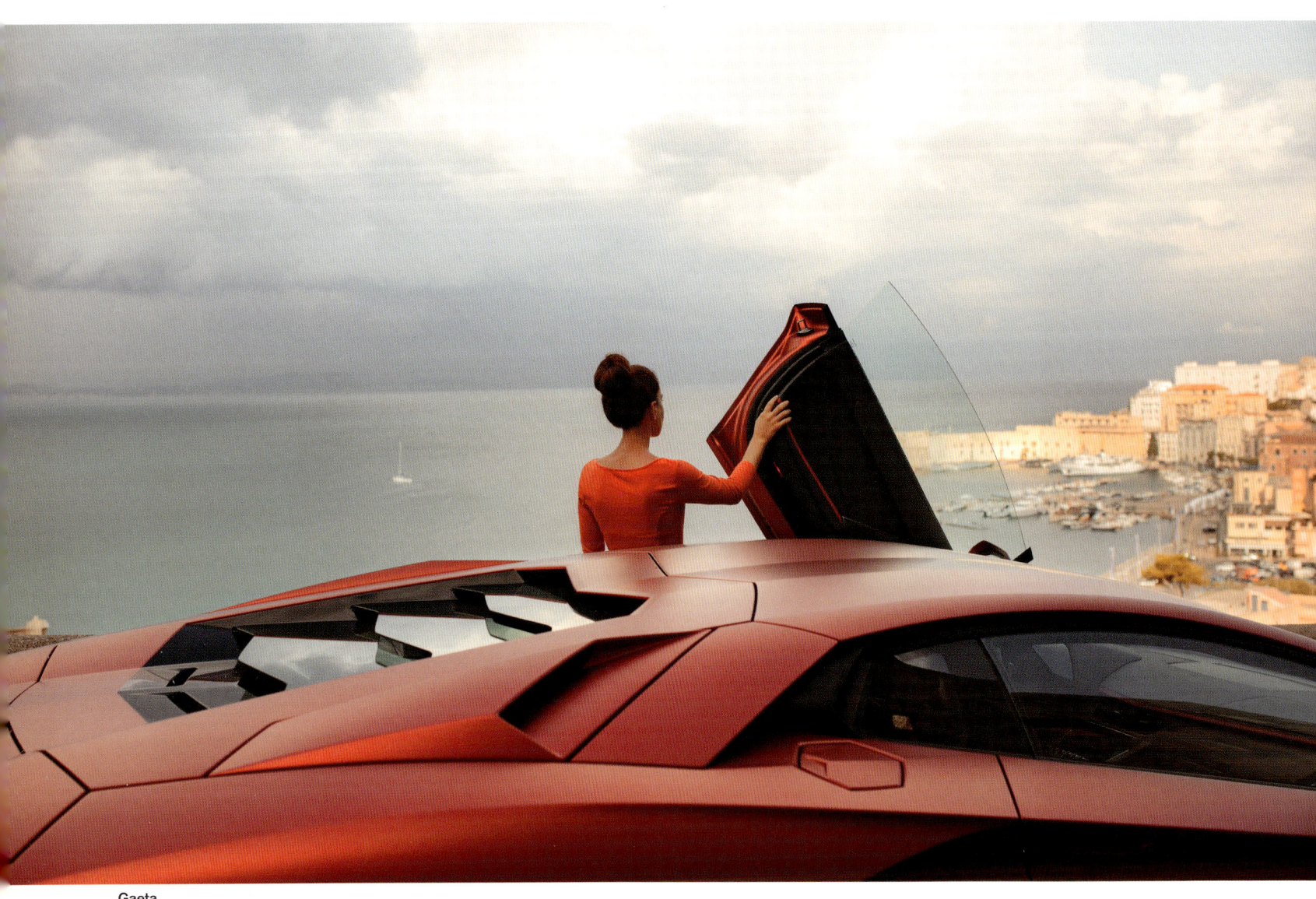
Gaeta

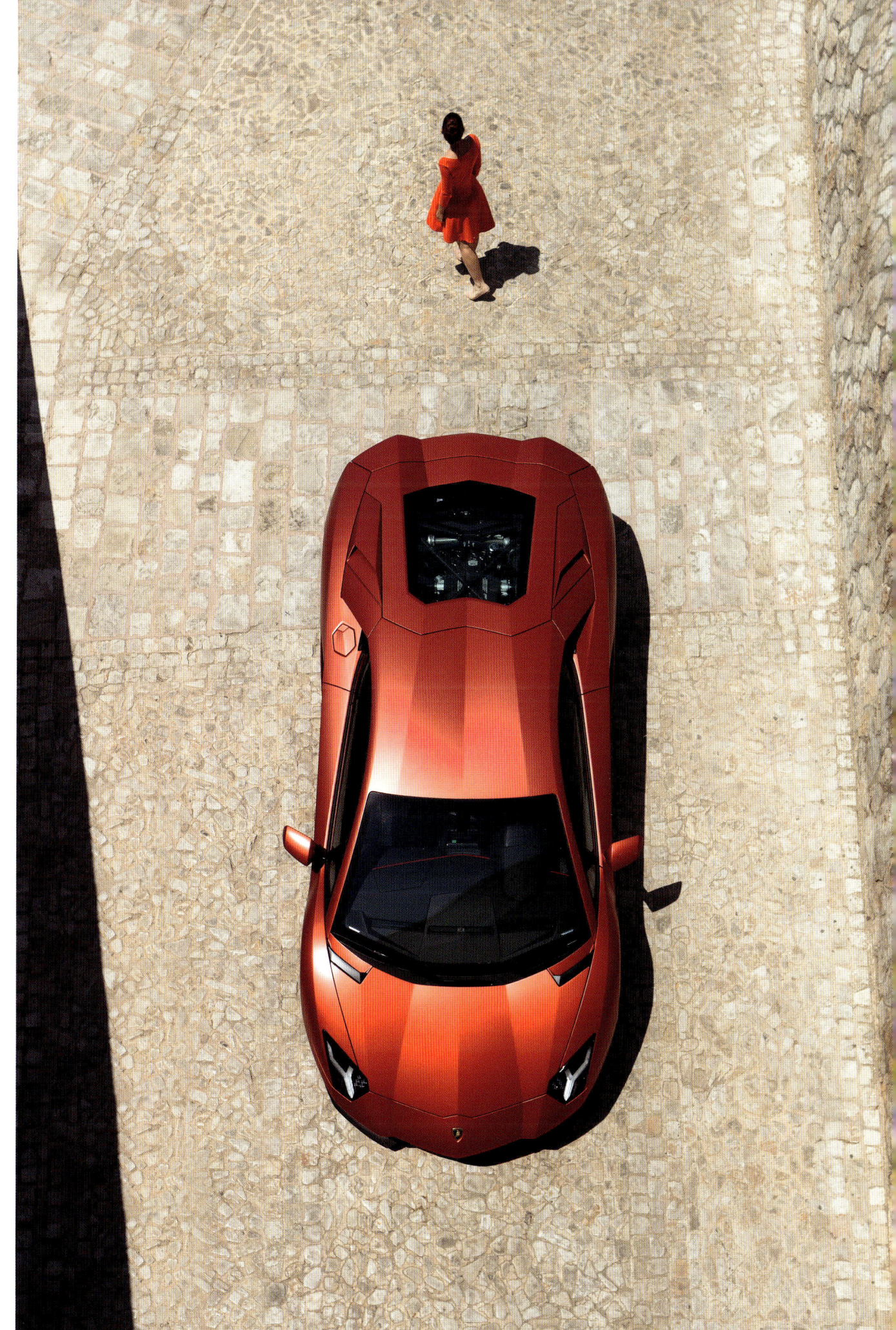

Sperlonga

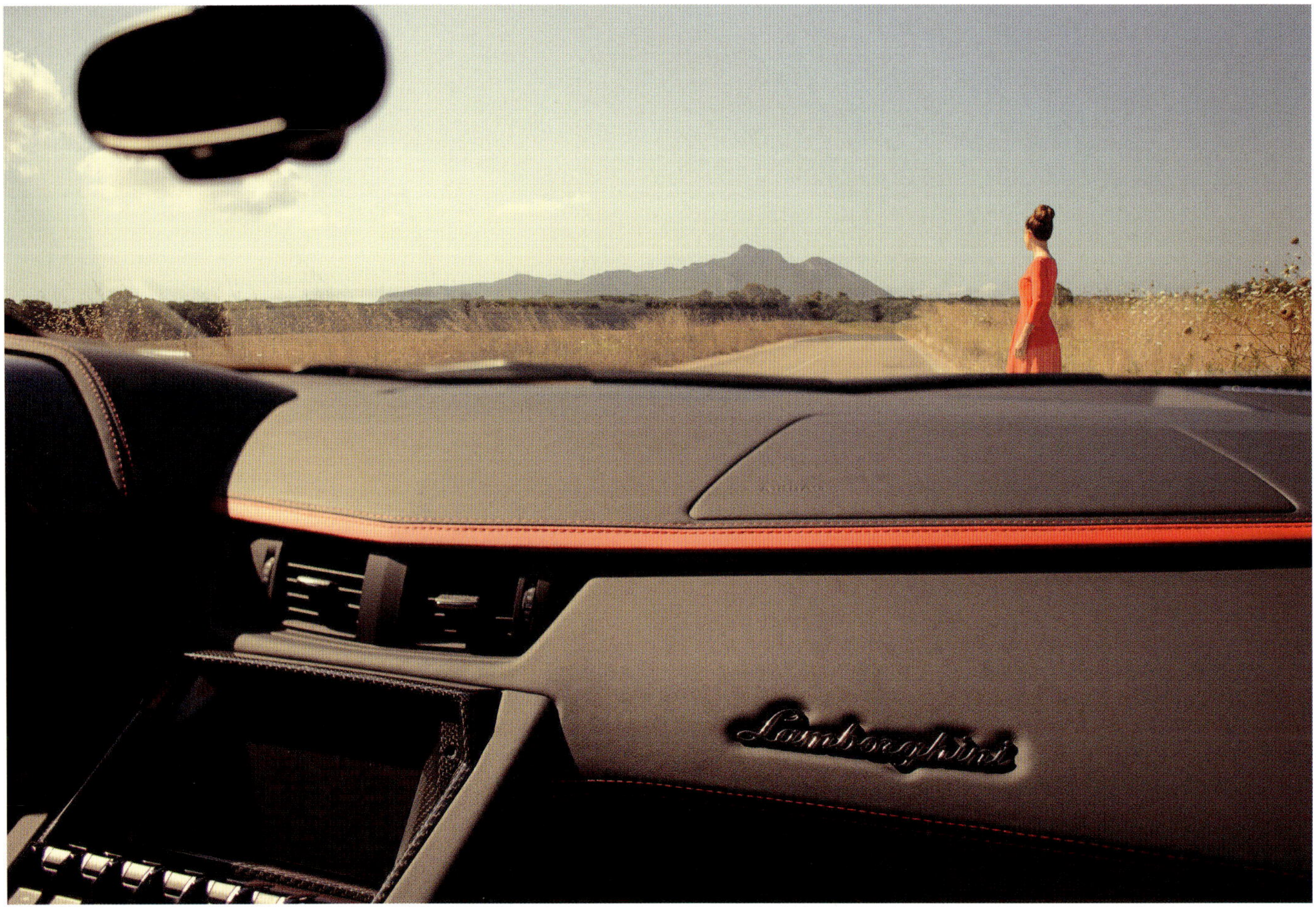

Sabaudia

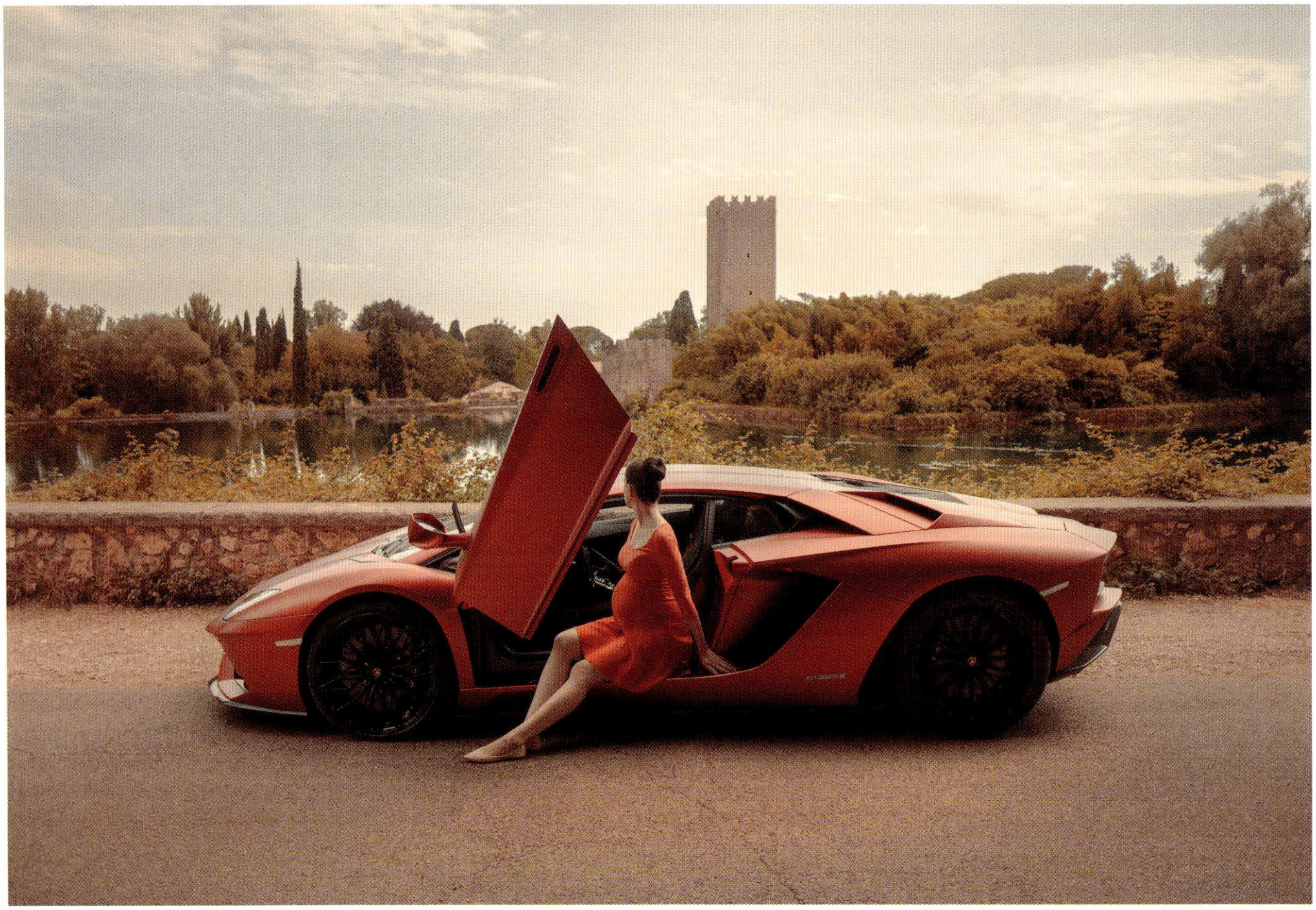

Ninfa Gardens,
Cisterna di Latina

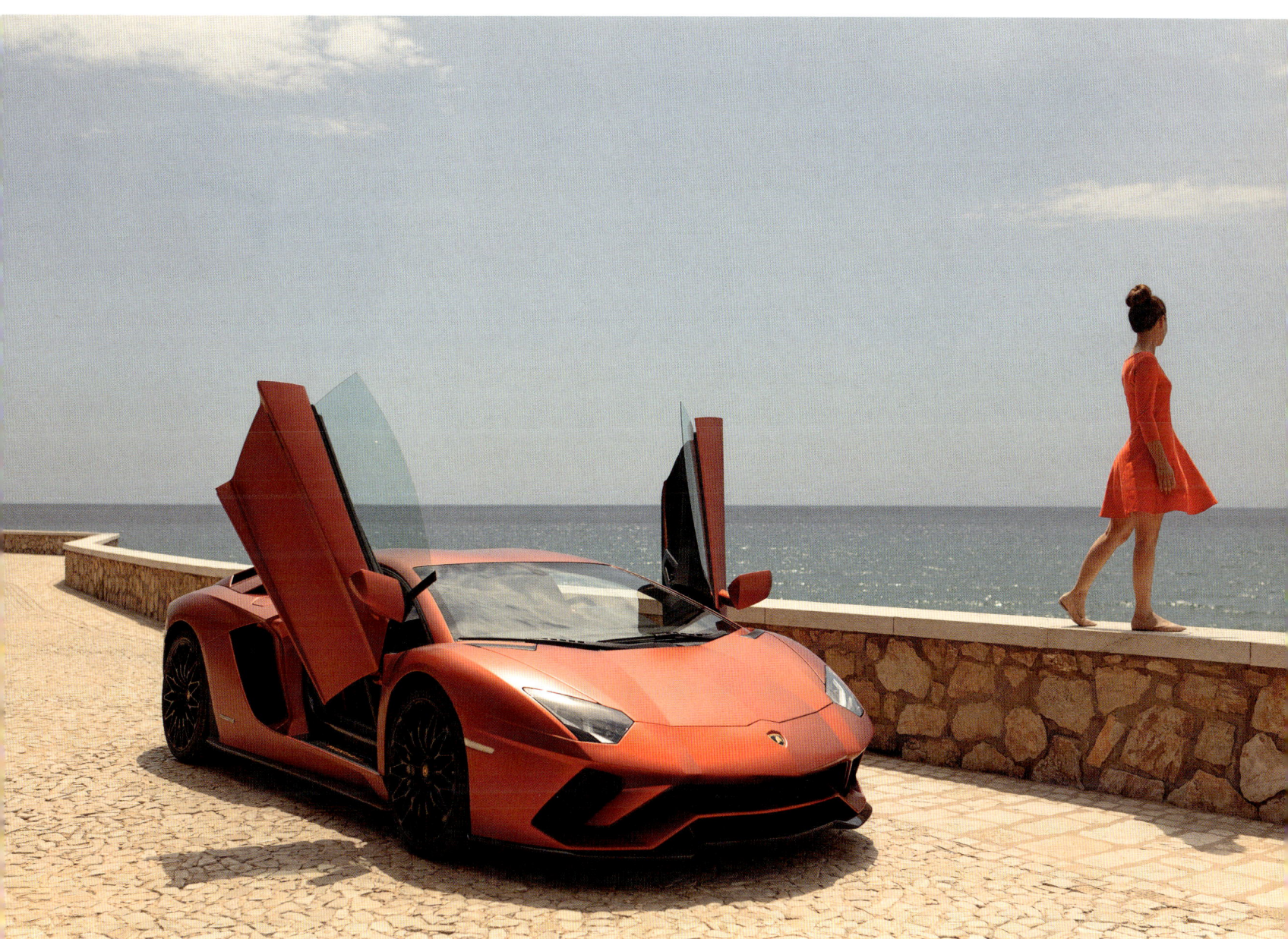

Sperlonga

"The Gulf of Gaeta, the Circeo promontory and the history of Rome provide the backdrop for a poetic dialogue between a woman and the Lamborghini's red profiles, studying the relationship between human beings and the environment"

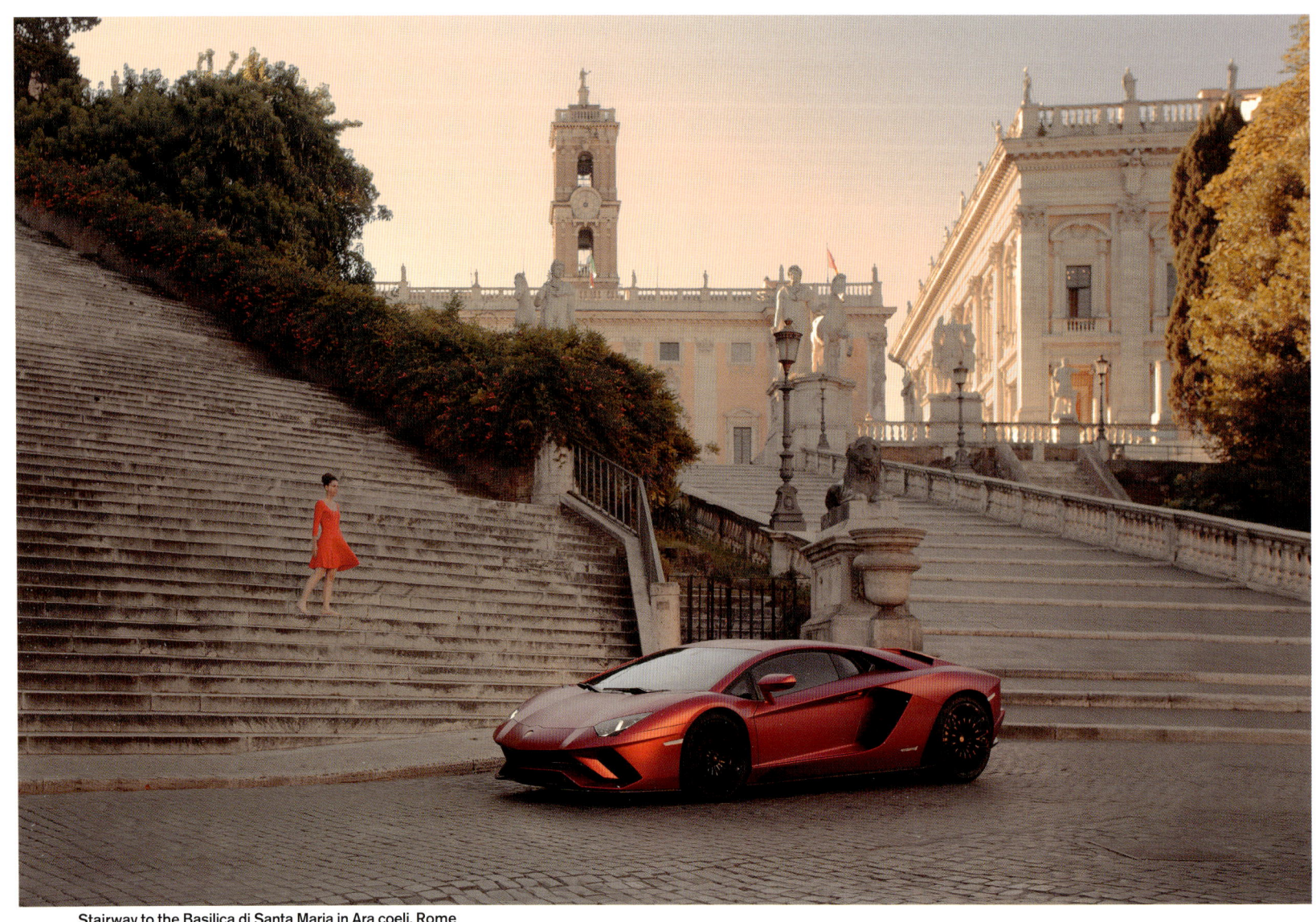
Stairway to the Basilica di Santa Maria in Ara coeli, Rome

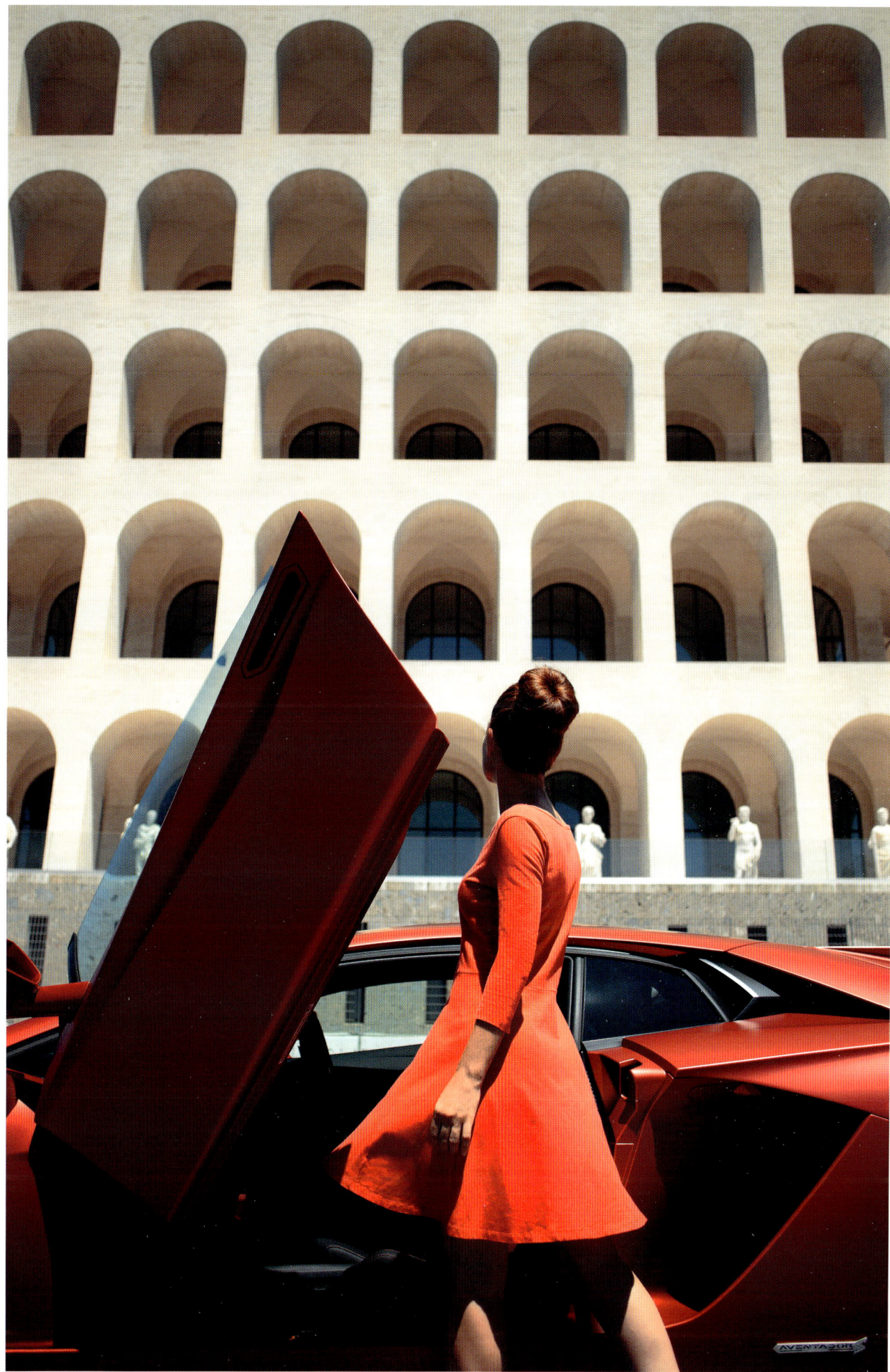

Palazzo della Civiltà
Italiana, Rome

Pages 128–29
Janiculum, Rome

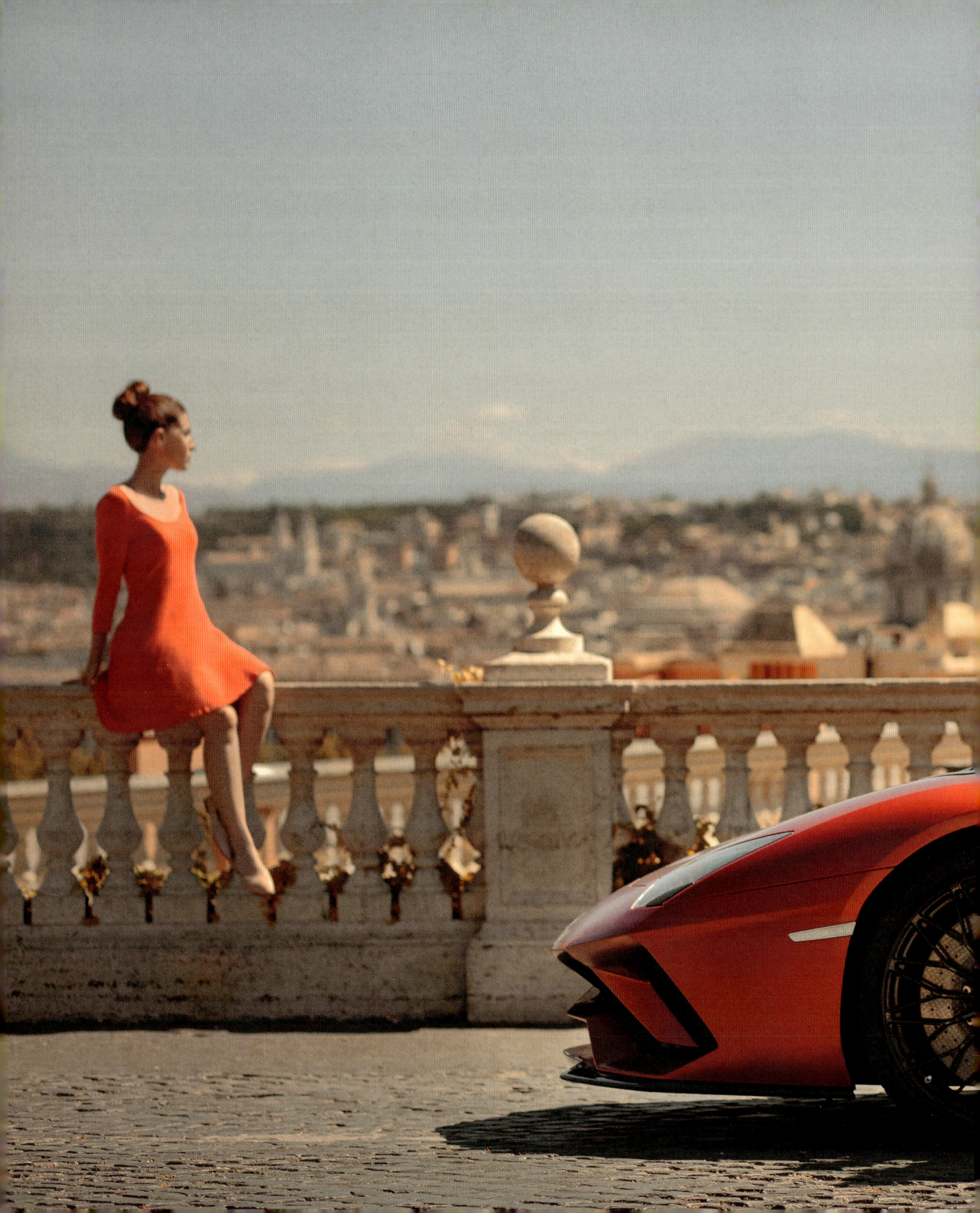

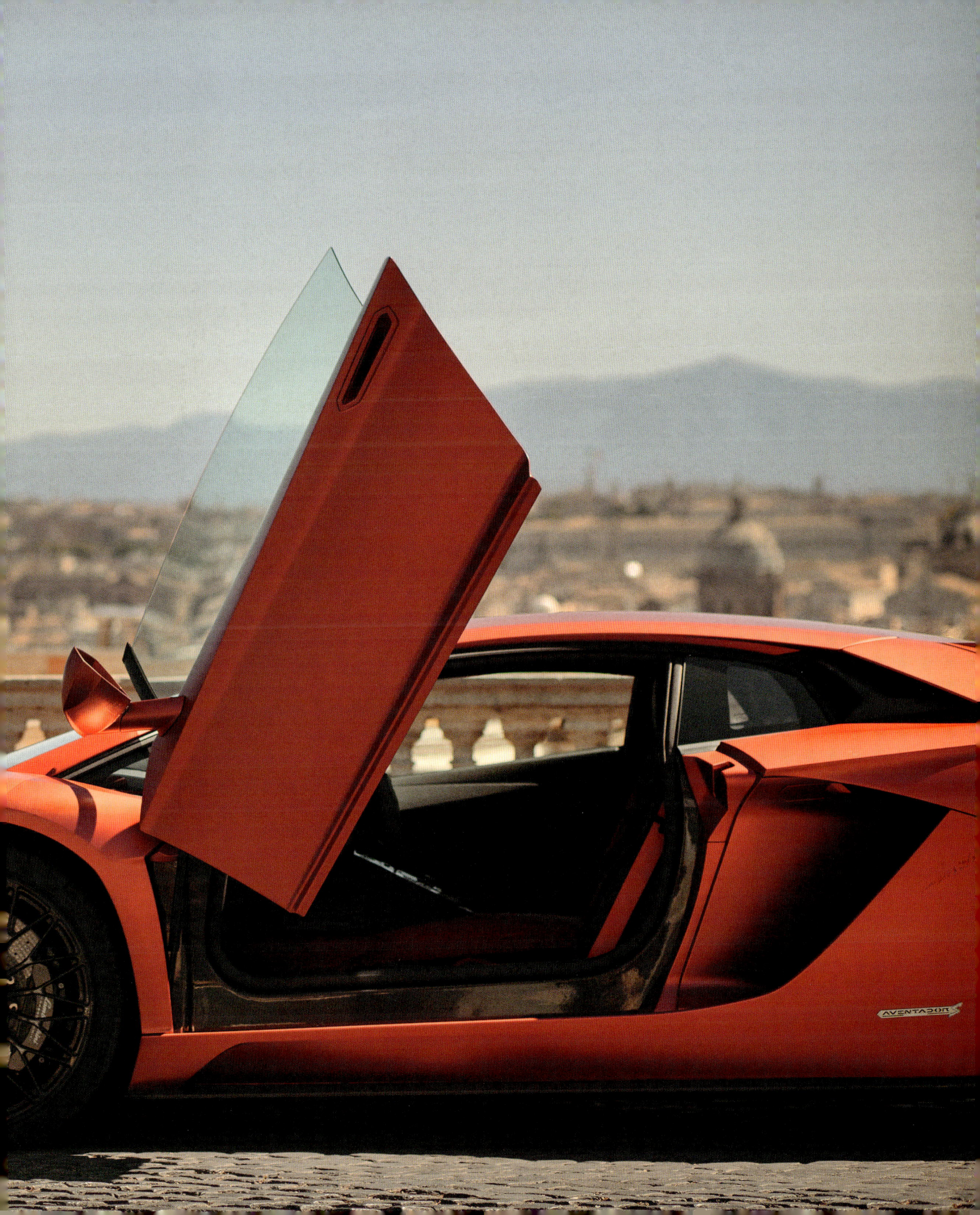

WOLFANGO SPACCARELLI
MARCHE

In his book *Viaggio in Italia* (Journey across Italy), Guido Piovene writes that the Marche is a distillate of our country, a harmonious whole in which cities of art, ancient small towns, the sea, hills and mountains blend together in a unique interplay of shapes and colours. Following this vision, I decided to interpret the Marche in a classical way, creating photos as if they were Renaissance paintings, forgoing movement and building a symbiotic relationship between the legendary Diablo 6.0 SE and the Marche region. The Lamborghini's shapes harmonize with the sea, the soft flowing hills and the Renaissance architecture. Its Oro Elios tint is just perfect for an area that itself is all about golden reflections.

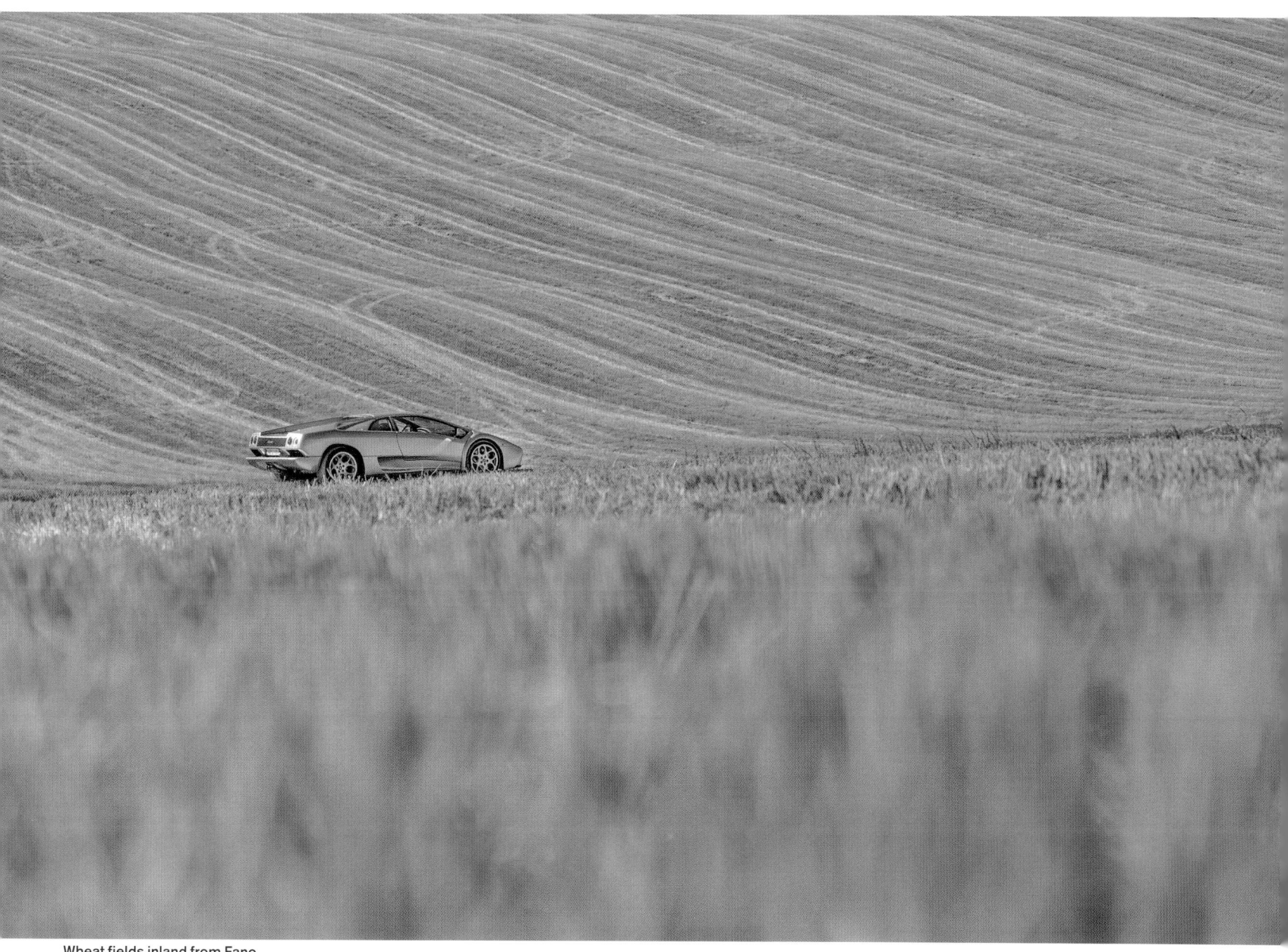
Wheat fields inland from Fano

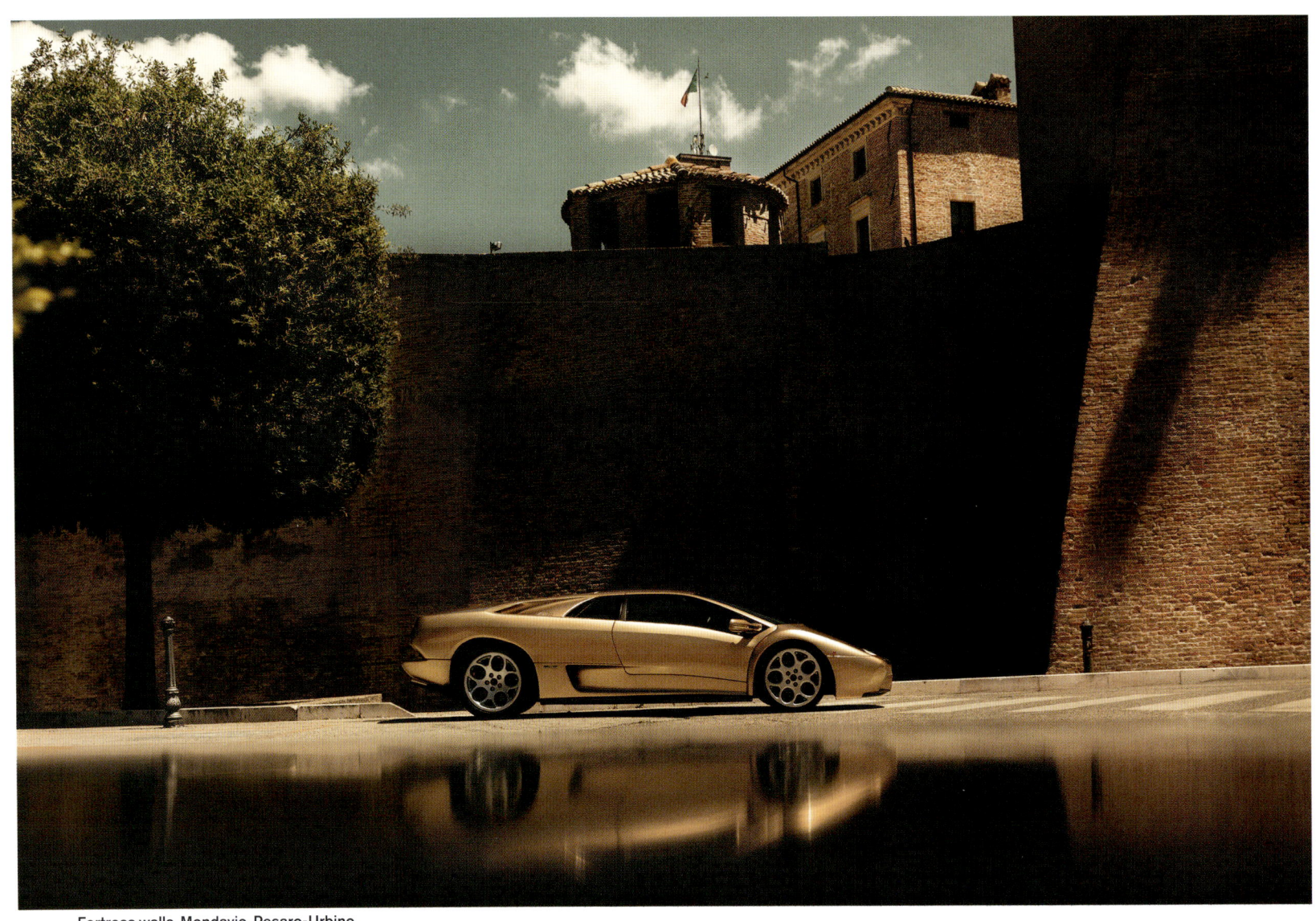
Fortress walls, Mondavio, Pesaro-Urbino

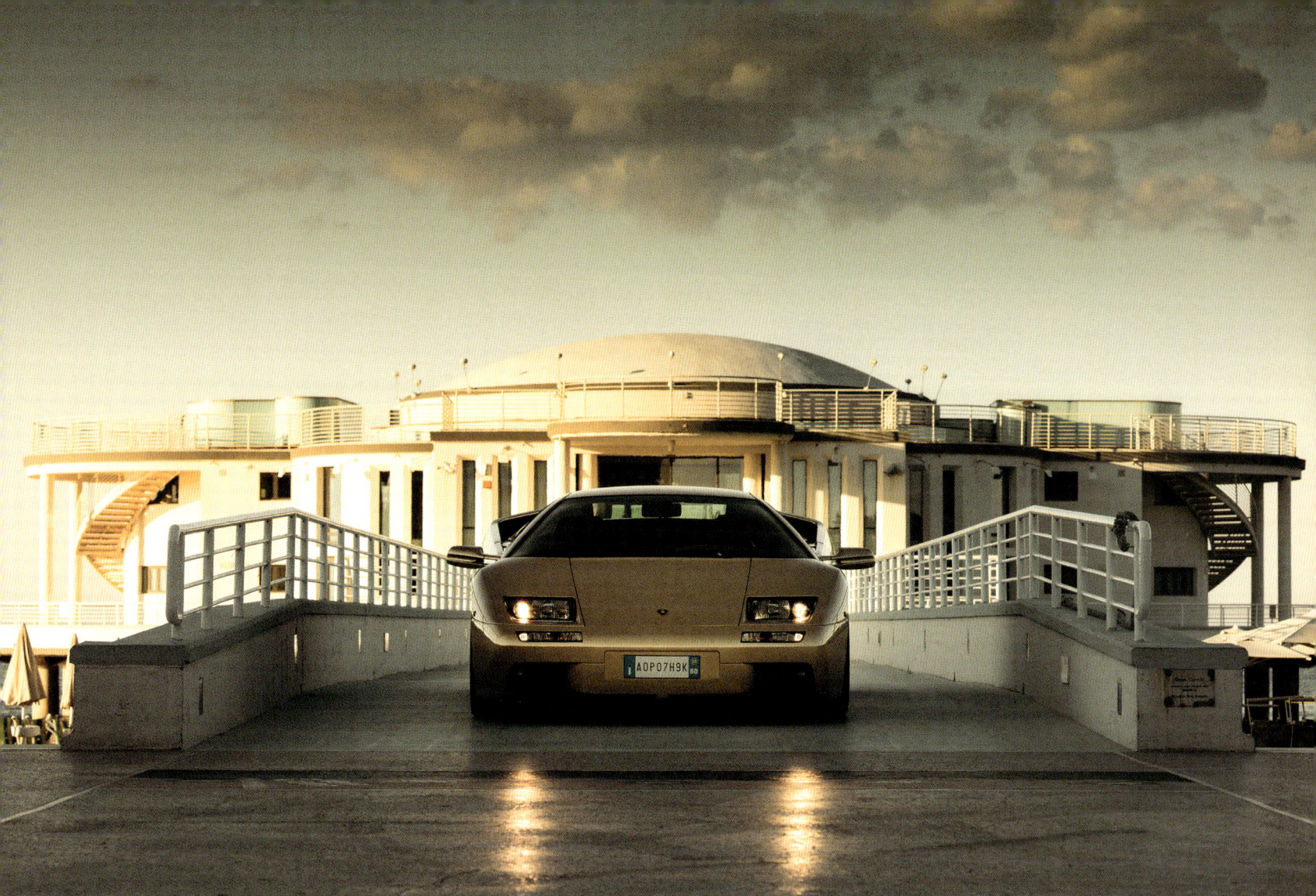

Rotonda a Mare, Senigallia

"The Lamborghini's shapes harmonize with the sea, the soft flowing hills and the Renaissance architecture. Its Oro Elios tint is just perfect for an area that itself is all about golden reflections"

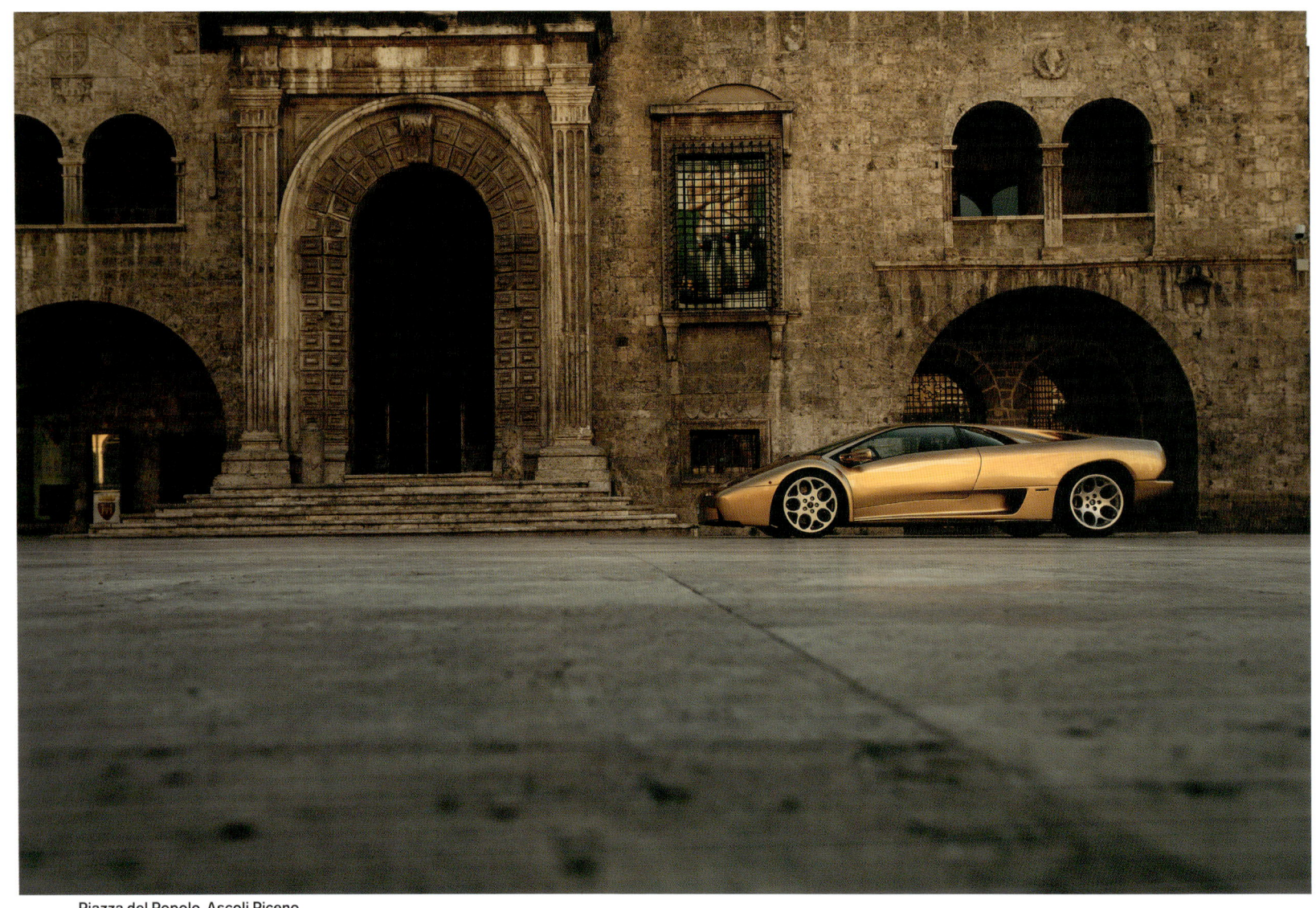
Piazza del Popolo, Ascoli Piceno

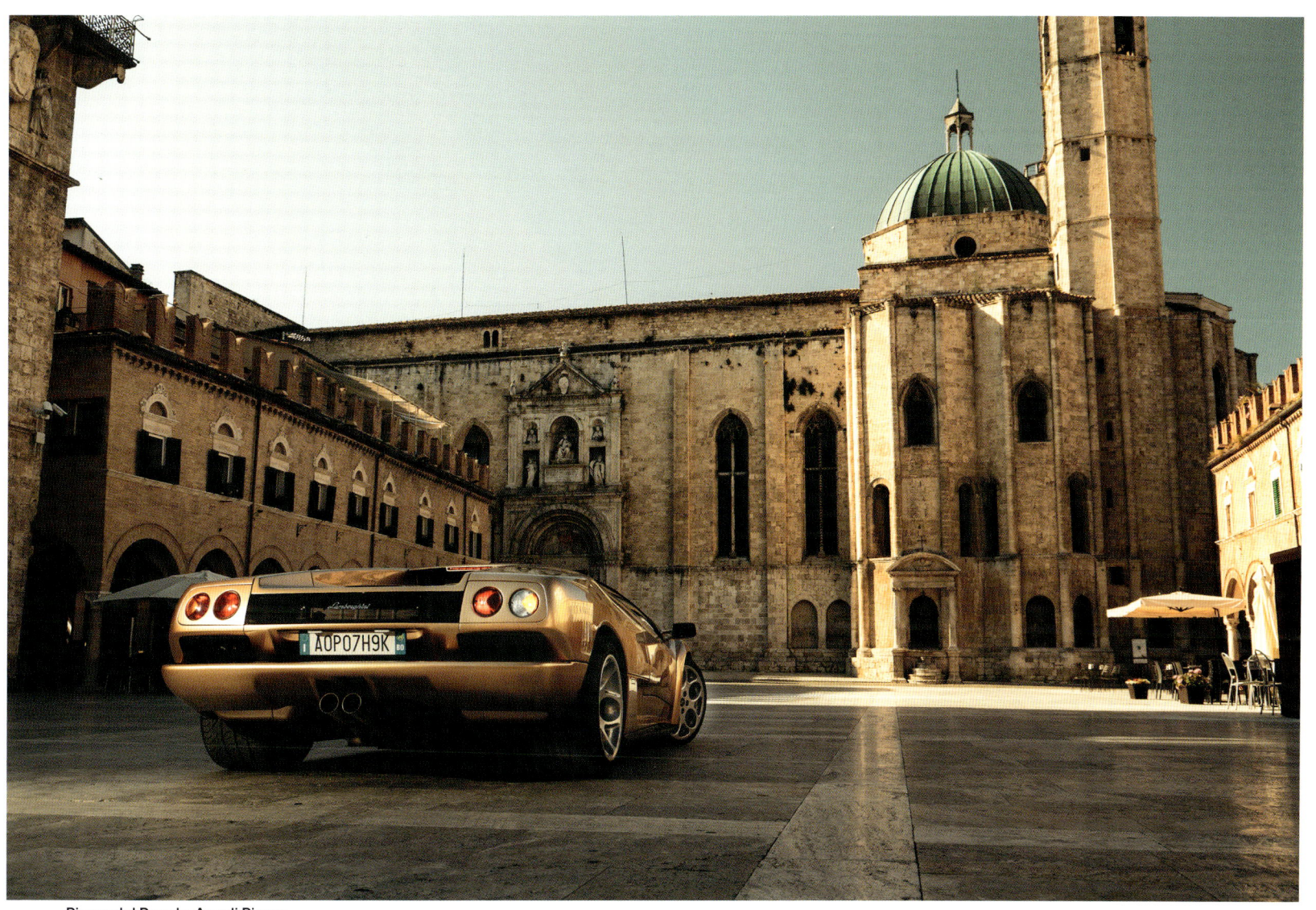
Piazza del Popolo, Ascoli Piceno

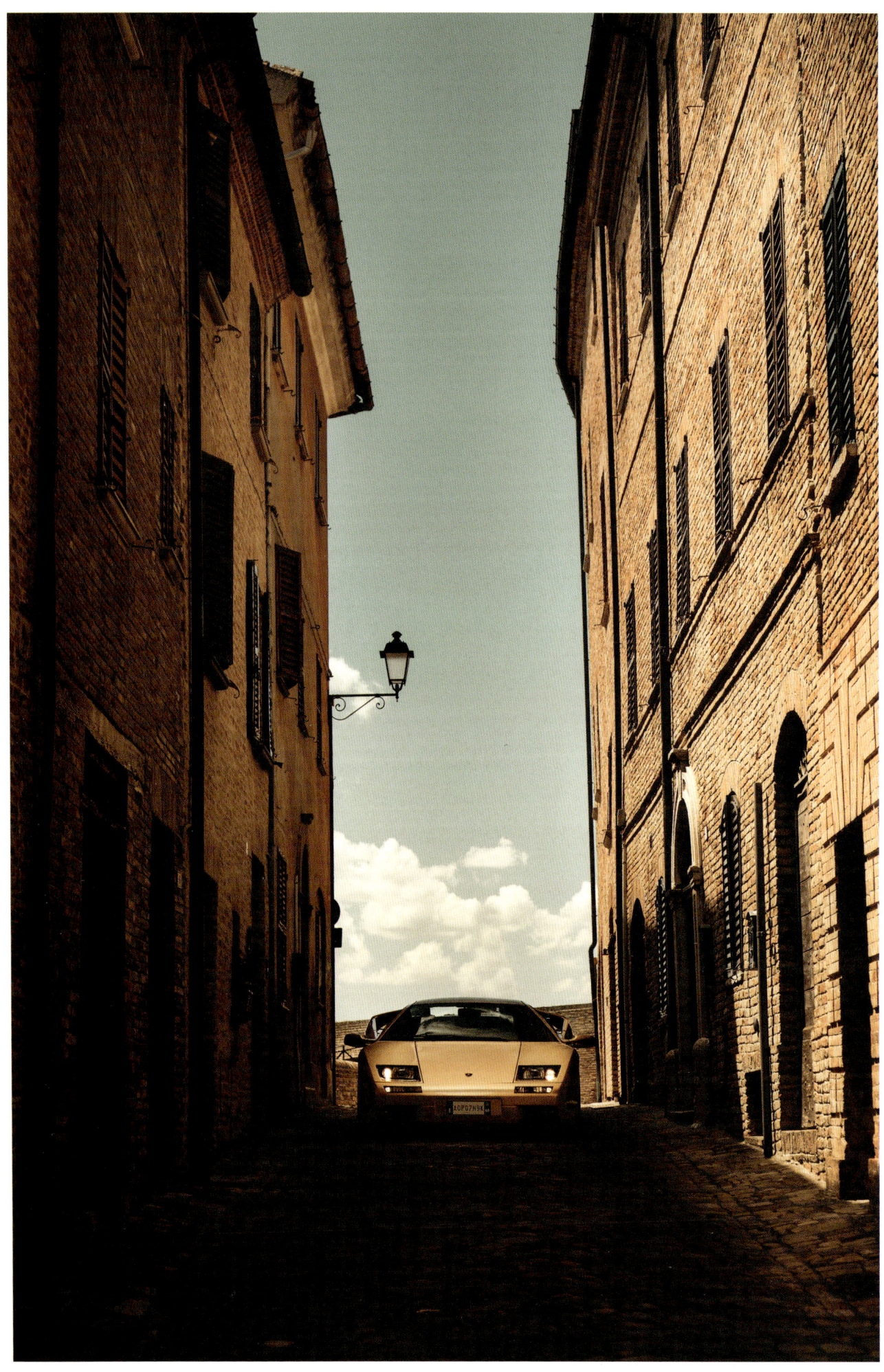

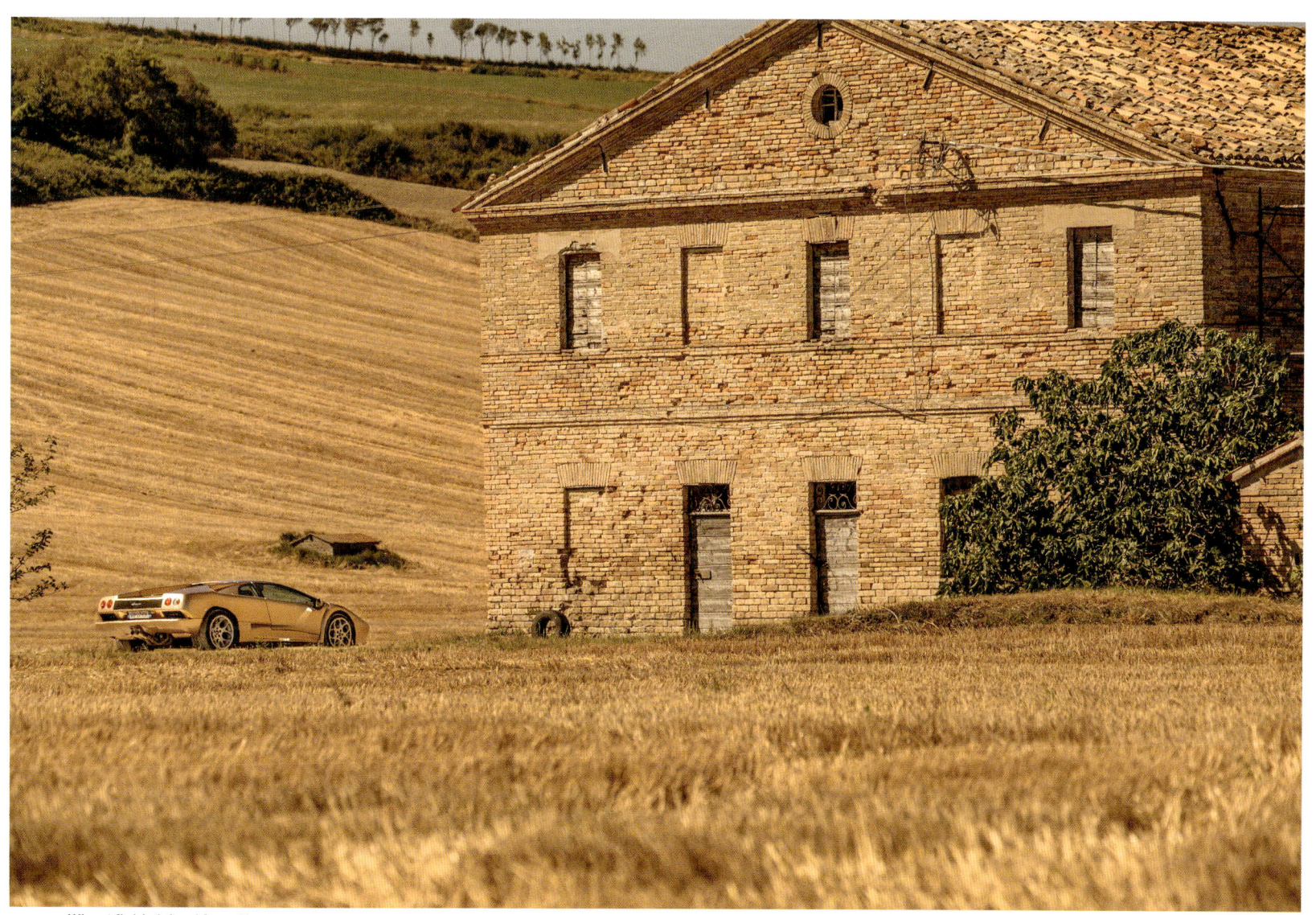

Wheat fields inland from Fano

Mondavio old town, Pesaro-Urbino

Pages 138–39 Mondavio old town, Pesaro-Urbino

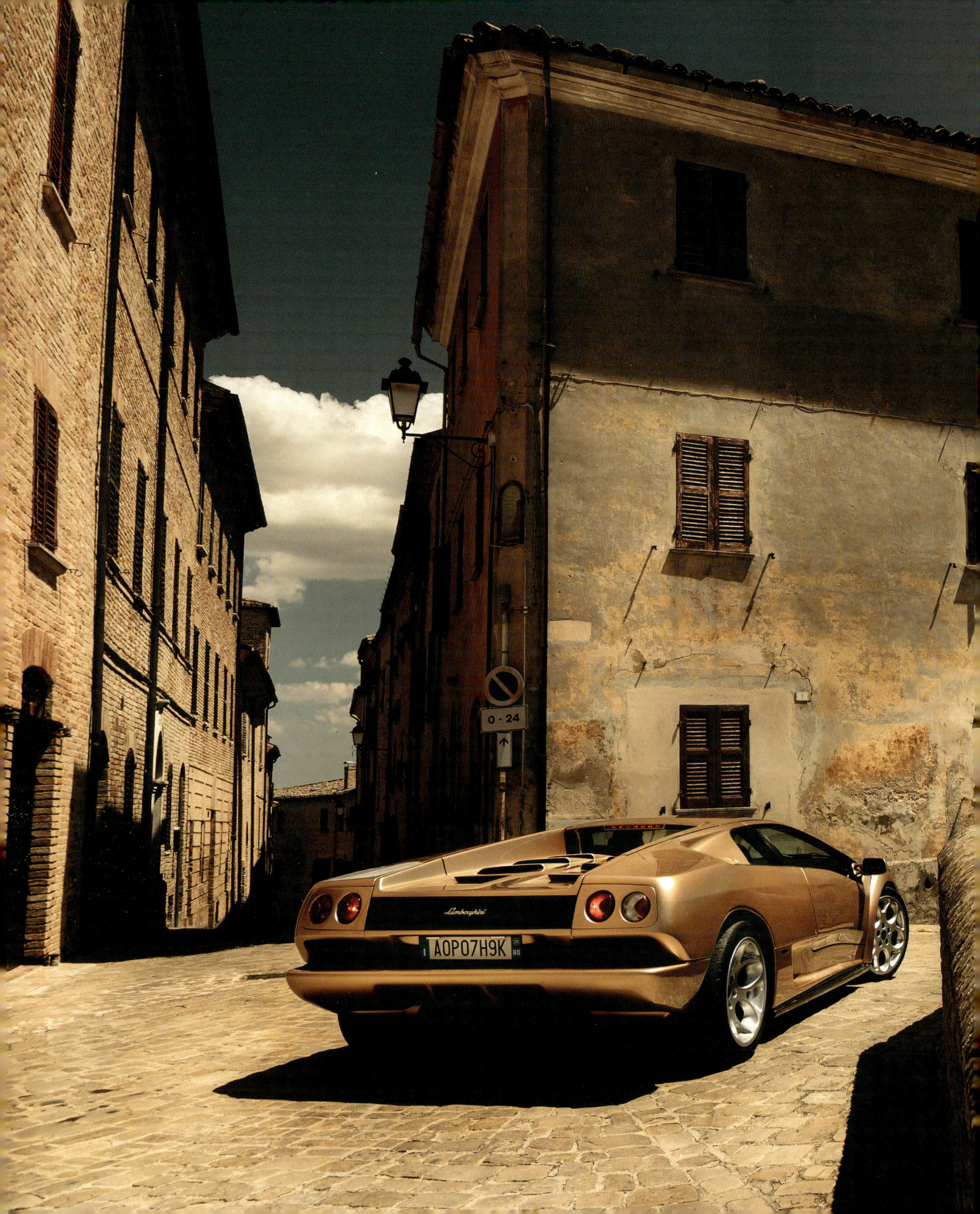

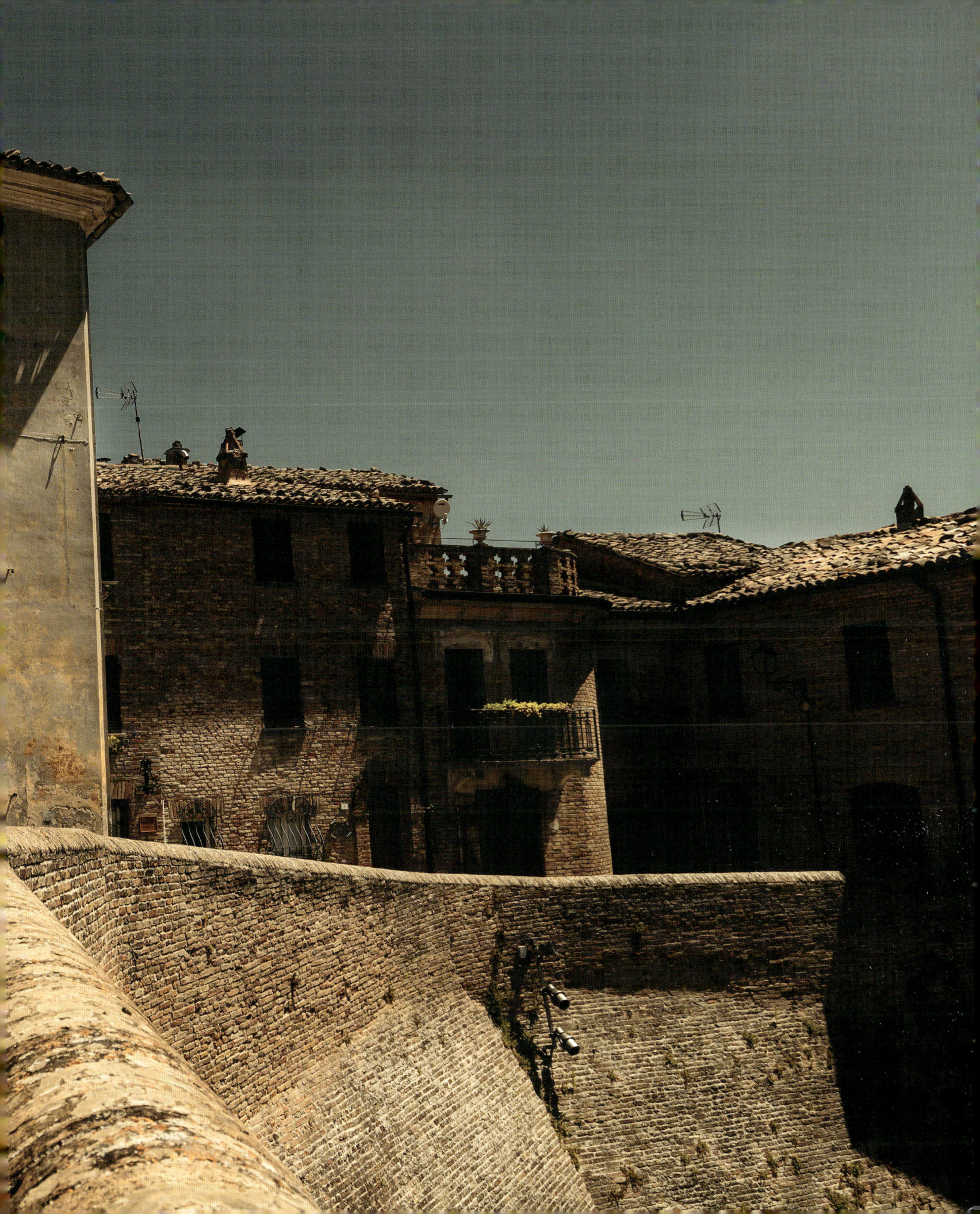

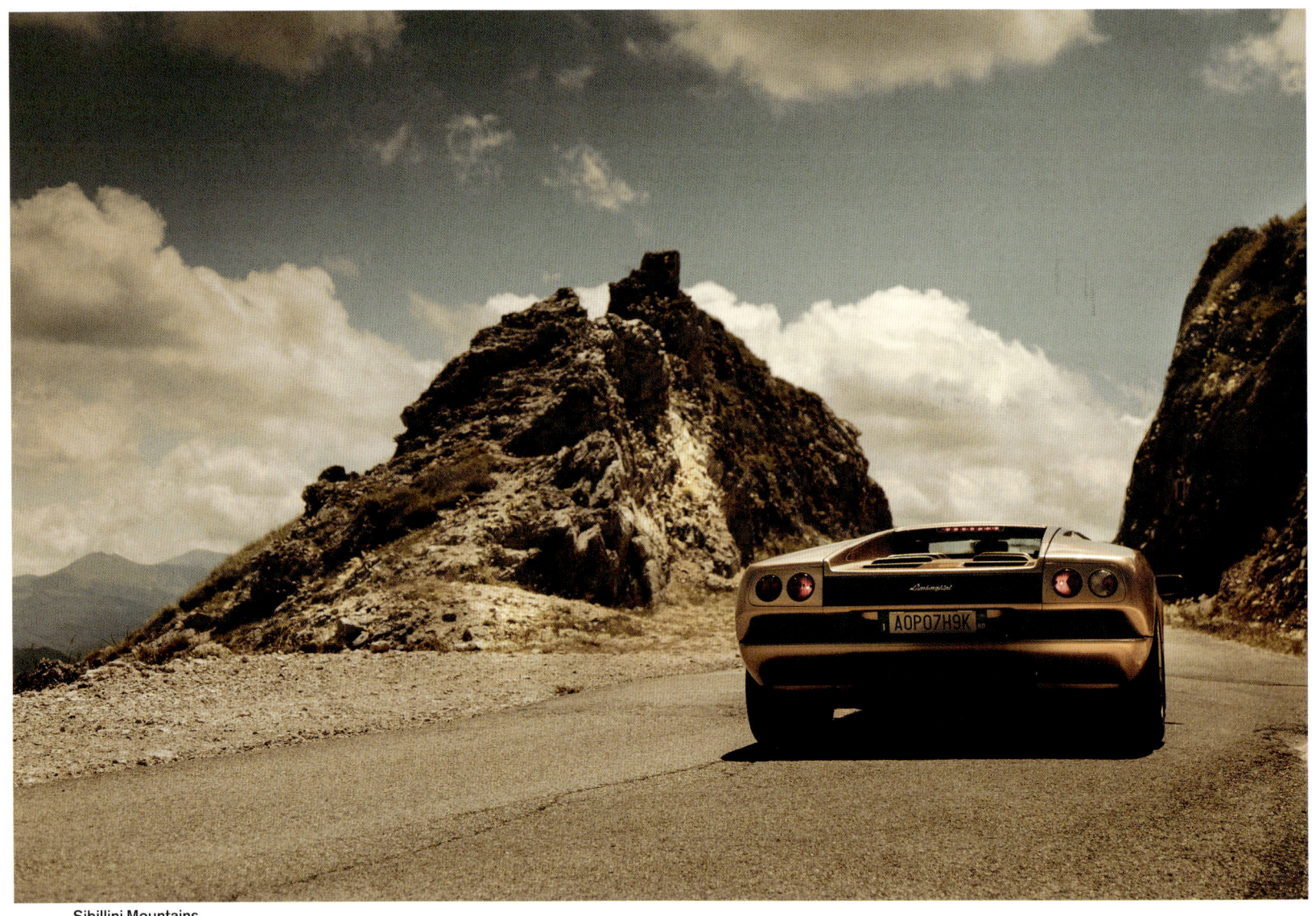
Sibillini Mountains

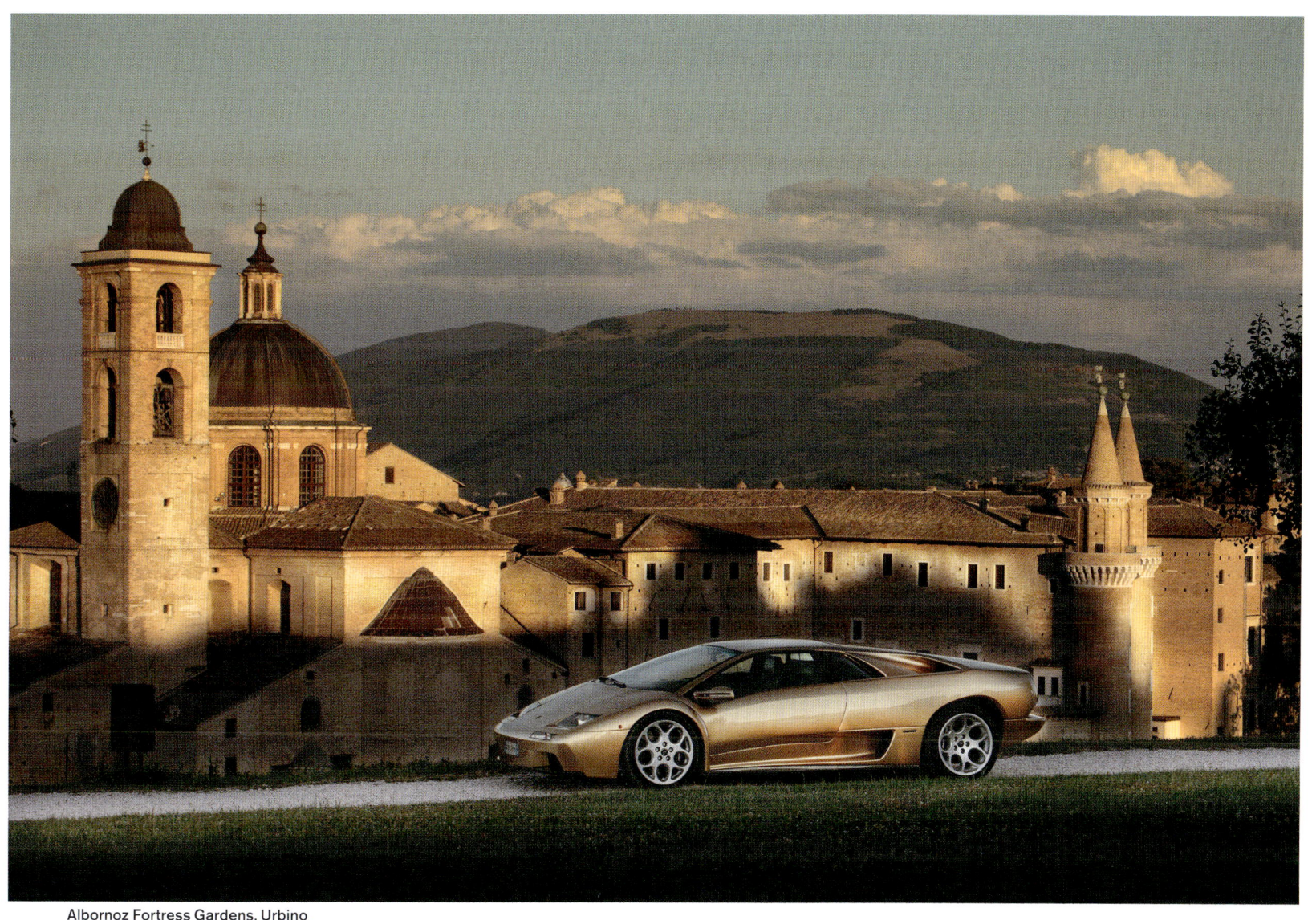
Albornoz Fortress Gardens, Urbino

# ALESSANDRO CINQUE
UMBRIA

My aesthetic is the result of an encounter between Italy and South America, where I decided to move. The neorealism of Rossellini, Visconti and Monicelli, recounting the everyday motions of a country just out of the war, is comparable to the magical realism of Peruvian writer Mario Vargas Llosa.

On this job, I wanted to portray the region as it gets back into gear after the first period of the Covid-19 pandemic. The amazement of meeting again in a square, in the fields, where I was born, a rural Umbria that is rediscovering its customs… The Lamborghini Aventador SVJ, an extraordinary element within the ordinary, catalyzing attention, breaking the routine, opening the doors to a magical realism that is a combination of the wonder of unexpected dissonance and the astonishment at rediscovering reality.

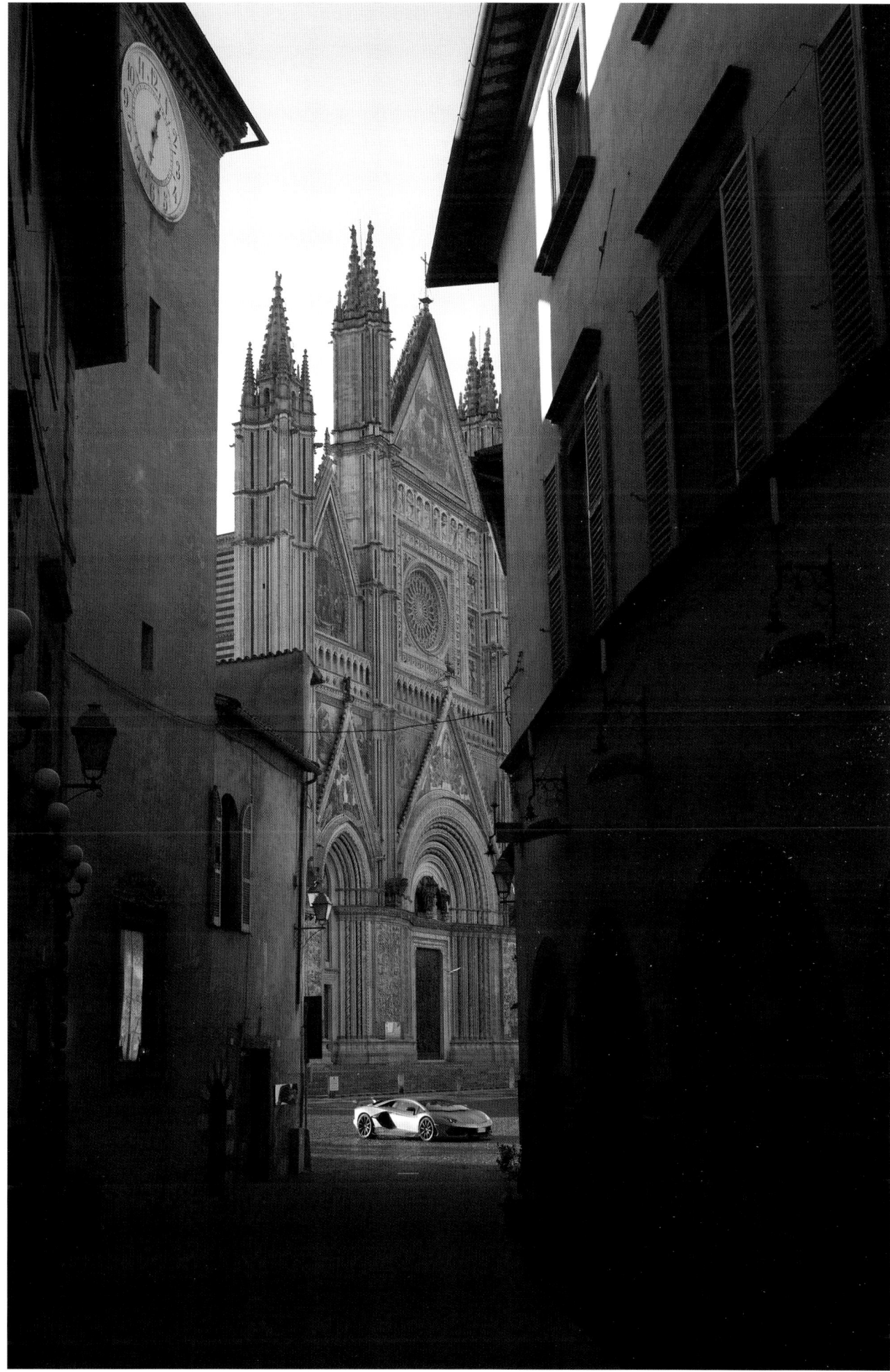

Orvieto Cathedral

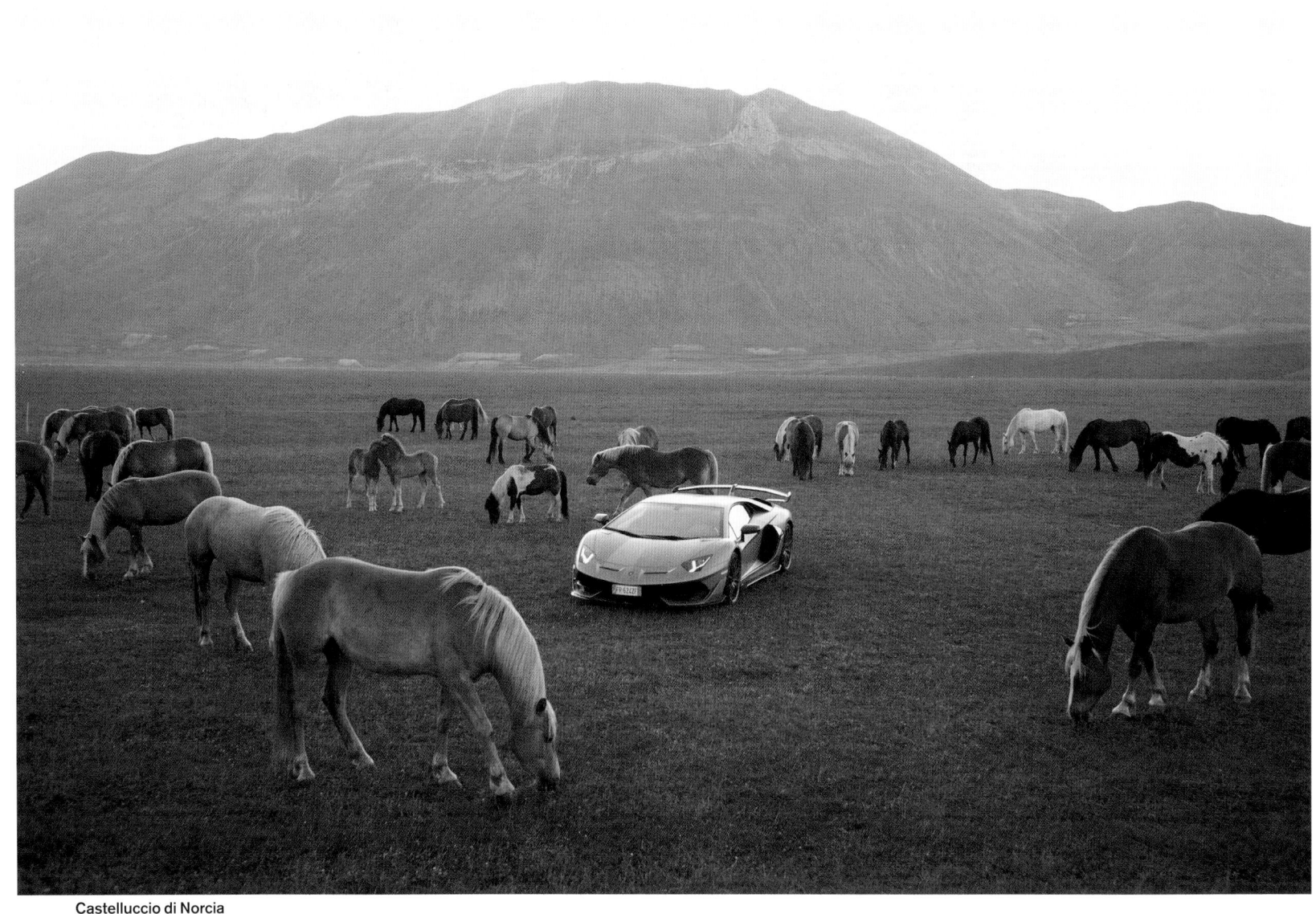
Castelluccio di Norcia

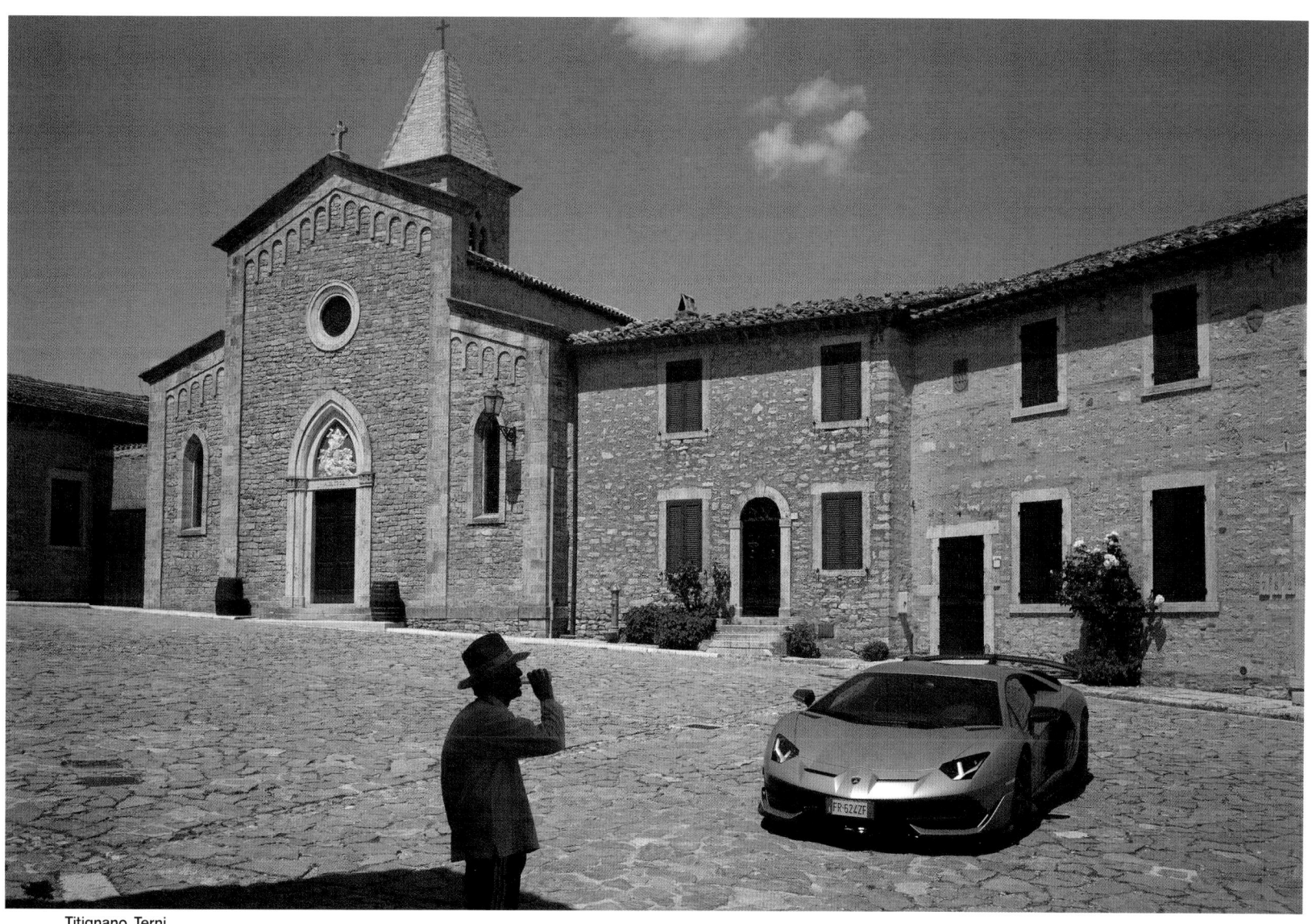
Titignano, Terni

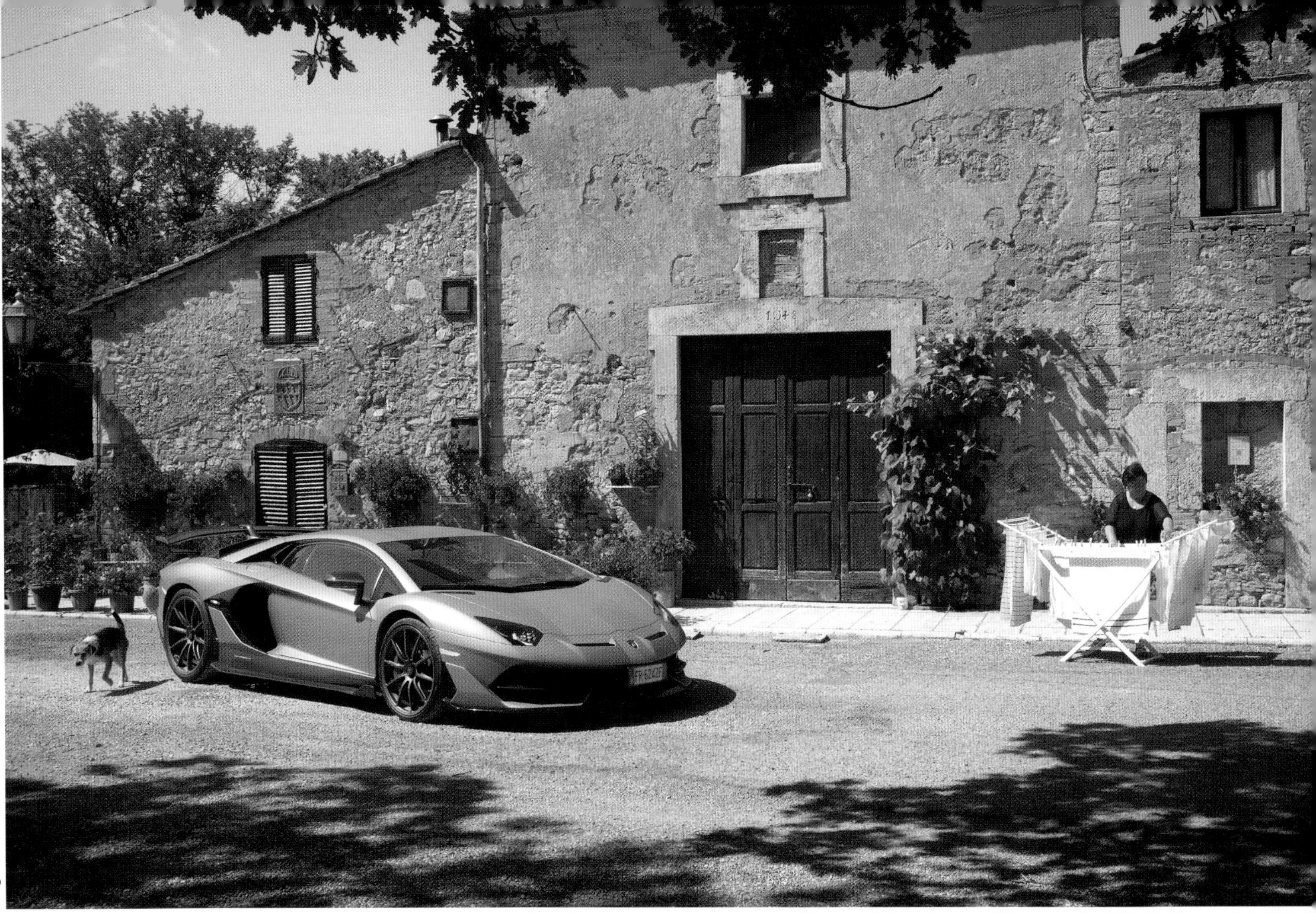

Titignano, Terni

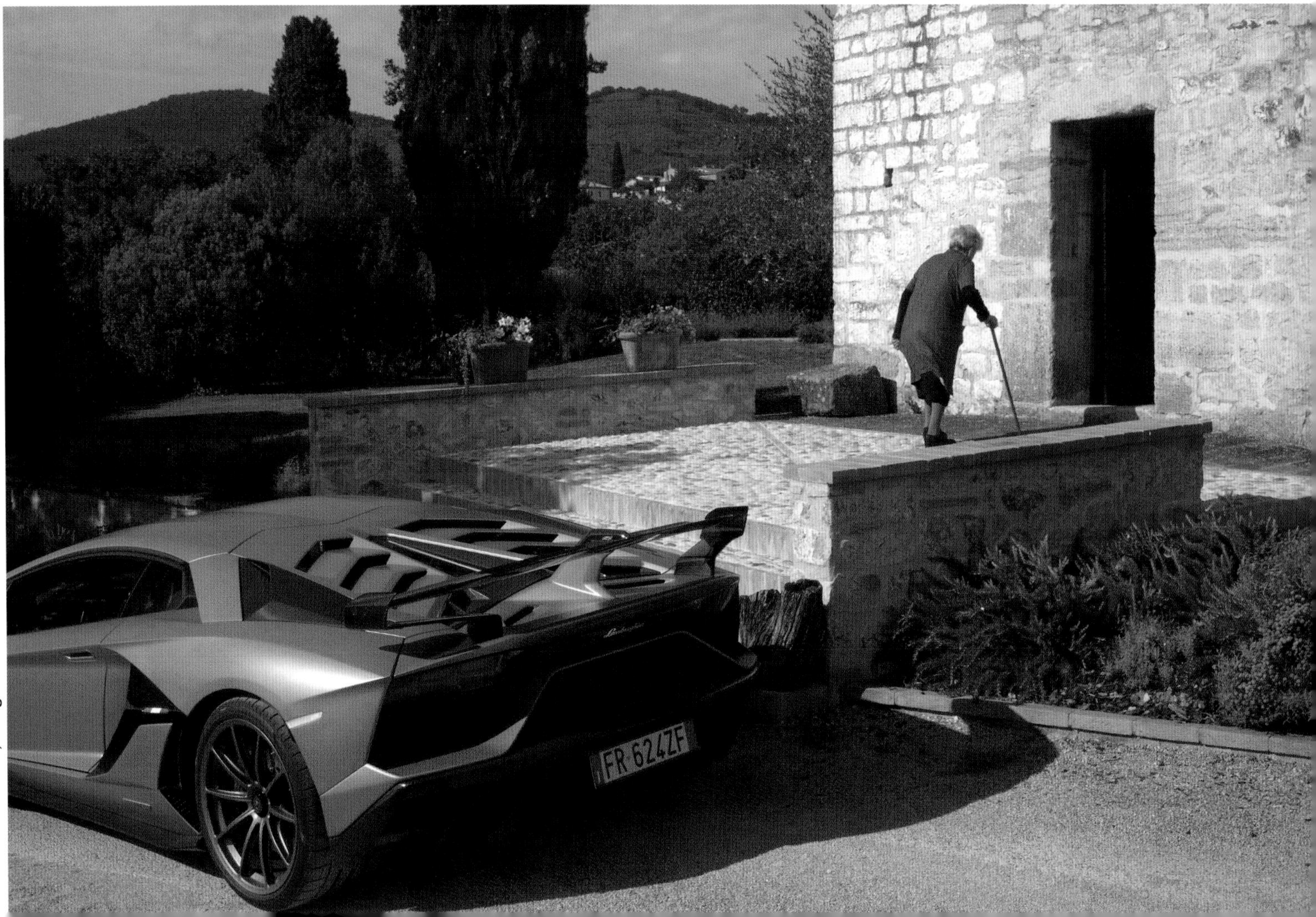

Trevi, Perugia

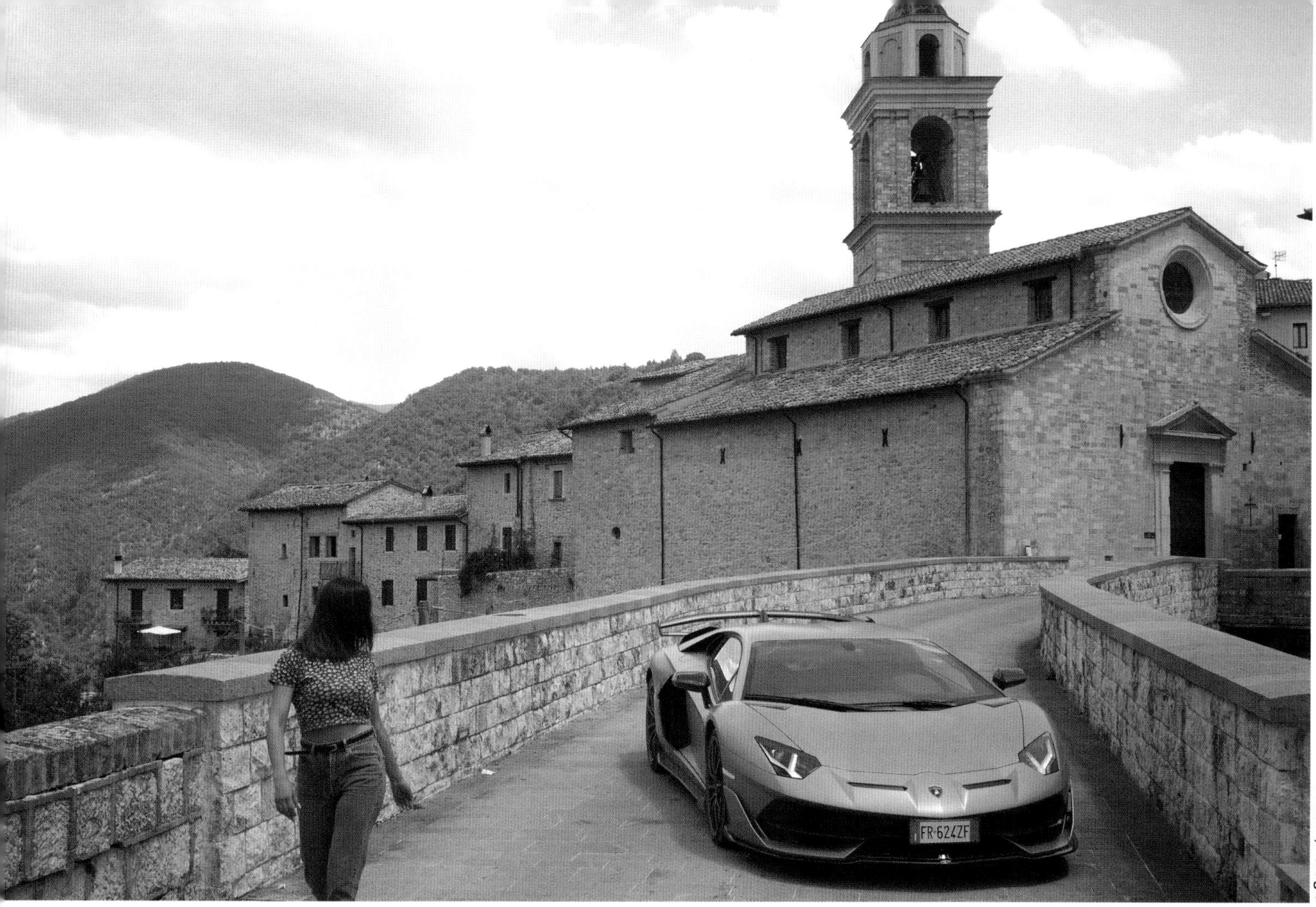

Sellano, Perugia

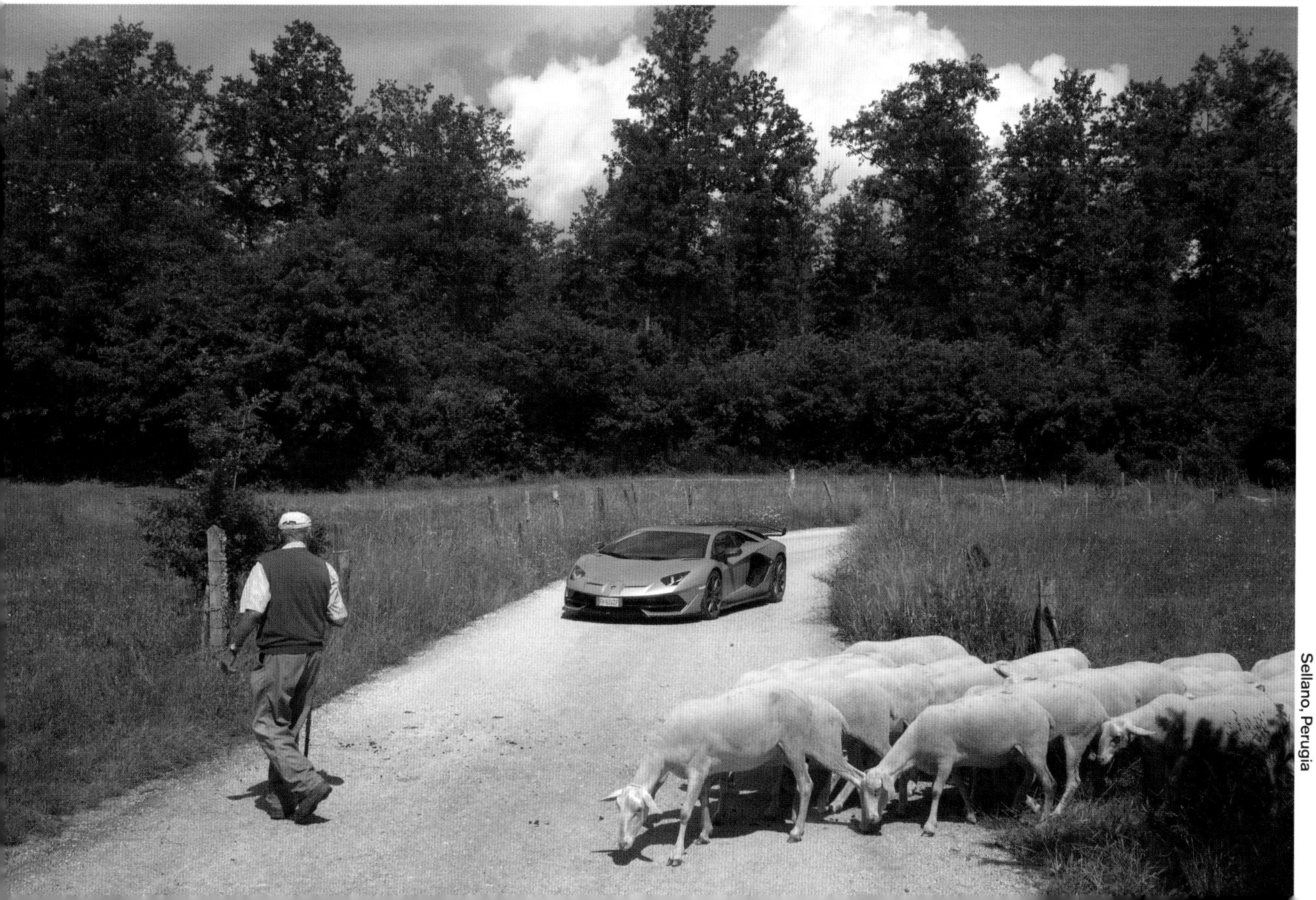

Sellano, Perugia

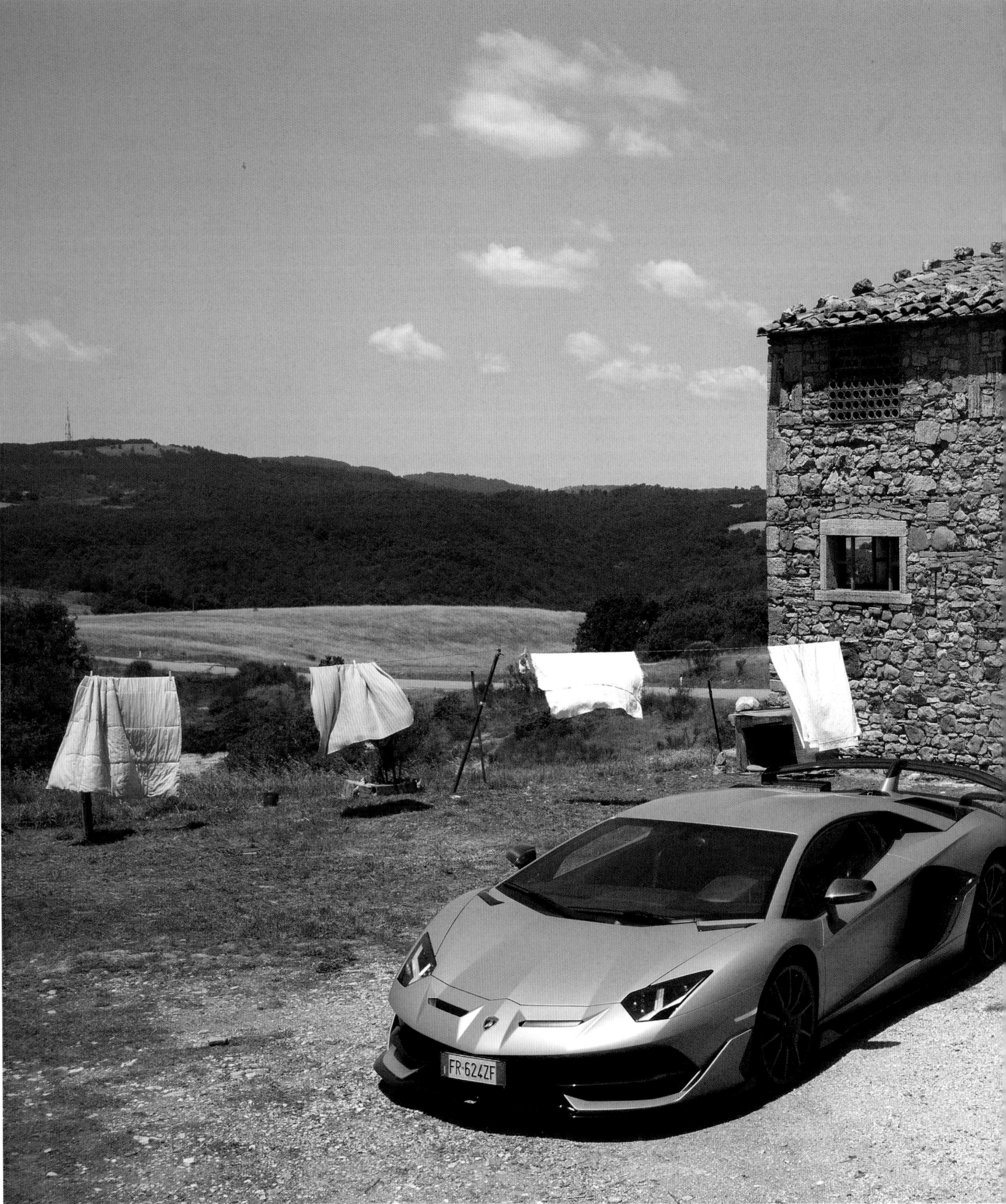

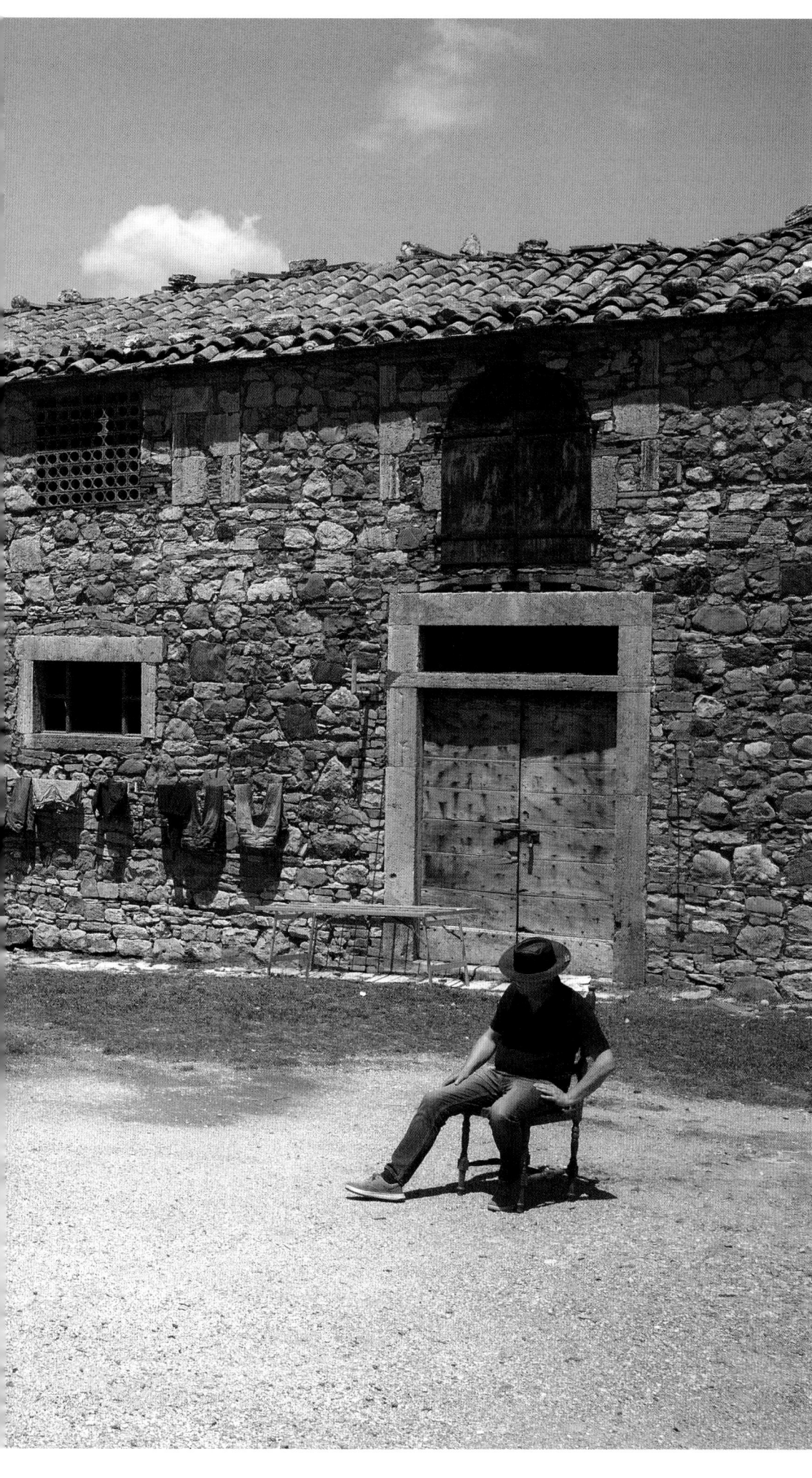

Titignano, Terni

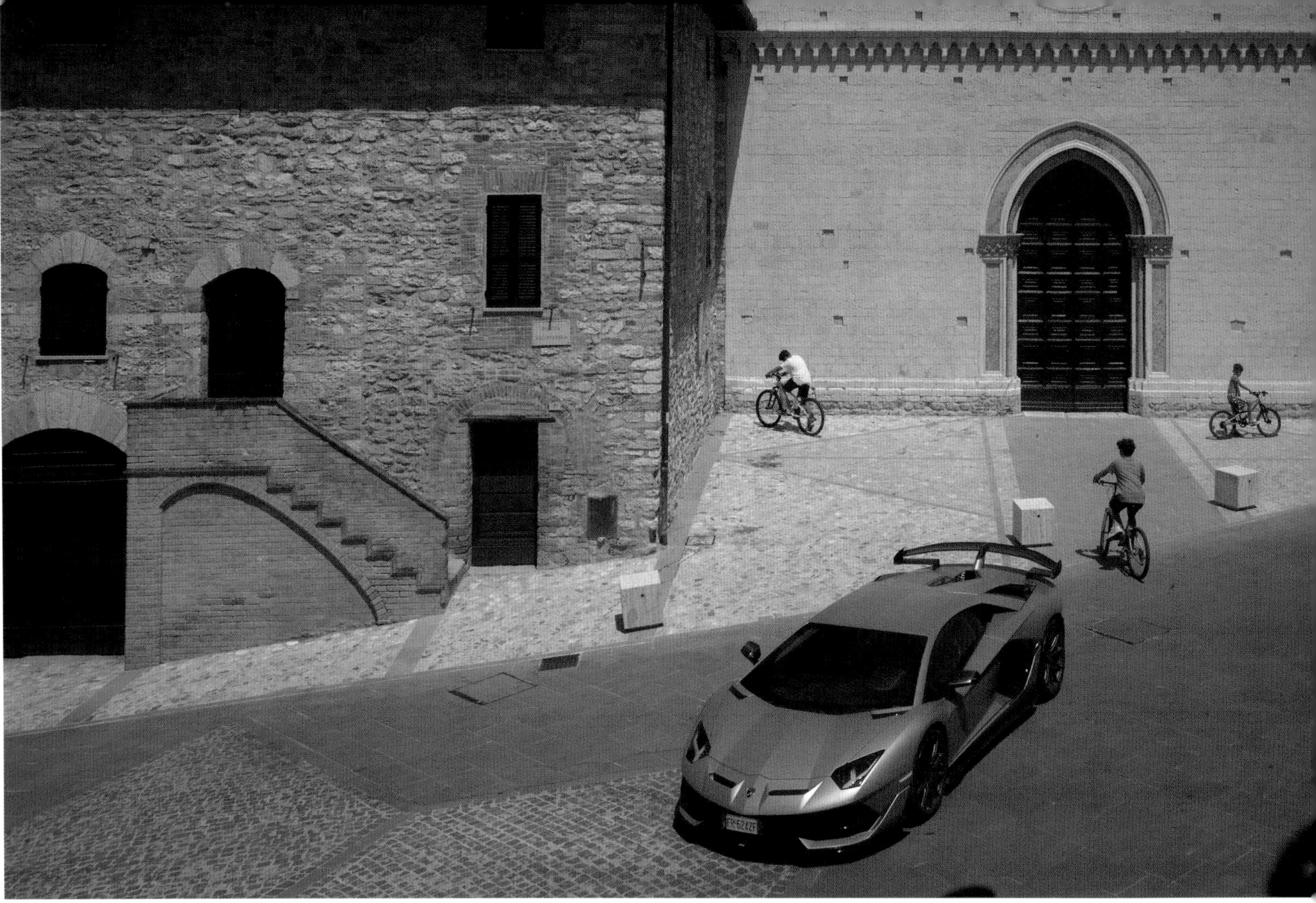

Spello, Perugia

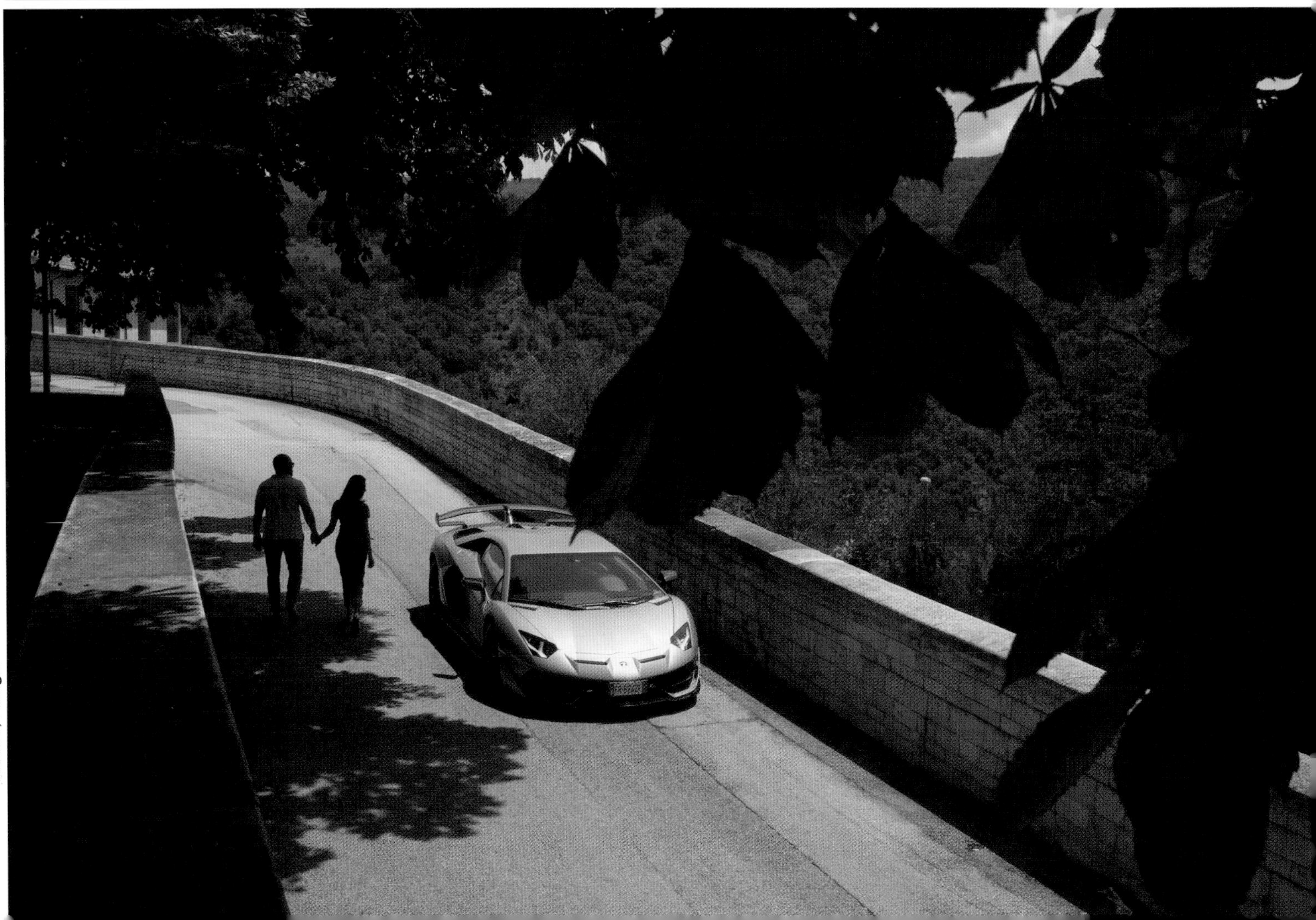

Sellano, Perugia

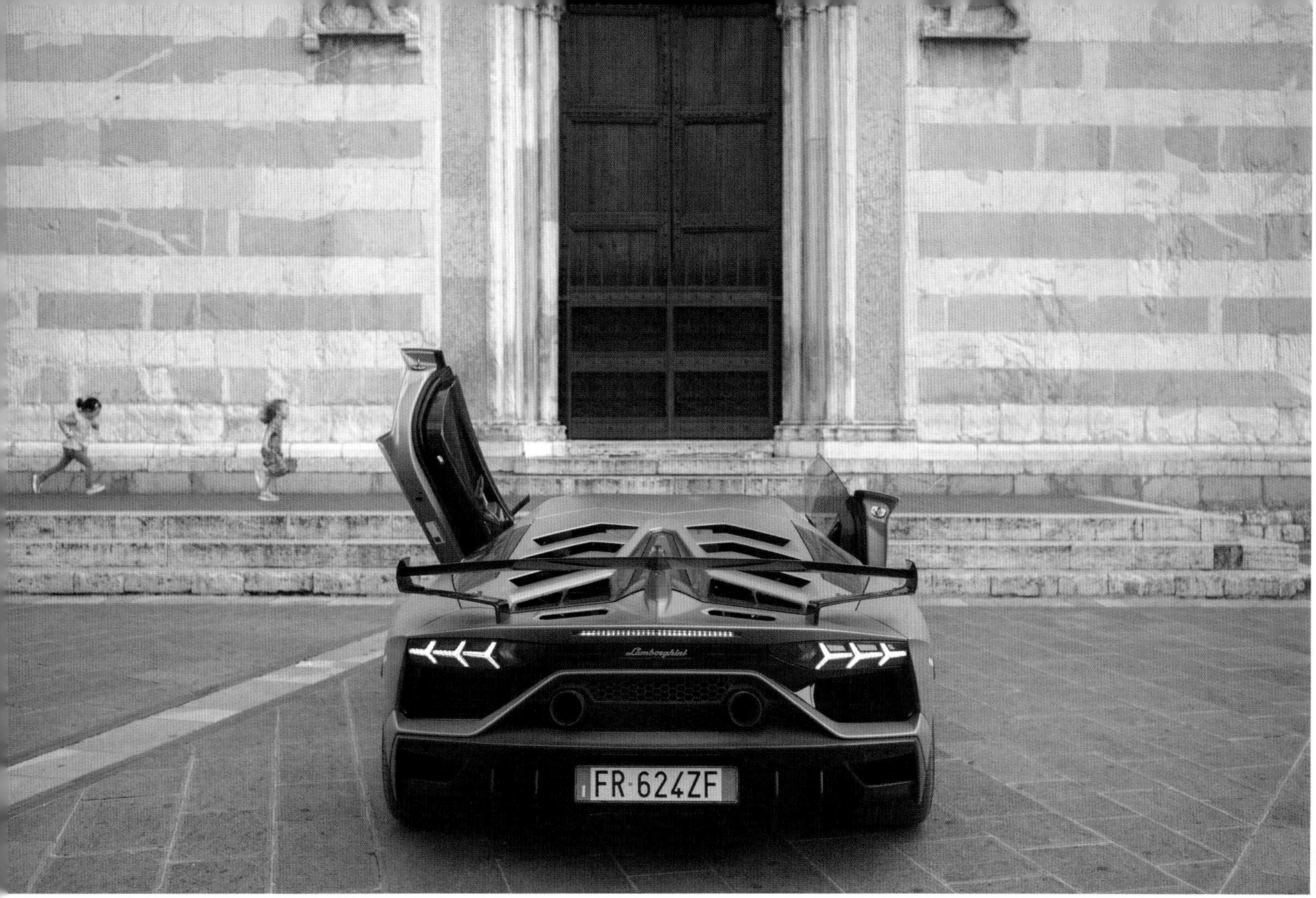

Assisi

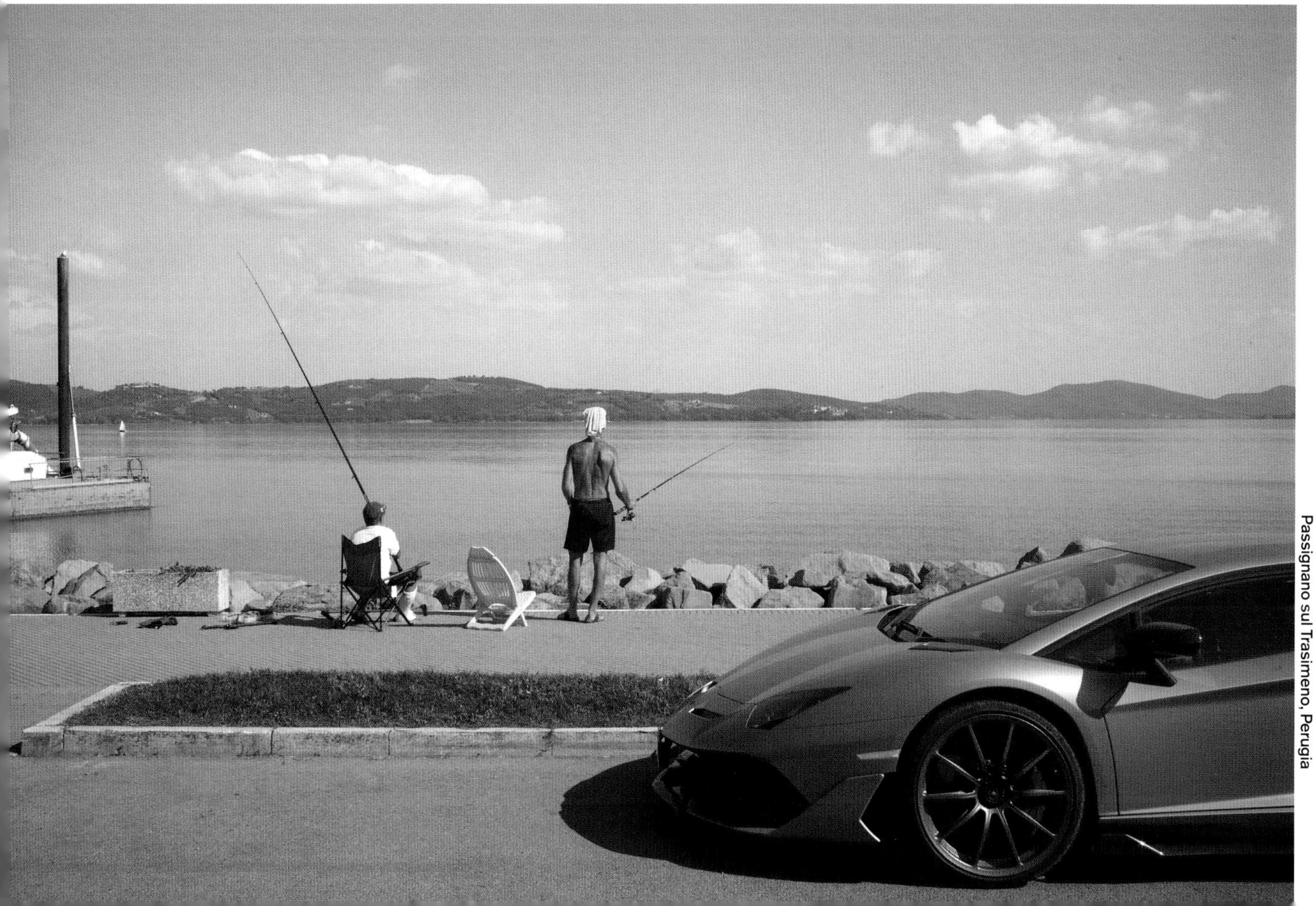

Passignano sul Trasimeno, Perugia

GABRIELE GALIMBERTI
TOSCANA

When I was a kid, I was a science fiction buff. I used to imagine my toy sports cars were spaceships. I had fun imagining a thousand possible worlds. To photograph the Huracán EVO RWD Spyder in my Tuscany was like going back in time, it pulled out those old passions in a kind of dreamlike narrative. I chose to involve children who have the same dreams today that I had back then, making images that evoke the purity and wonder of childhood, recolouring the Val di Chiana, its villages, hills, wheat fields and vineyards. I transformed this region into a science fiction film set in which the past resurfaces to join the future where, just like a Lamborghini, dreams are born in colour.

All photos taken in Castiglion Fiorentino and Cortona, Arezzo
In order of appearance: Pierle (opposite), Church of Santa Margherita in Cortona (p. 154), Villa di Brolio (p. 155), Montecchio Castle (p. 160), and the centuries-old plane tree at Villa Passerini in Pergo (p. 161)

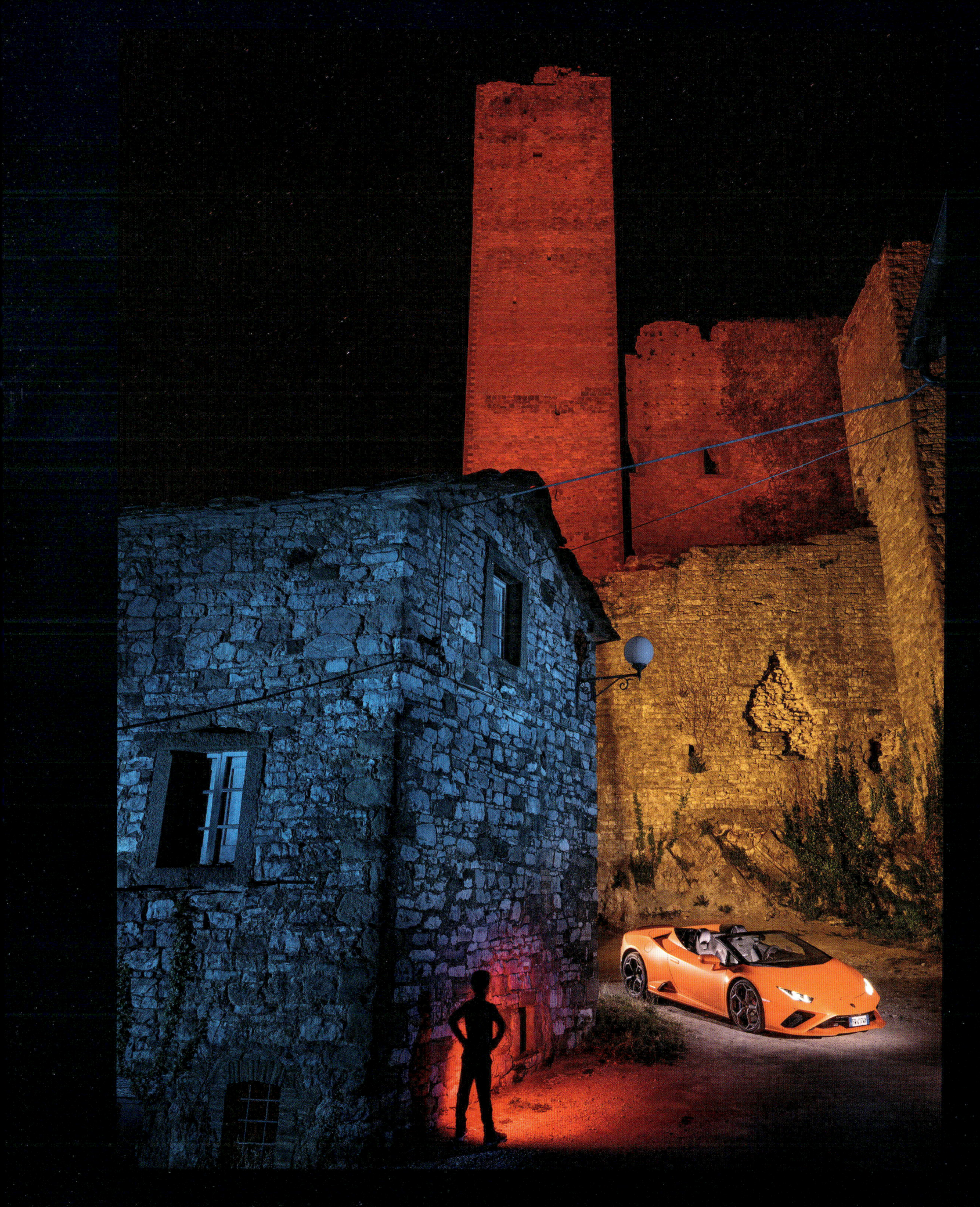

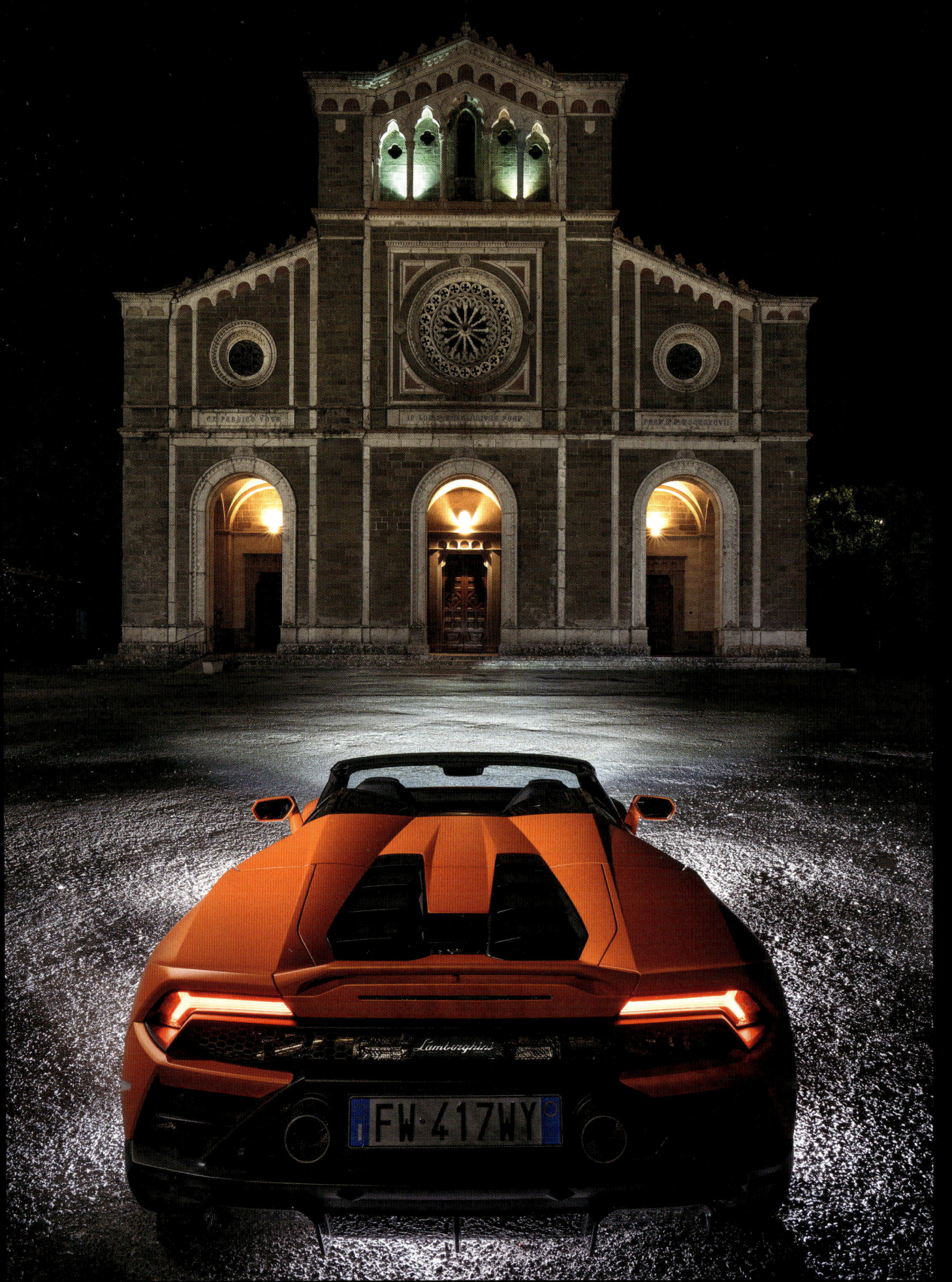

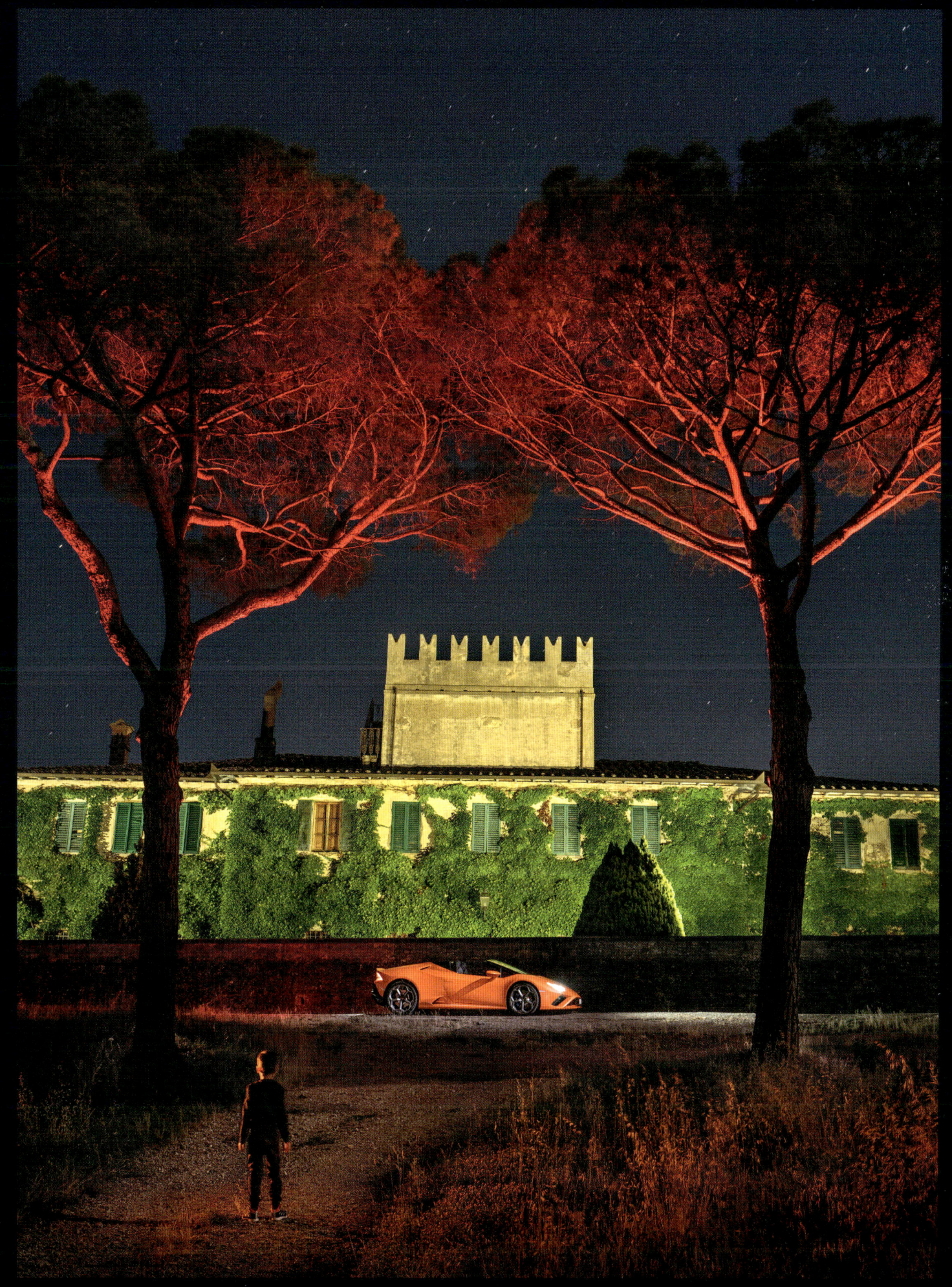

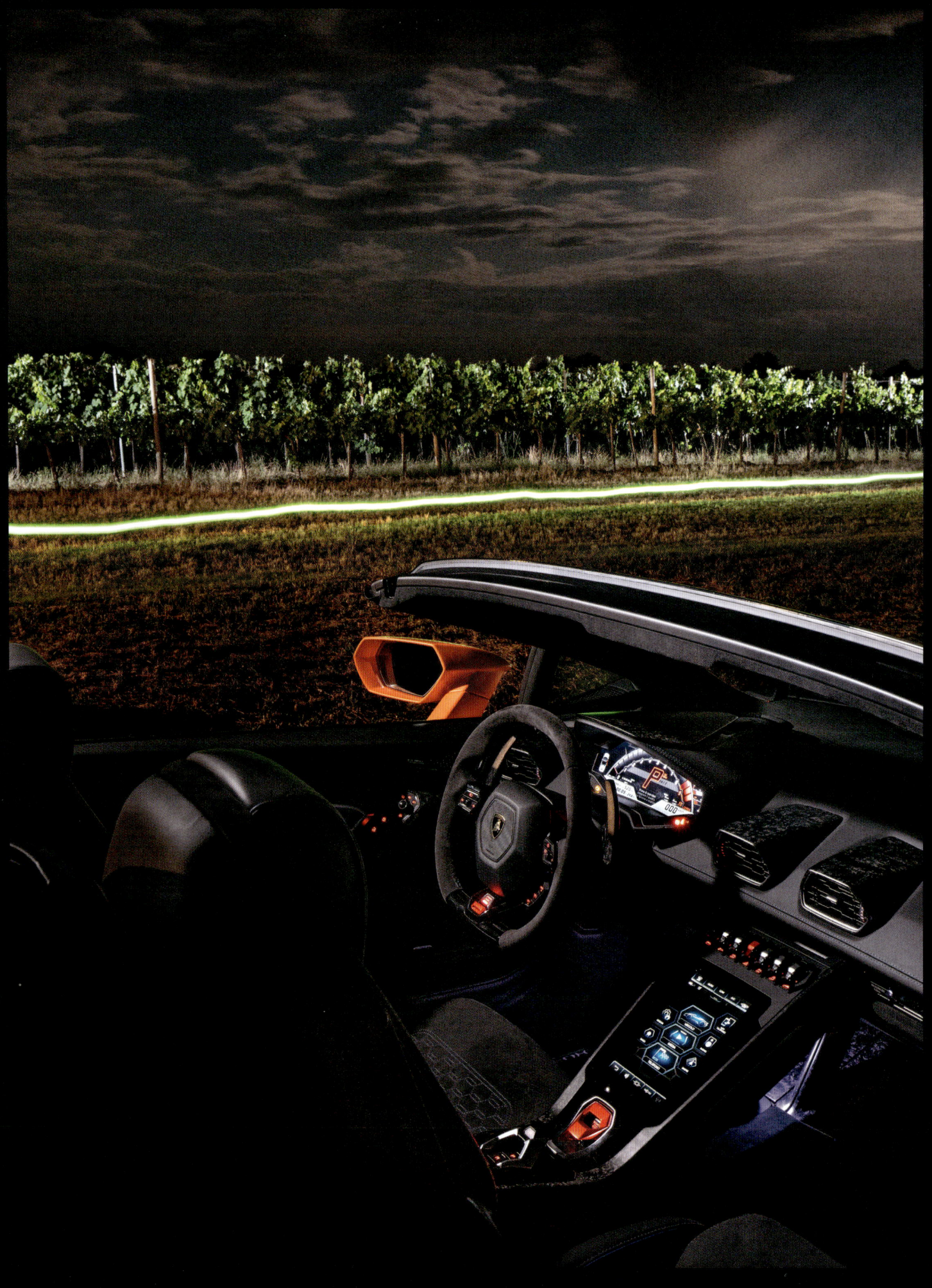

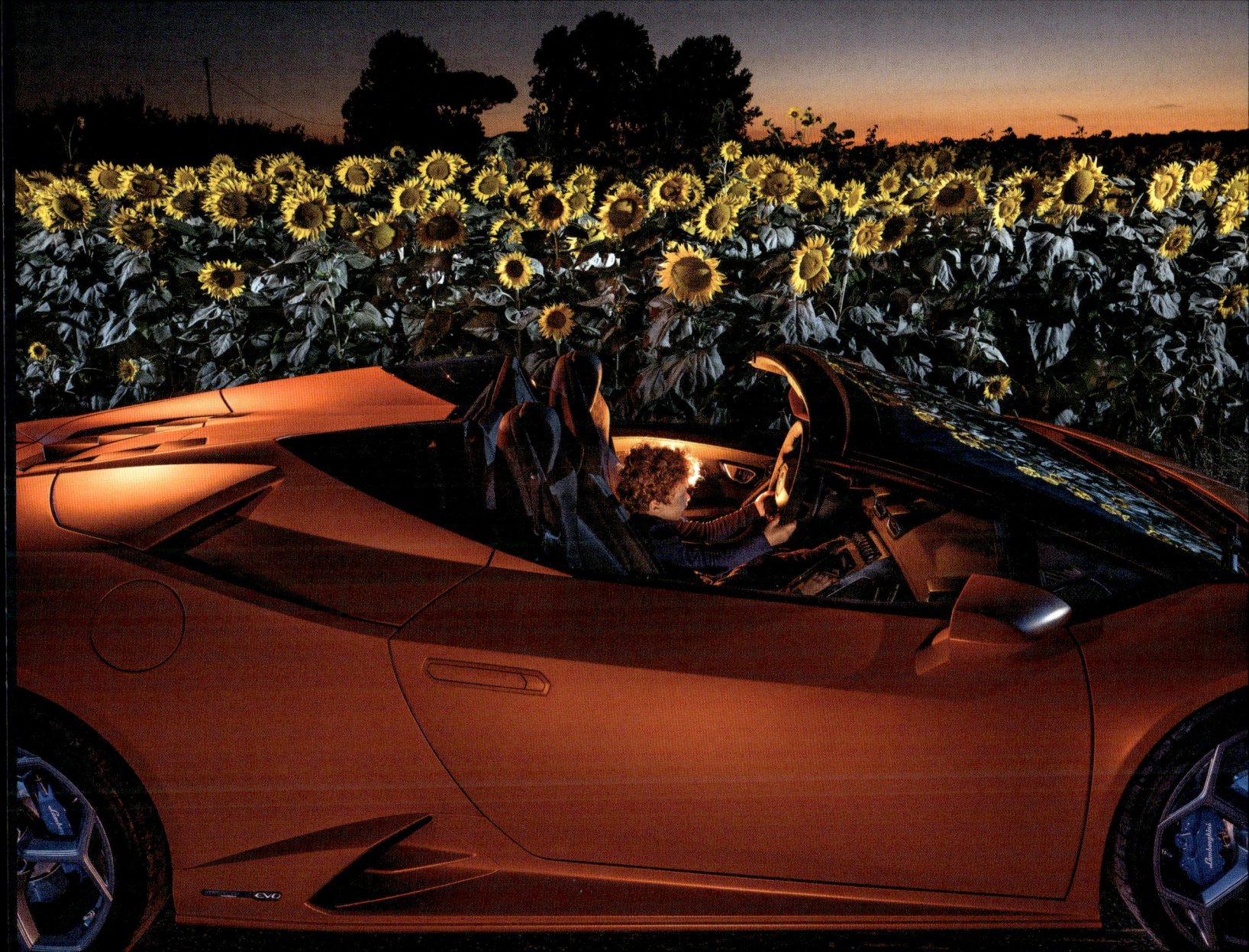

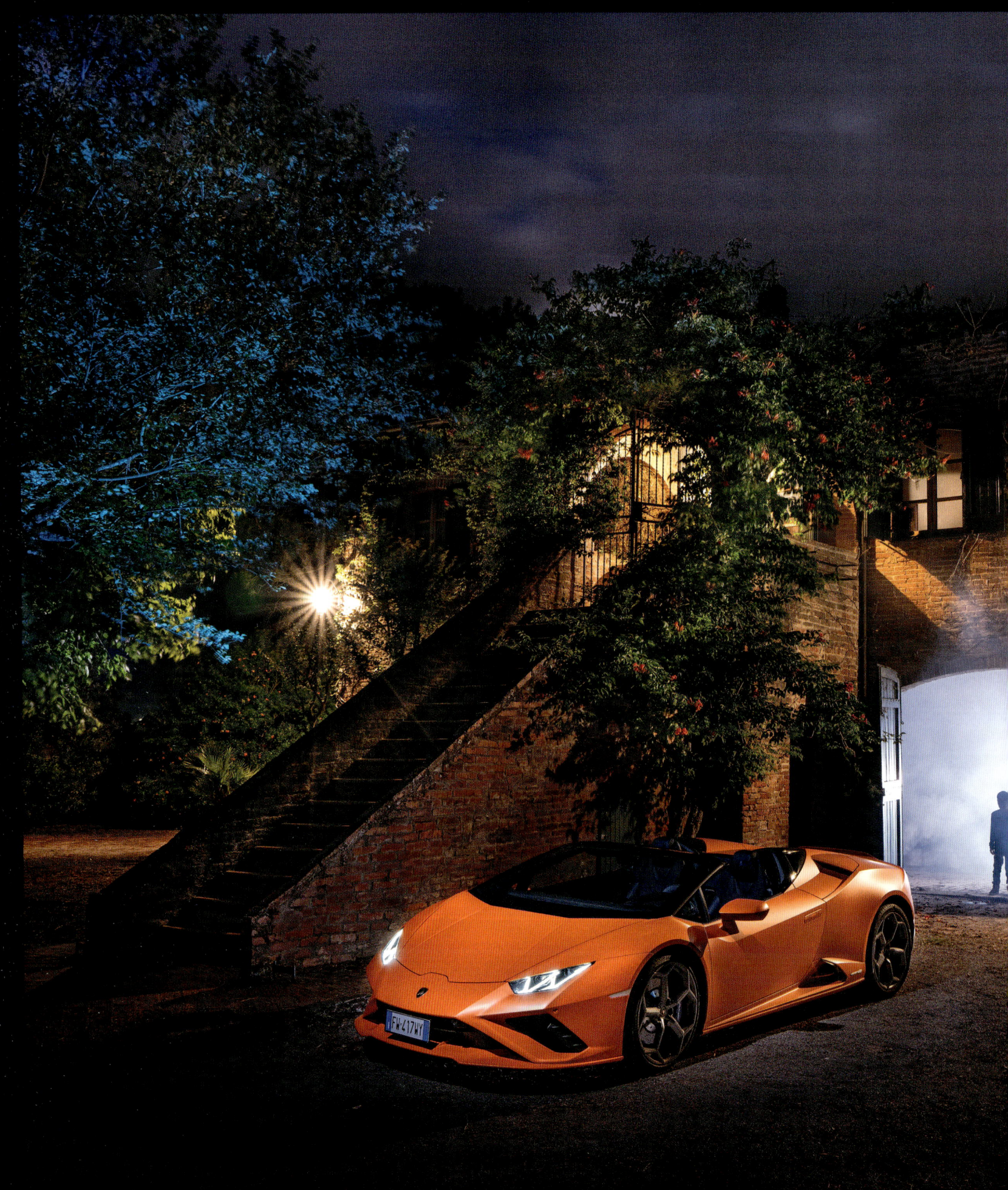

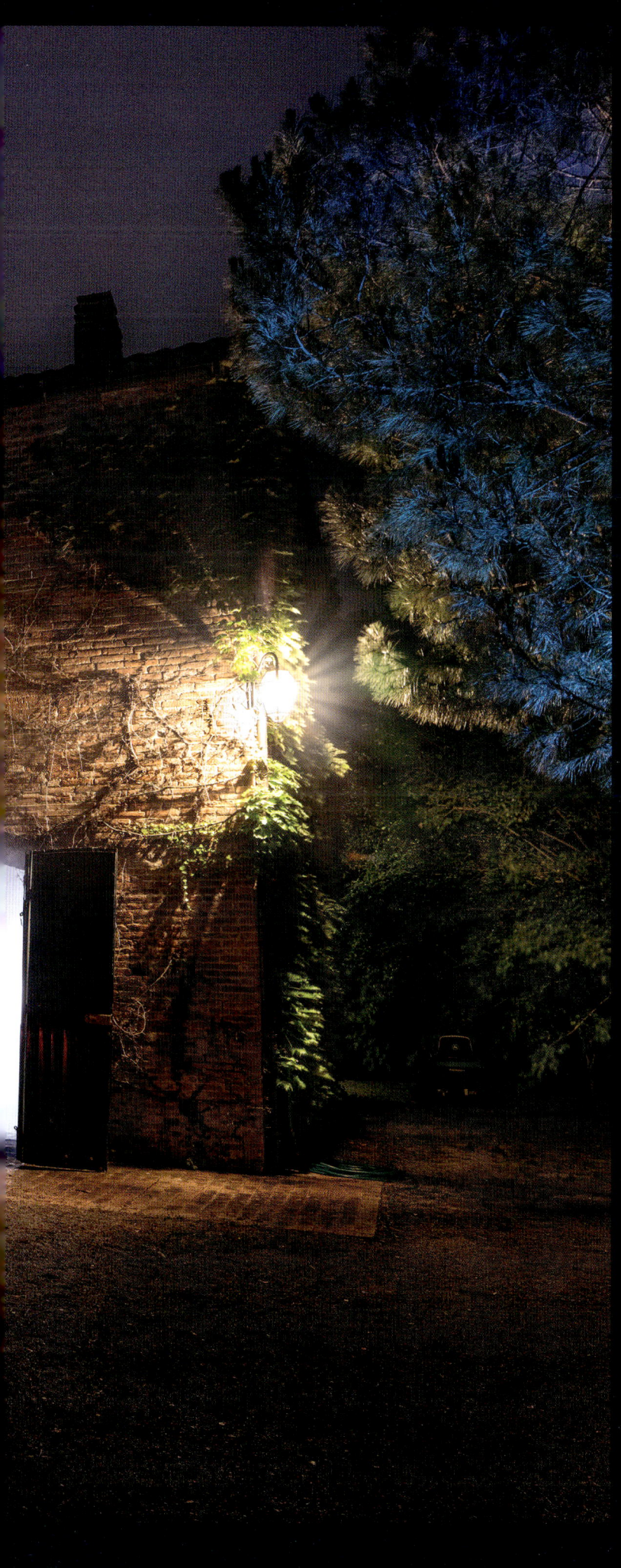

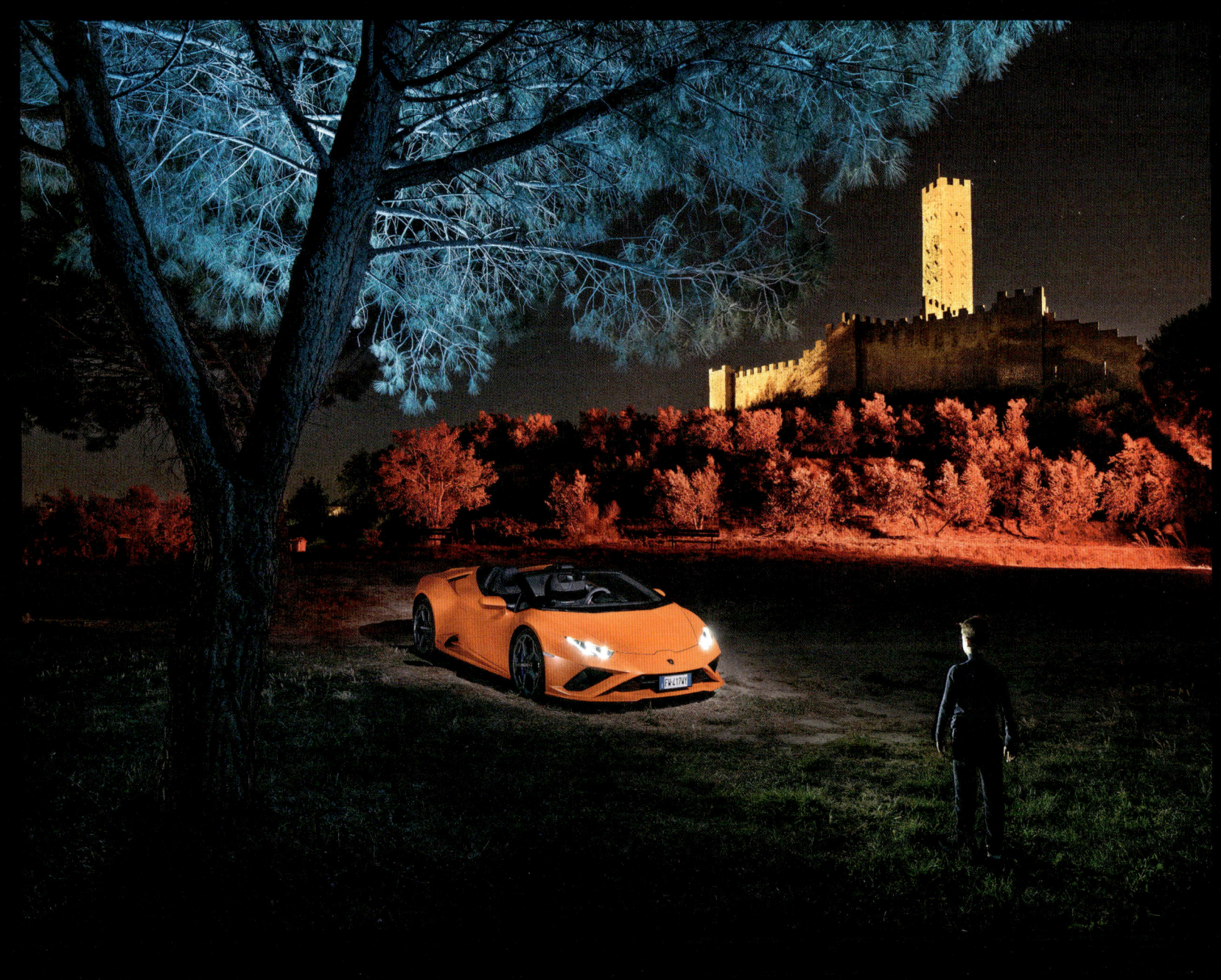

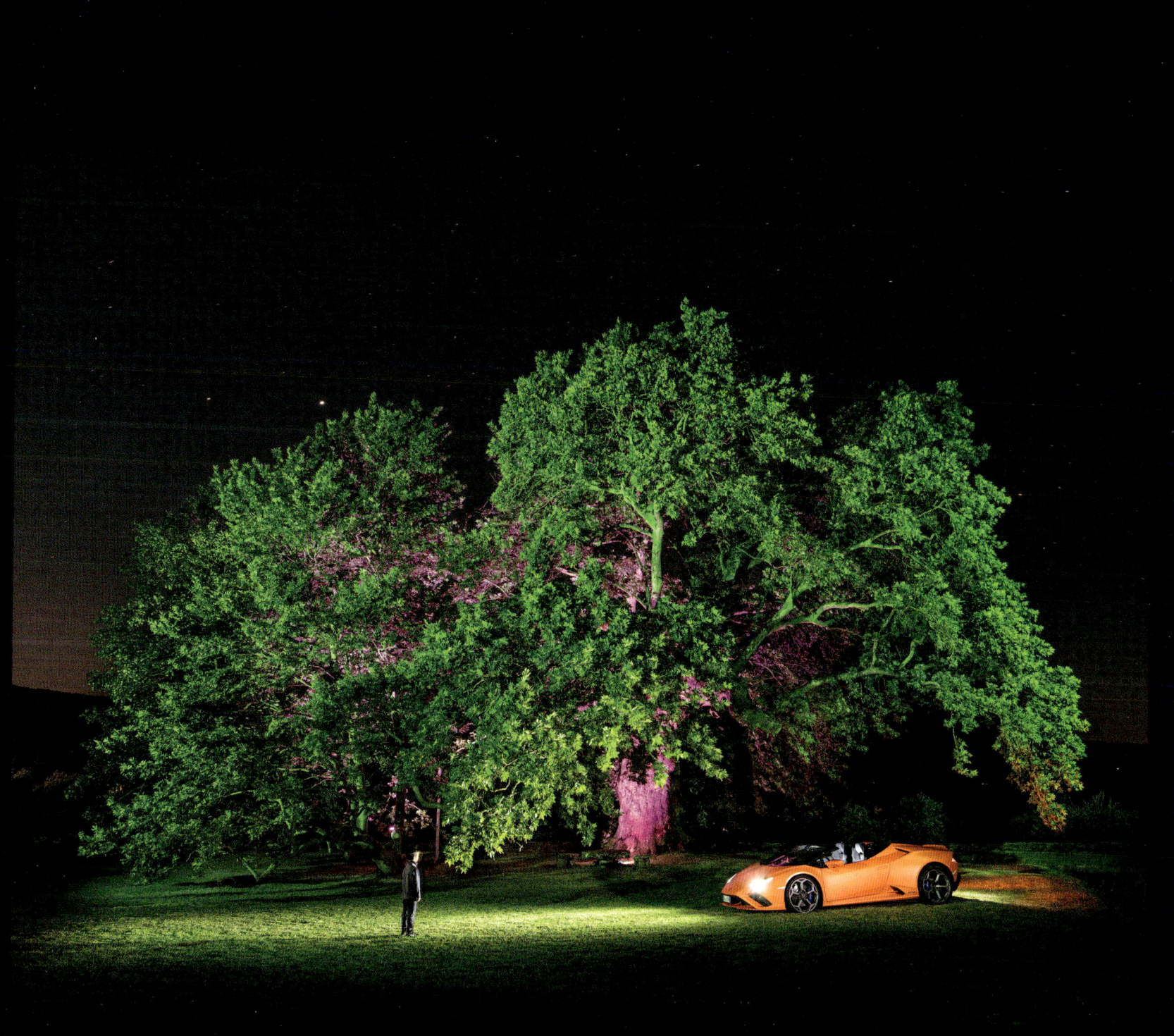

"Italy's speciality is that all these different and contrasting beauties, which unfold one by one in a single swathe of land, are represented in a compendium that, like a prism, in a limited space reflects all the rest of the world in its many faces"

Guido Piovene

PIERO GEMELLI
EMILIA-ROMAGNA

Only those who know, love and control speed can appreciate slowness. I organized my photographic project around this concept in Emilia-Romagna, on a purely contemplative journey aboard an Aventador S Roadster. I took shots conceived as "Postcards of Italy", from the most famous cities to hidden spots off the beaten track. I stopped in places that would allow my gaze to slow down and take the time to calmly and silently admire a magnificence that is, in fact, an expression of who we are, of Italian culture as identity. Places far from the road, where silence amplifies the roar of the engine, where the car, a symbol of technological development and the quality of high-end Italian manufacturing, dovetails with the wonders of a territory that welcomes it and speaks of us.

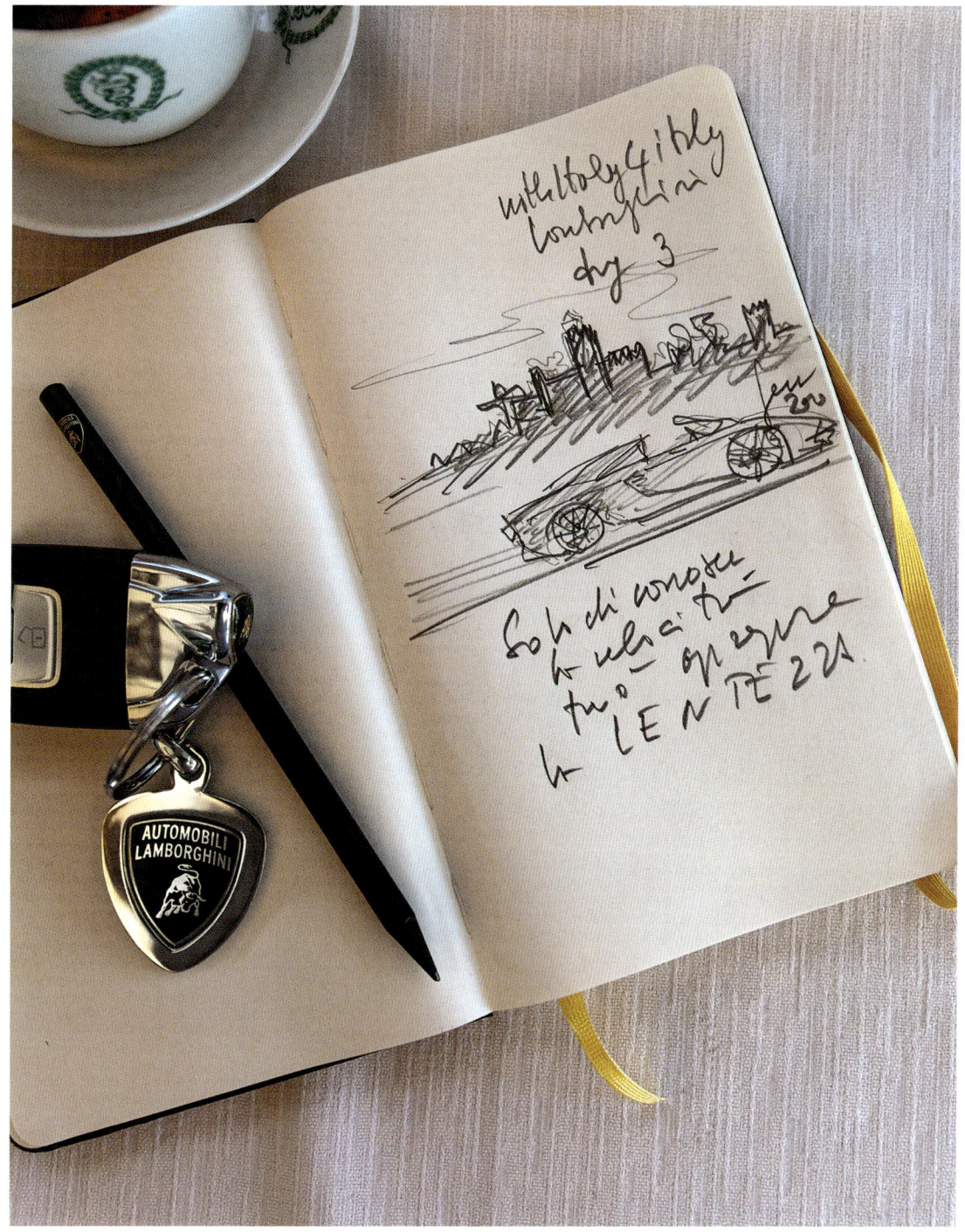

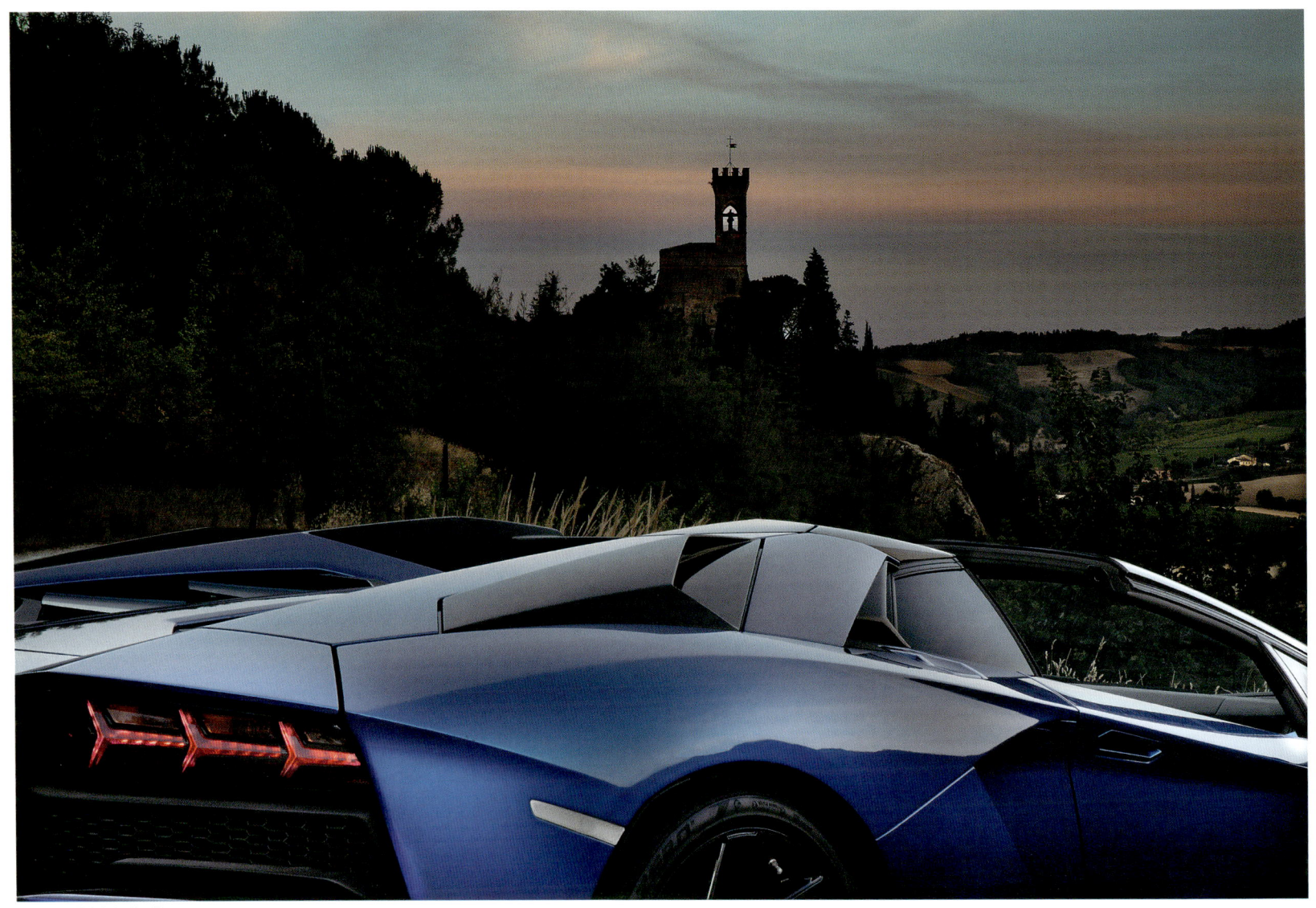
Brisighella, Ravenna

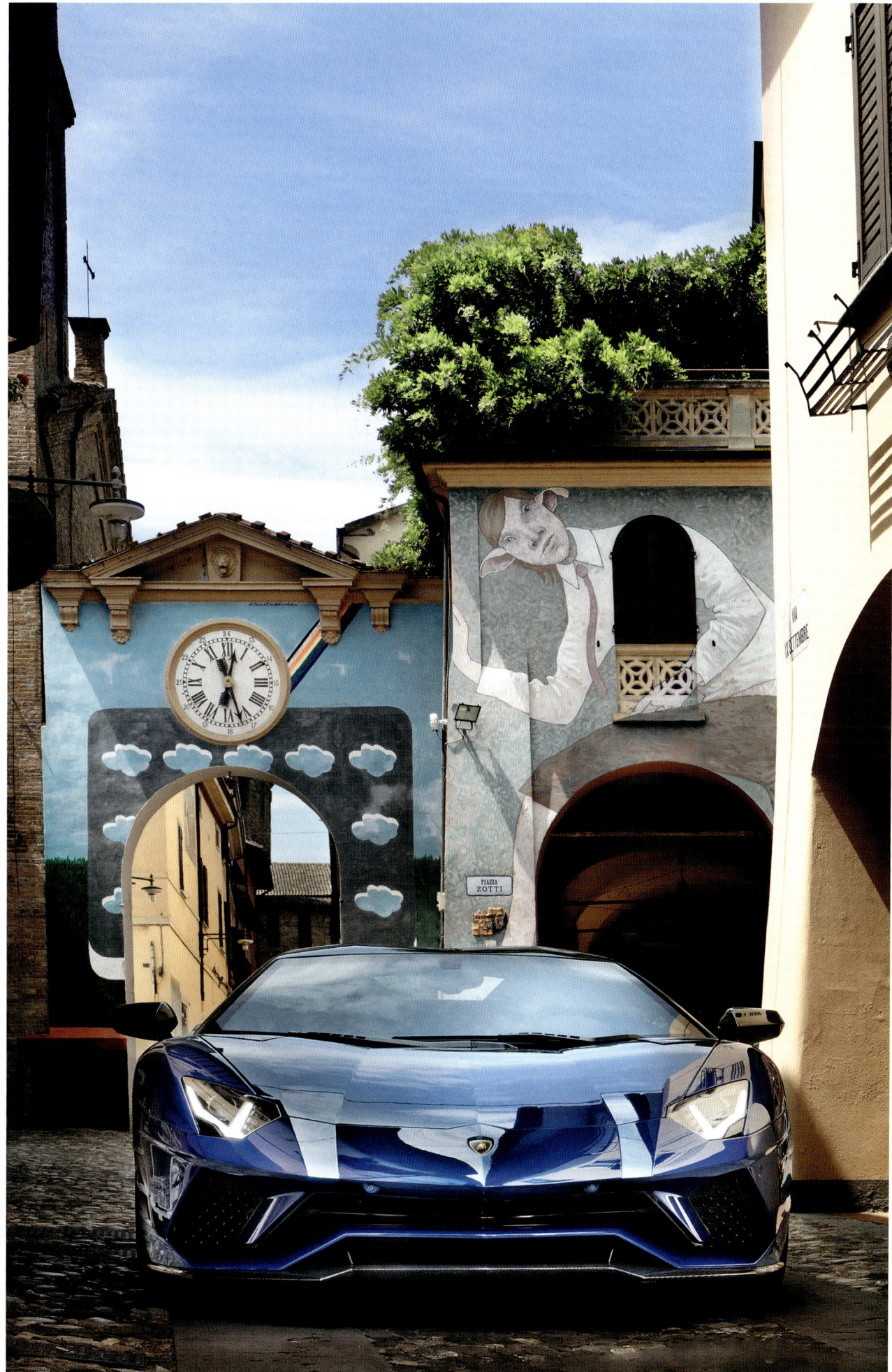

Dozza, Bologna

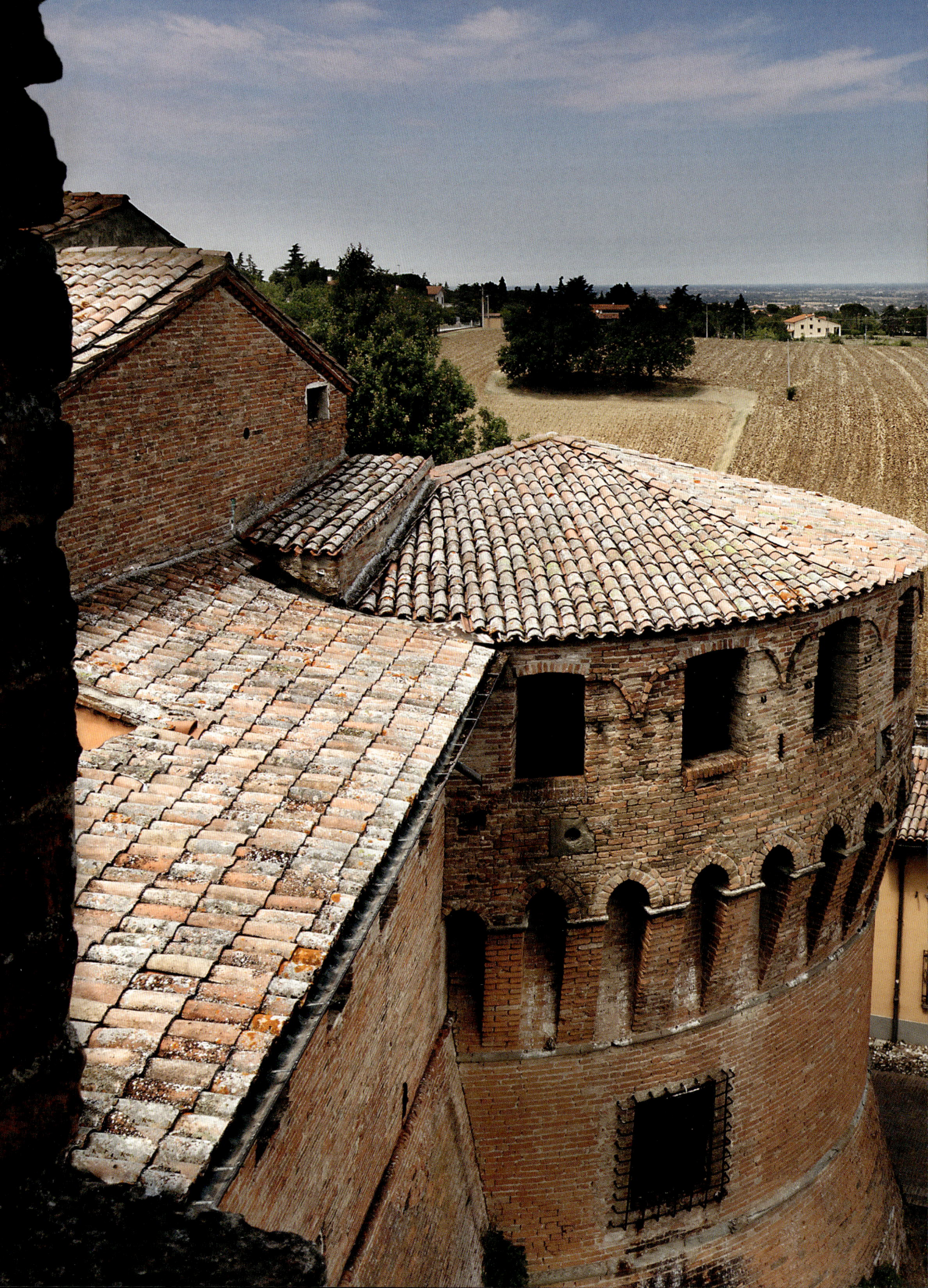

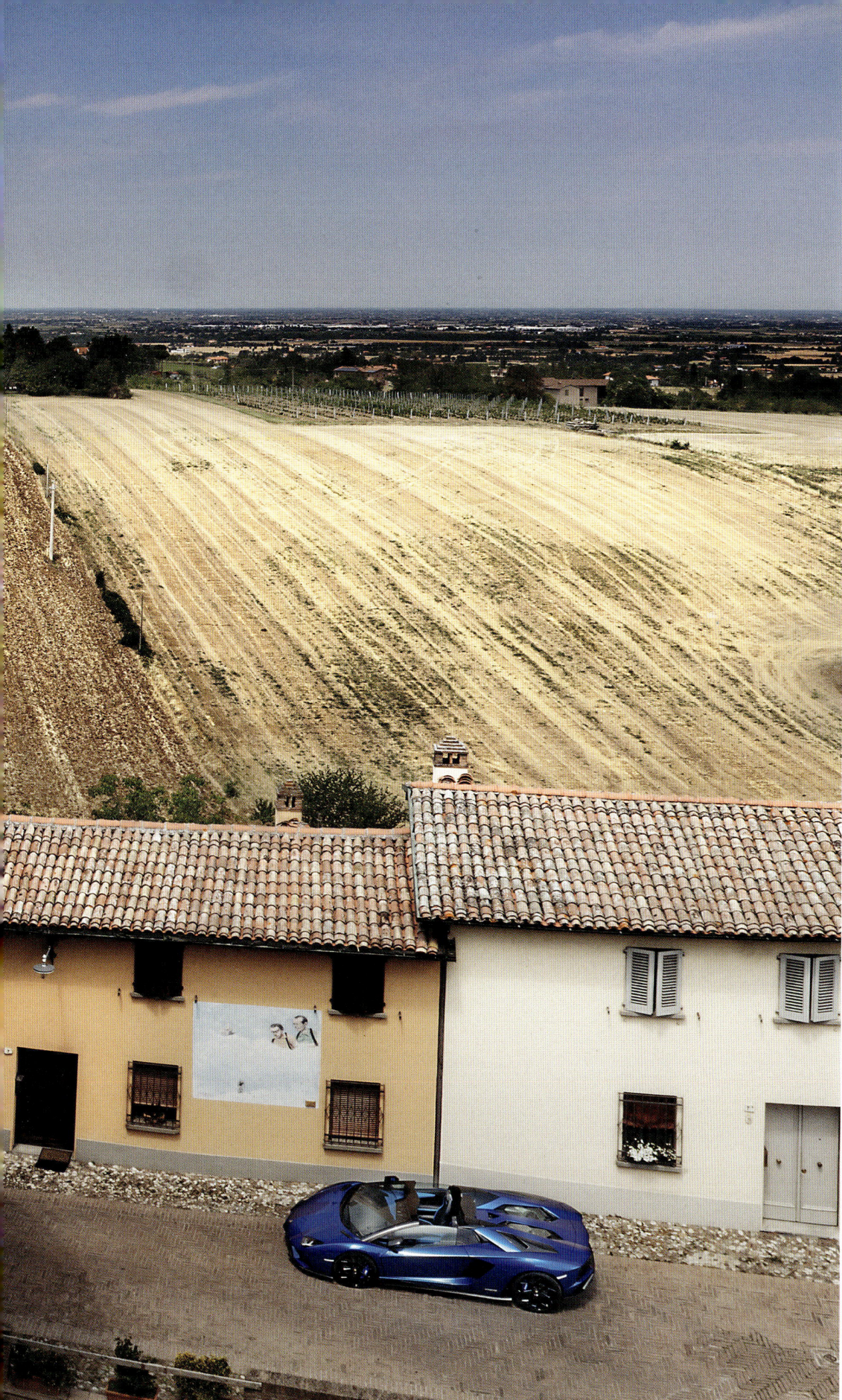

Rocca Sforzesca at Dozza, Bologna, view from the tower

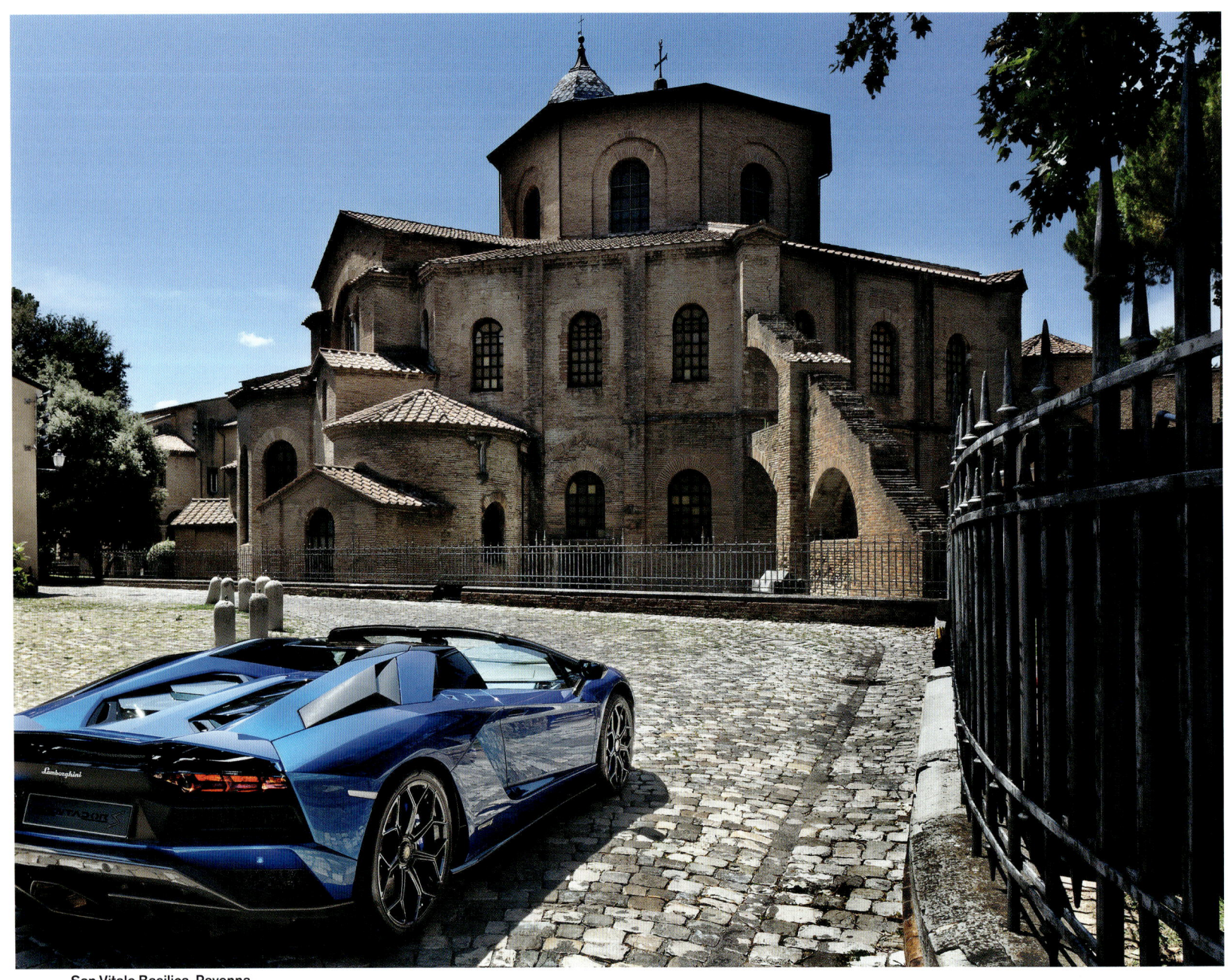

San Vitale Basilica, Ravenna

View of Torrechiara Castle, Langhirano, Parma

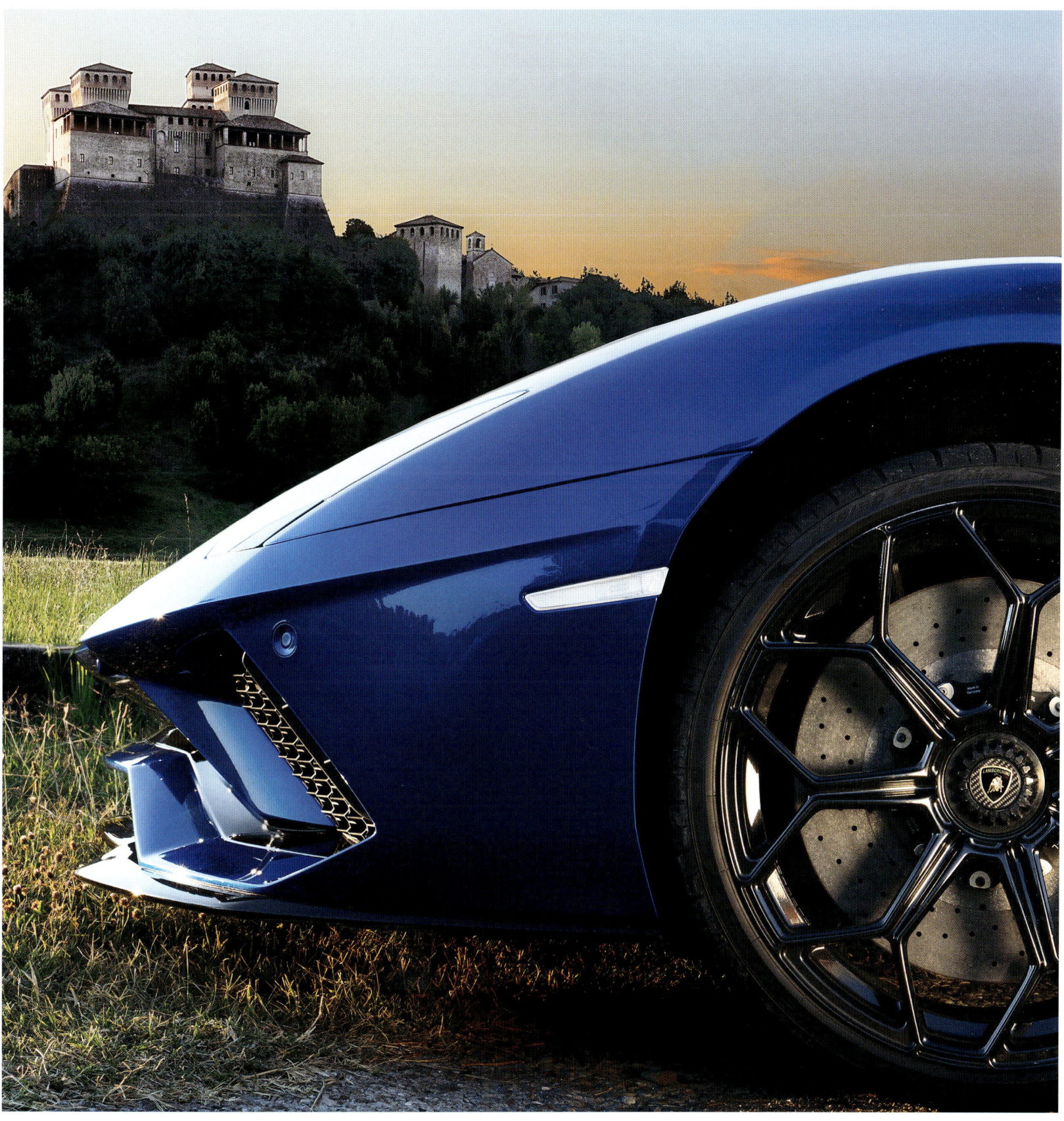

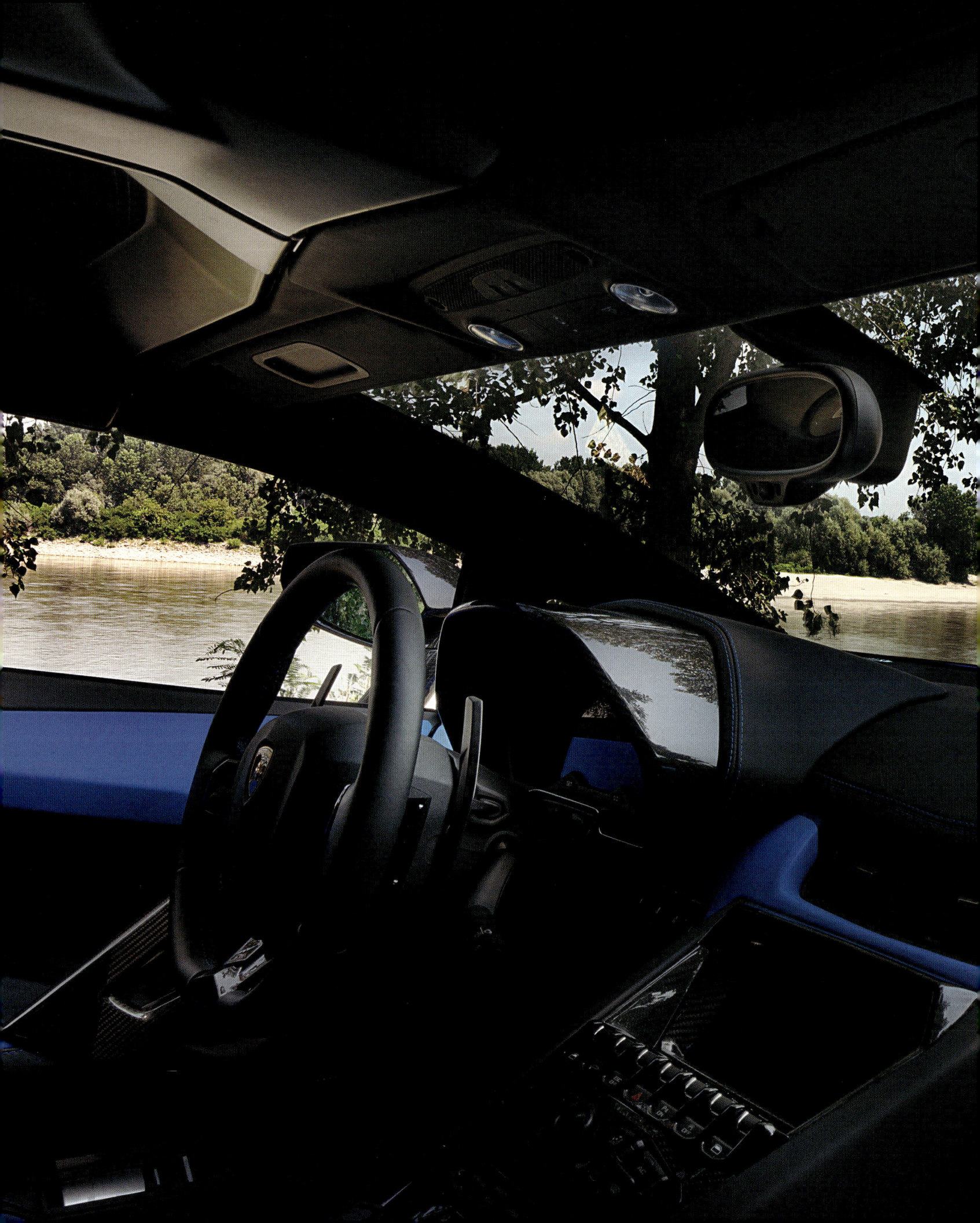

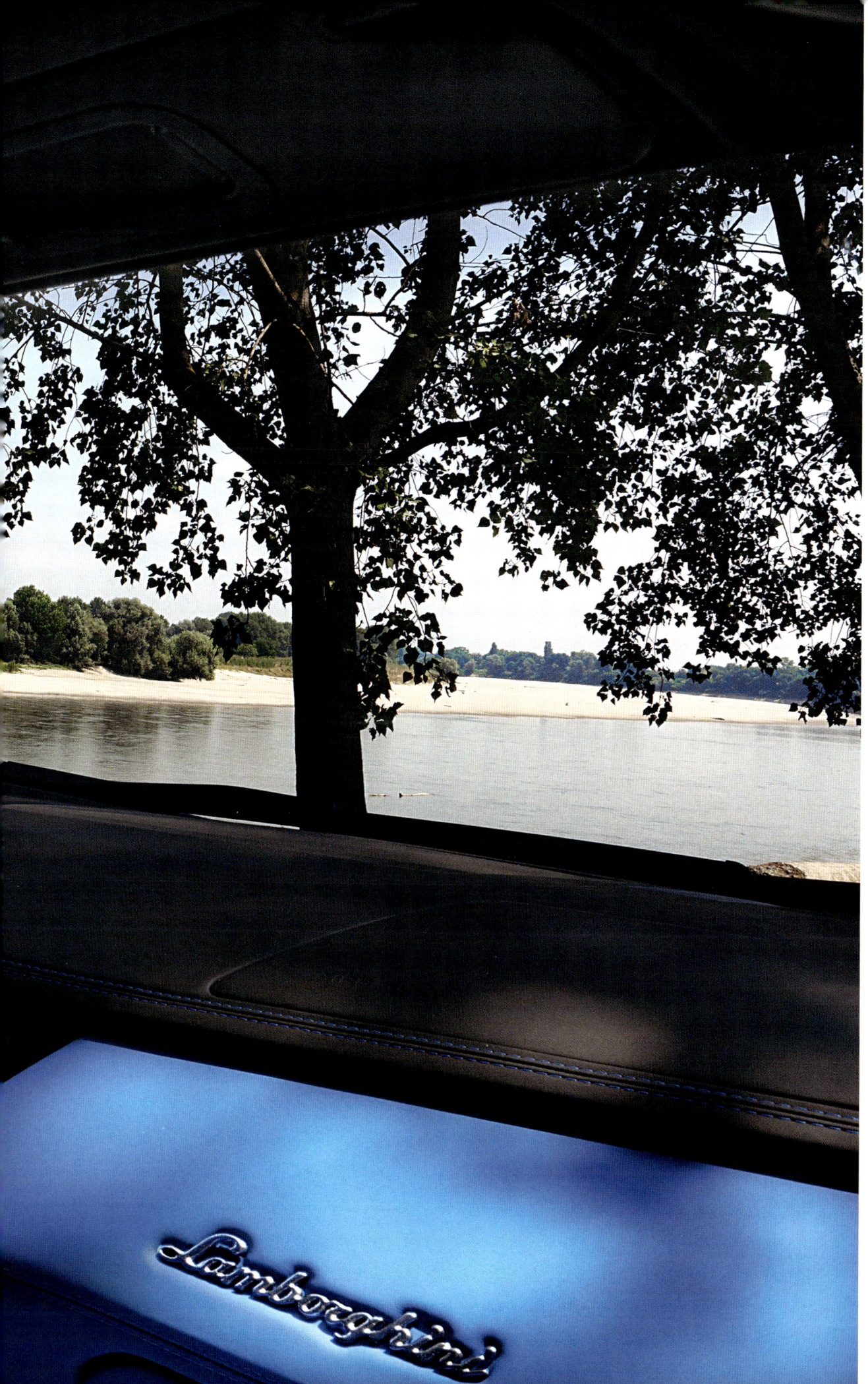

Polesine parmense,
Riva del Po

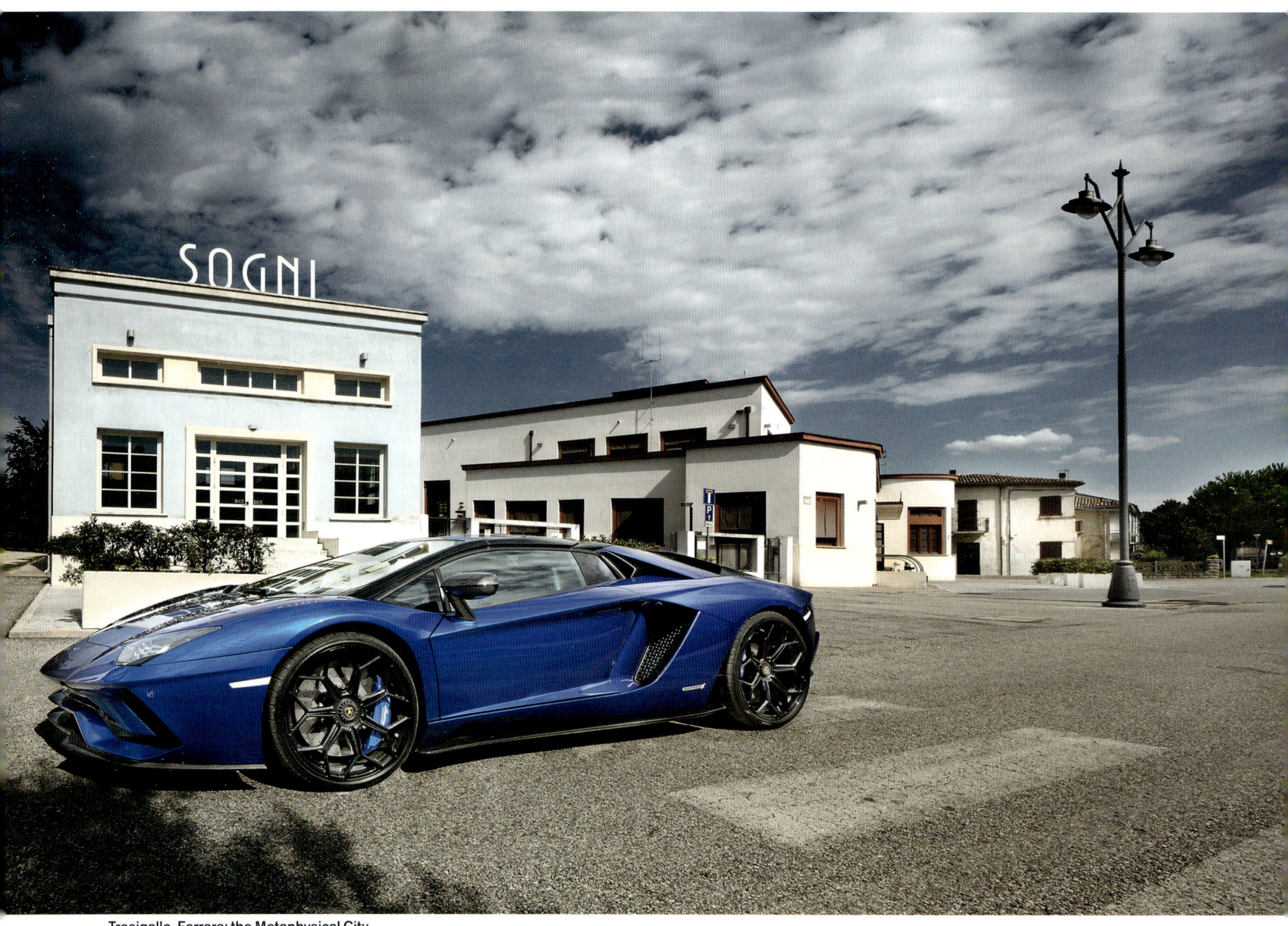
Tresigallo, Ferrara: the Metaphysical City

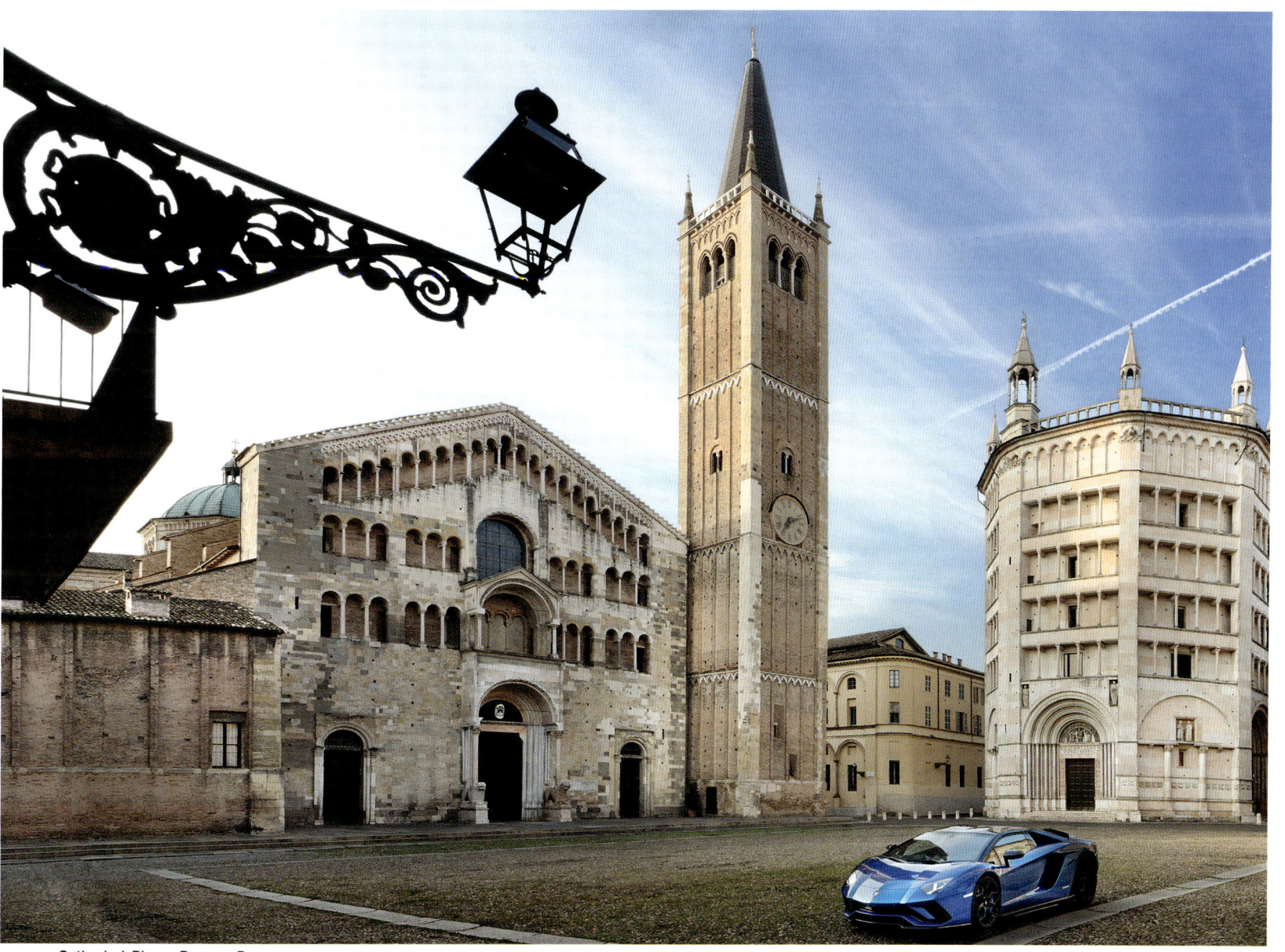

Cathedral, Piazza Duomo, Parma

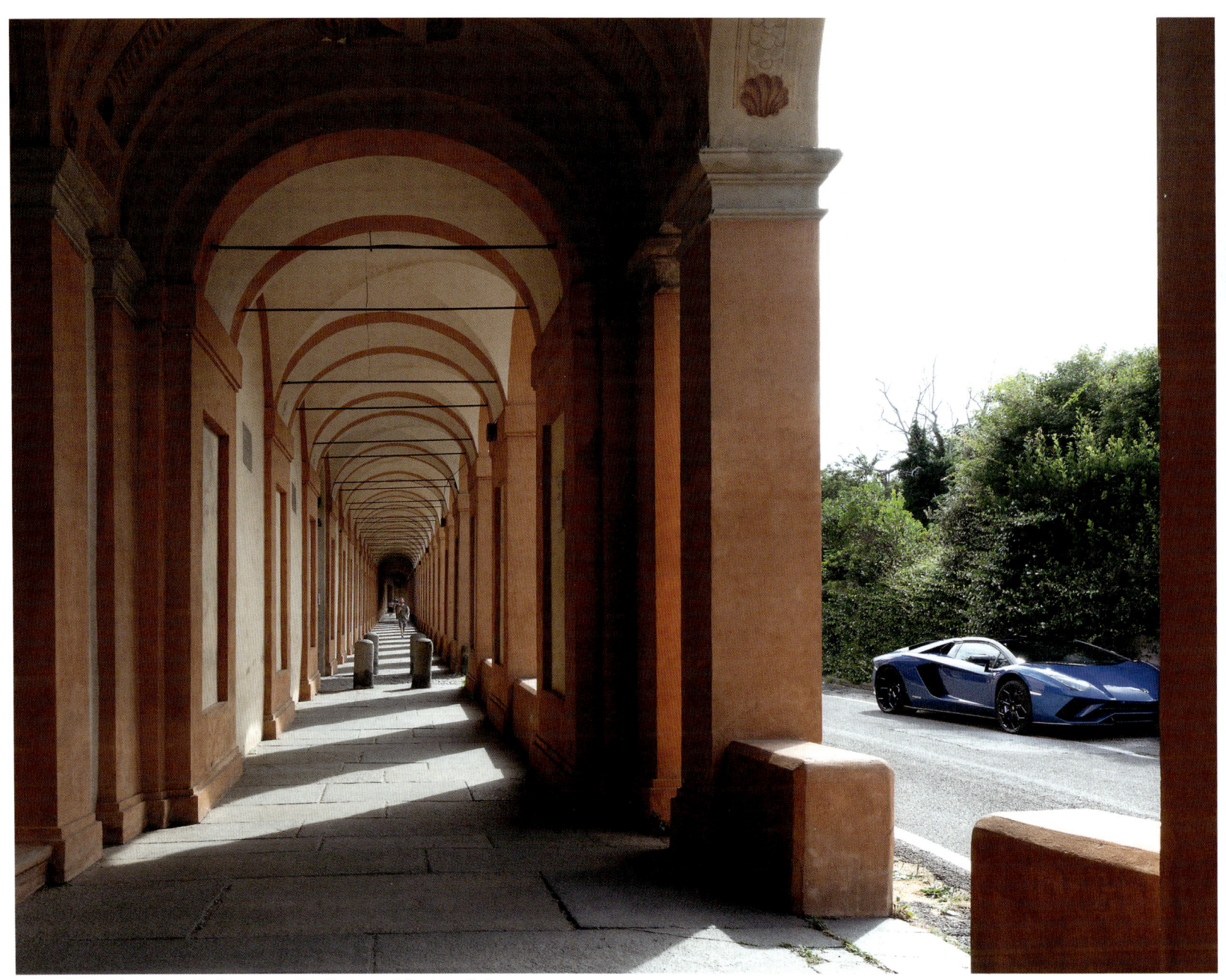

Pedestrian route to the Sanctuary of the Madonna di San Luca, Bologna

Terme Berzieri, Salsomaggiore Terme

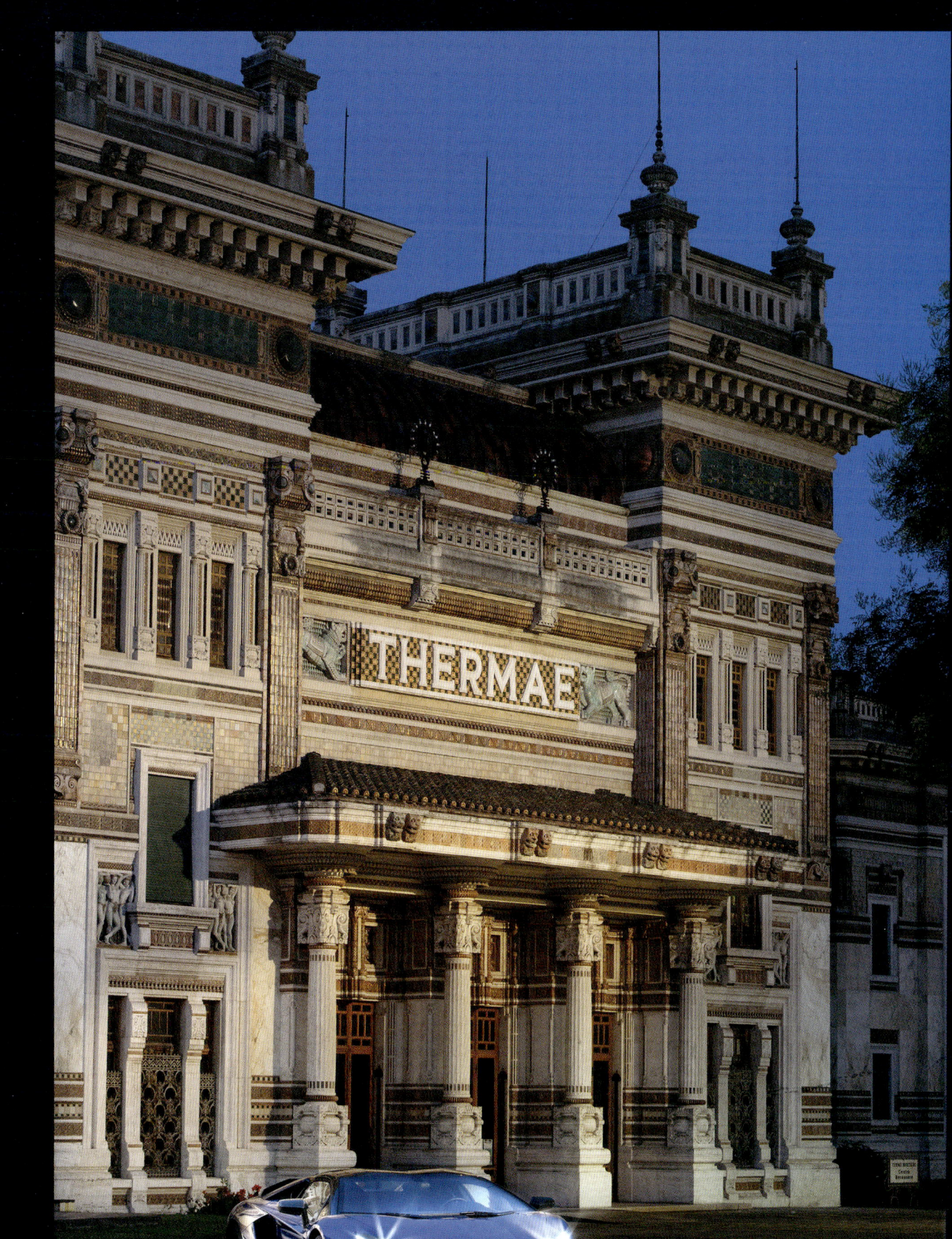

MARCO VALMARANA
VENETO

I wanted to fully leverage the stylistic essence and aristocratic elegance of my Veneto to tell its story, in a synthesis between perfect natural landscapes, architecture and the timeless design of the Countach 25th Anniversary. A marriage of style celebrated between Venetian villas, with their priceless artistic heritage, and the breathtaking views of the Dolomite passes at Giau. Finally, my beloved lagoon, its islands in the background and, away in the distance, Venice. A journey back in time, a visual evocation of the value, richness and superiority of our past.

Asolo

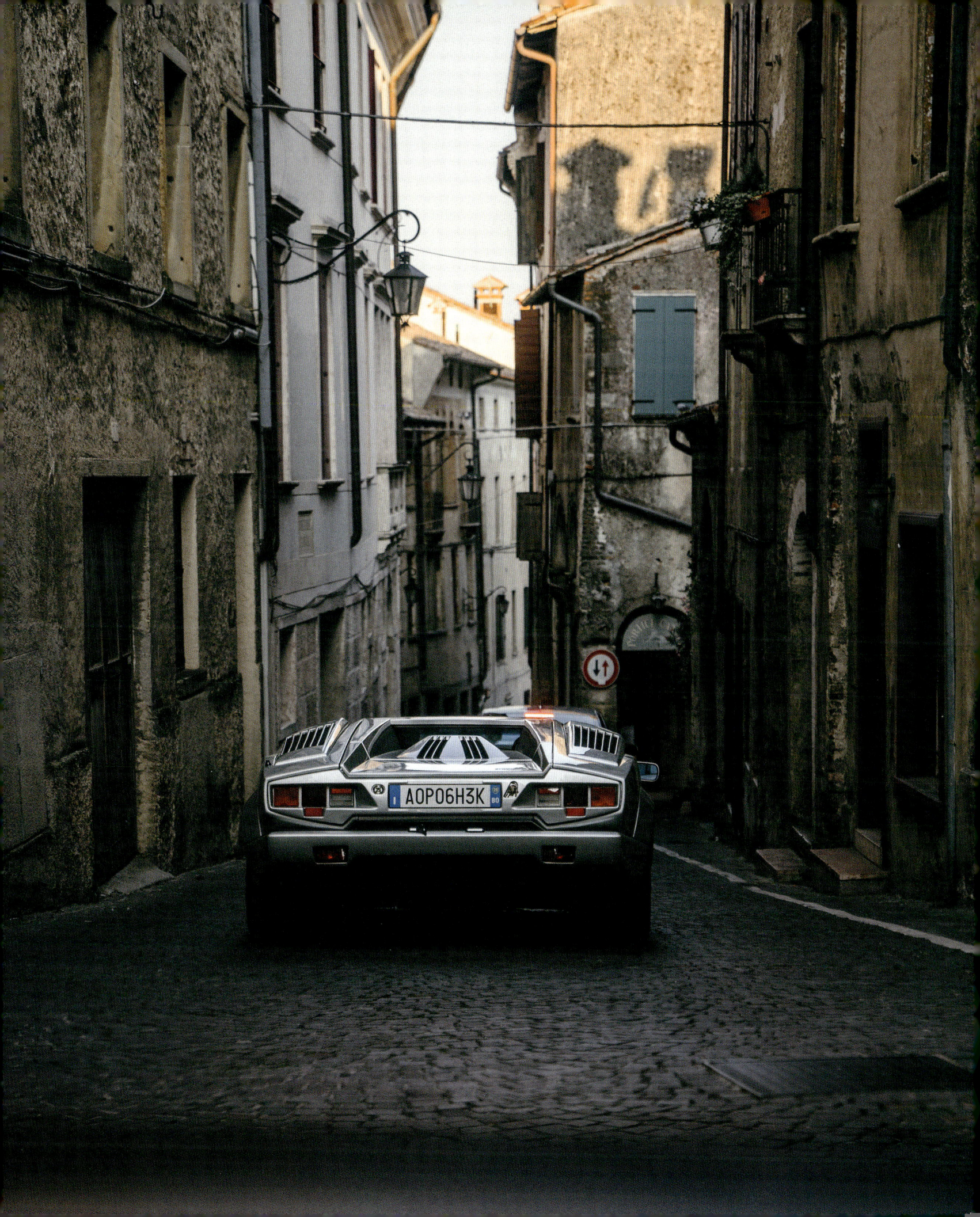

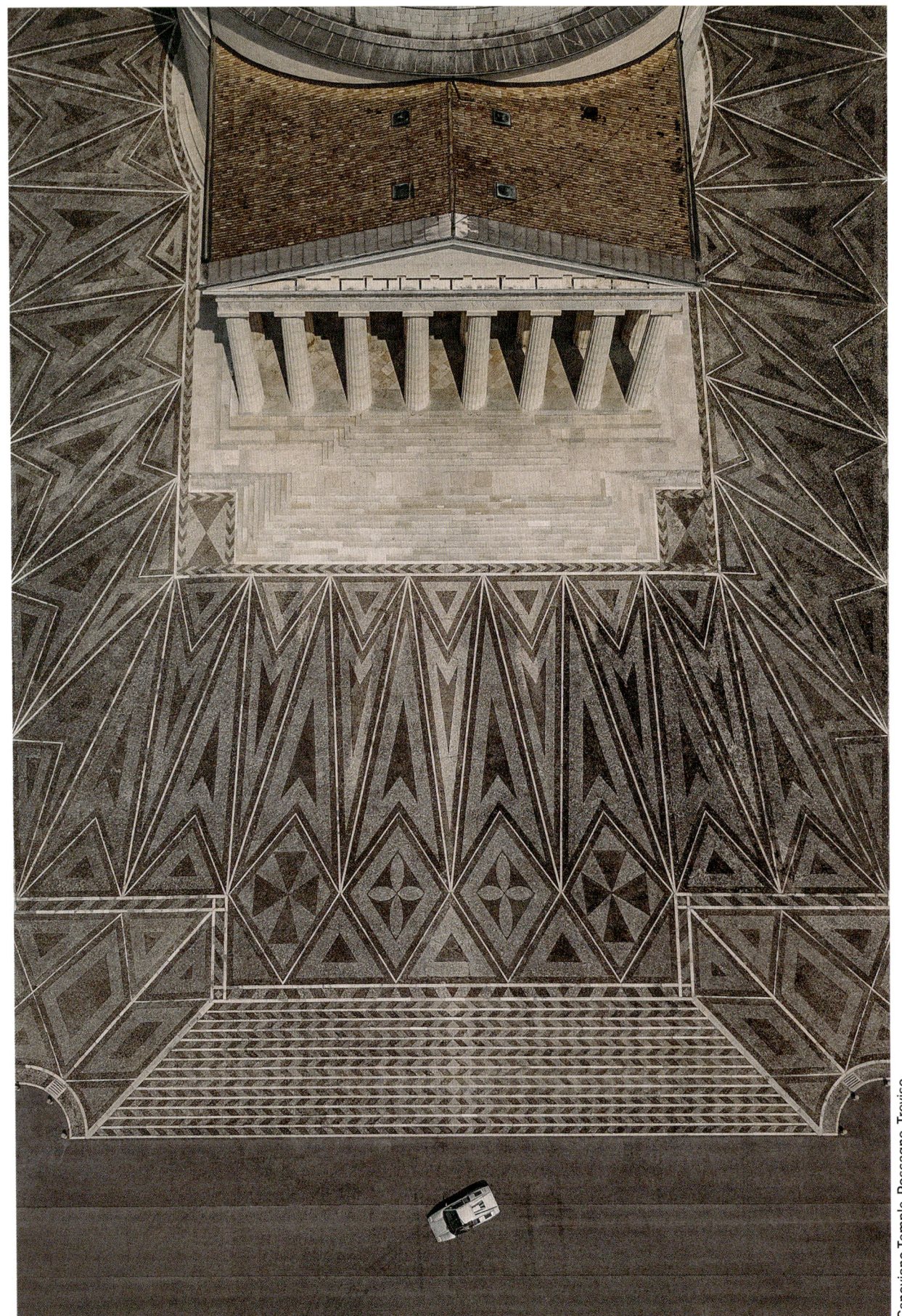

Canoviano Temple, Possagno, Treviso

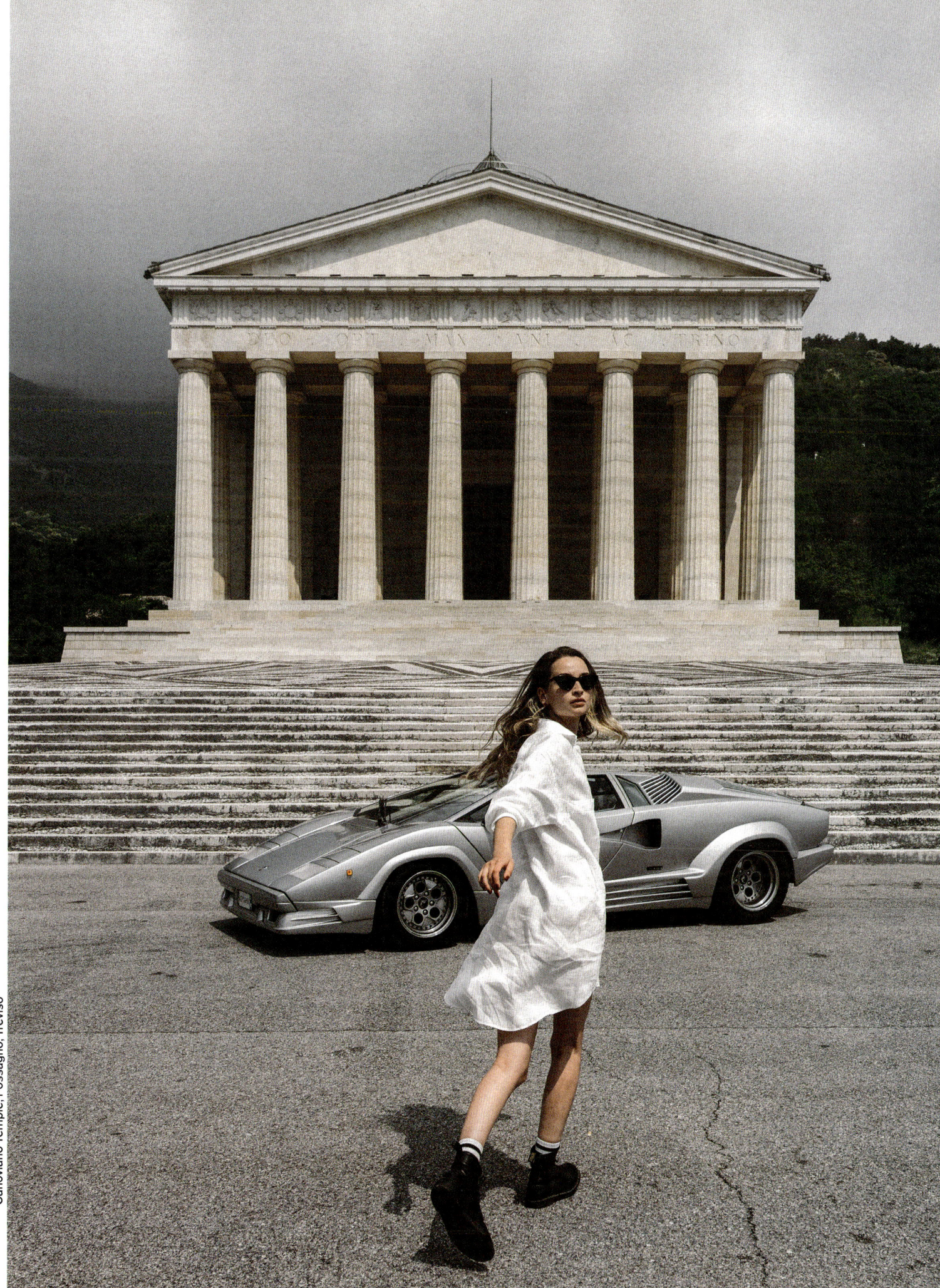

Canoviano Temple, Possagno, Treviso

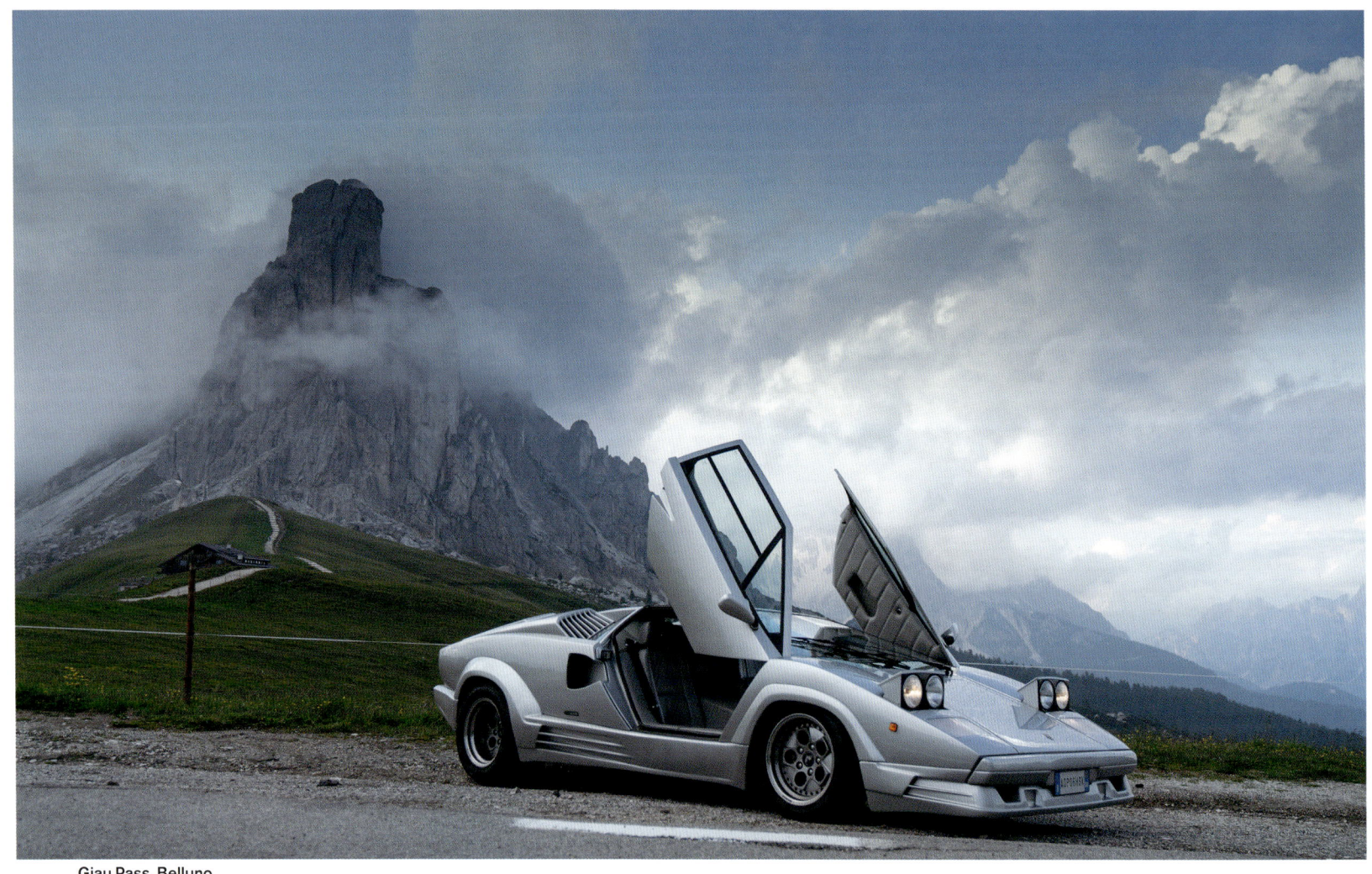

Giau Pass, Belluno

Villa Barbarigo, Valsanzibio, Padua

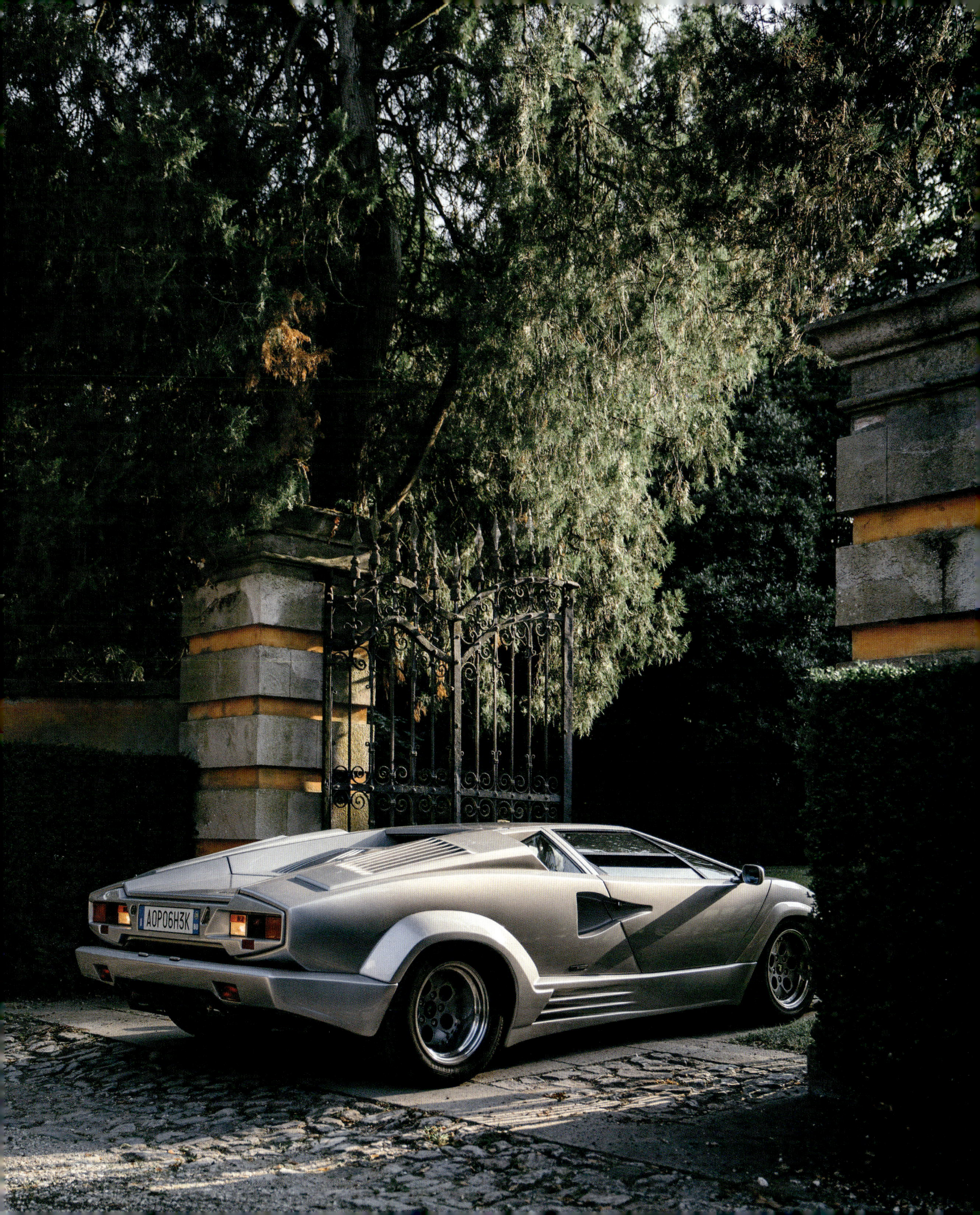

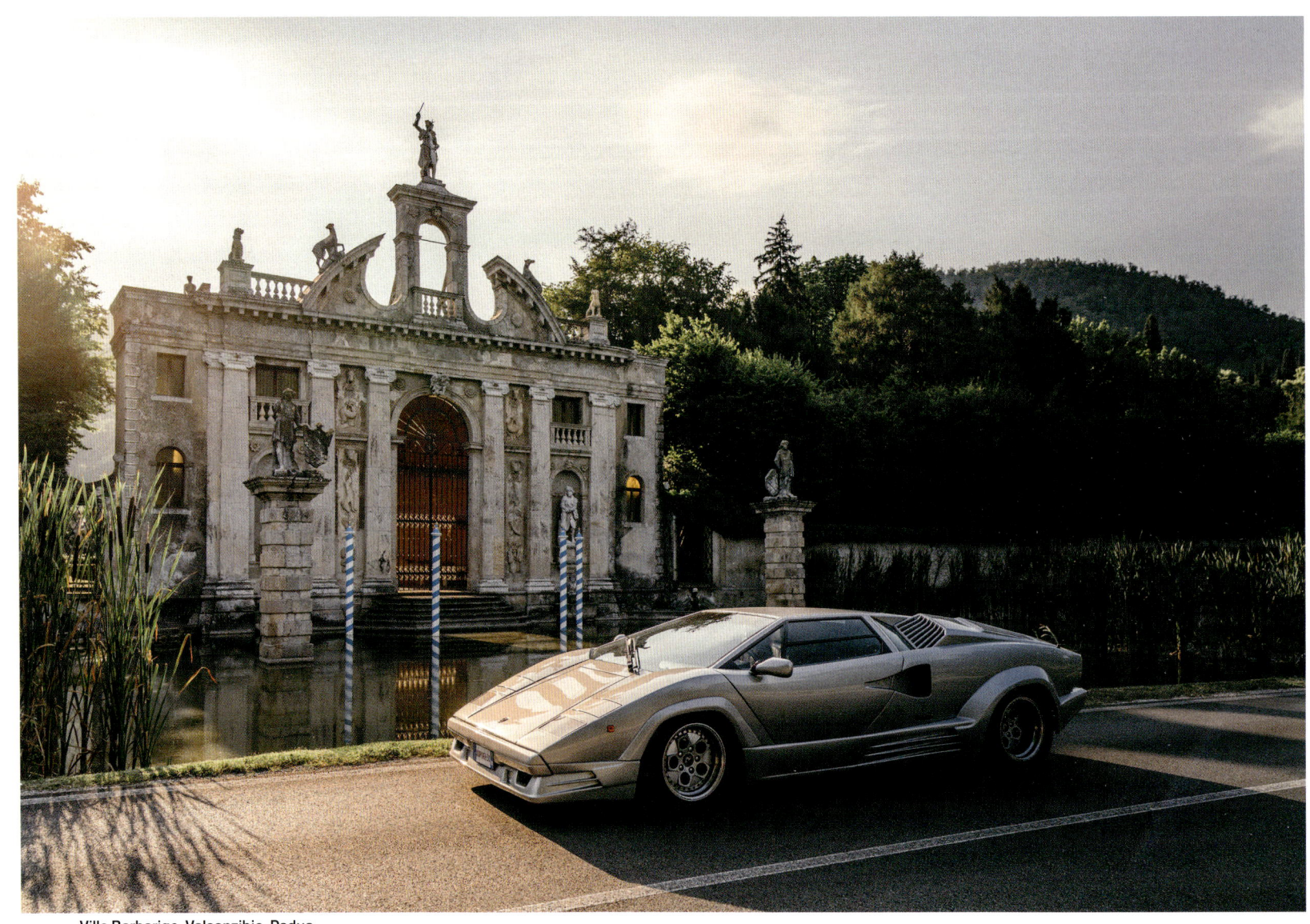
Villa Barbarigo, Valsanzibio, Padua

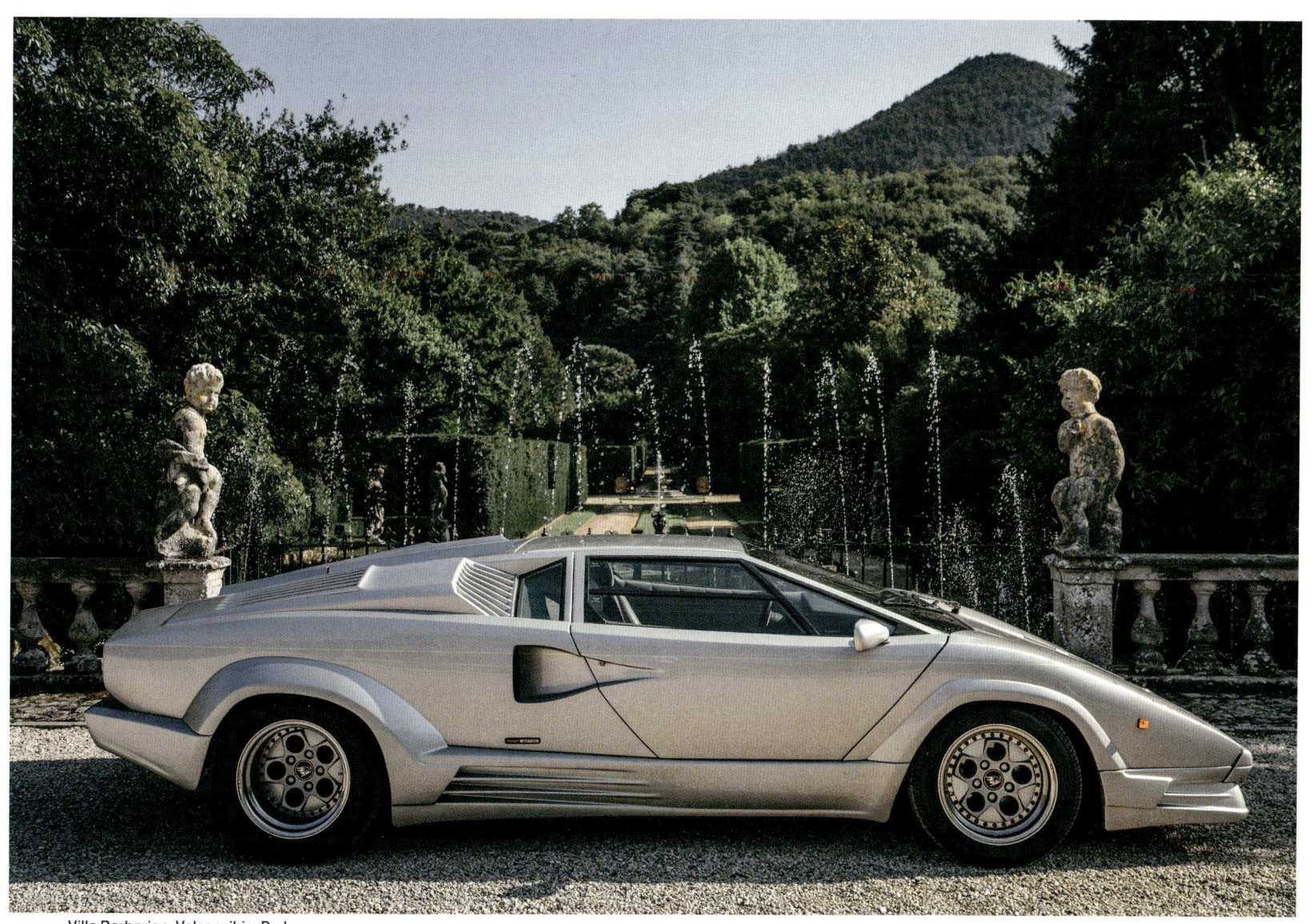

Villa Barbarigo, Valsanzibio, Padua

Pages 186–87 Lio Piccolo, Venice

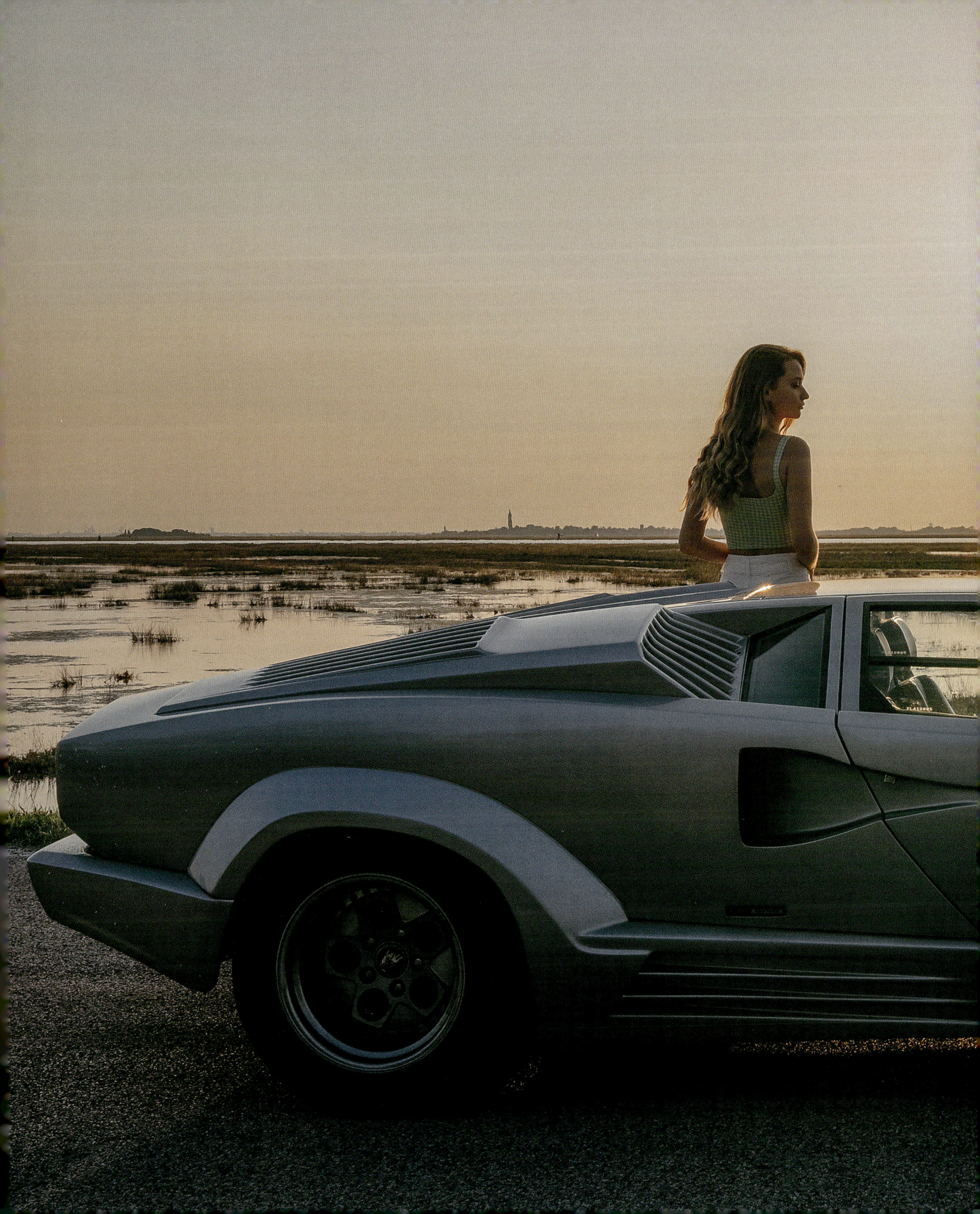

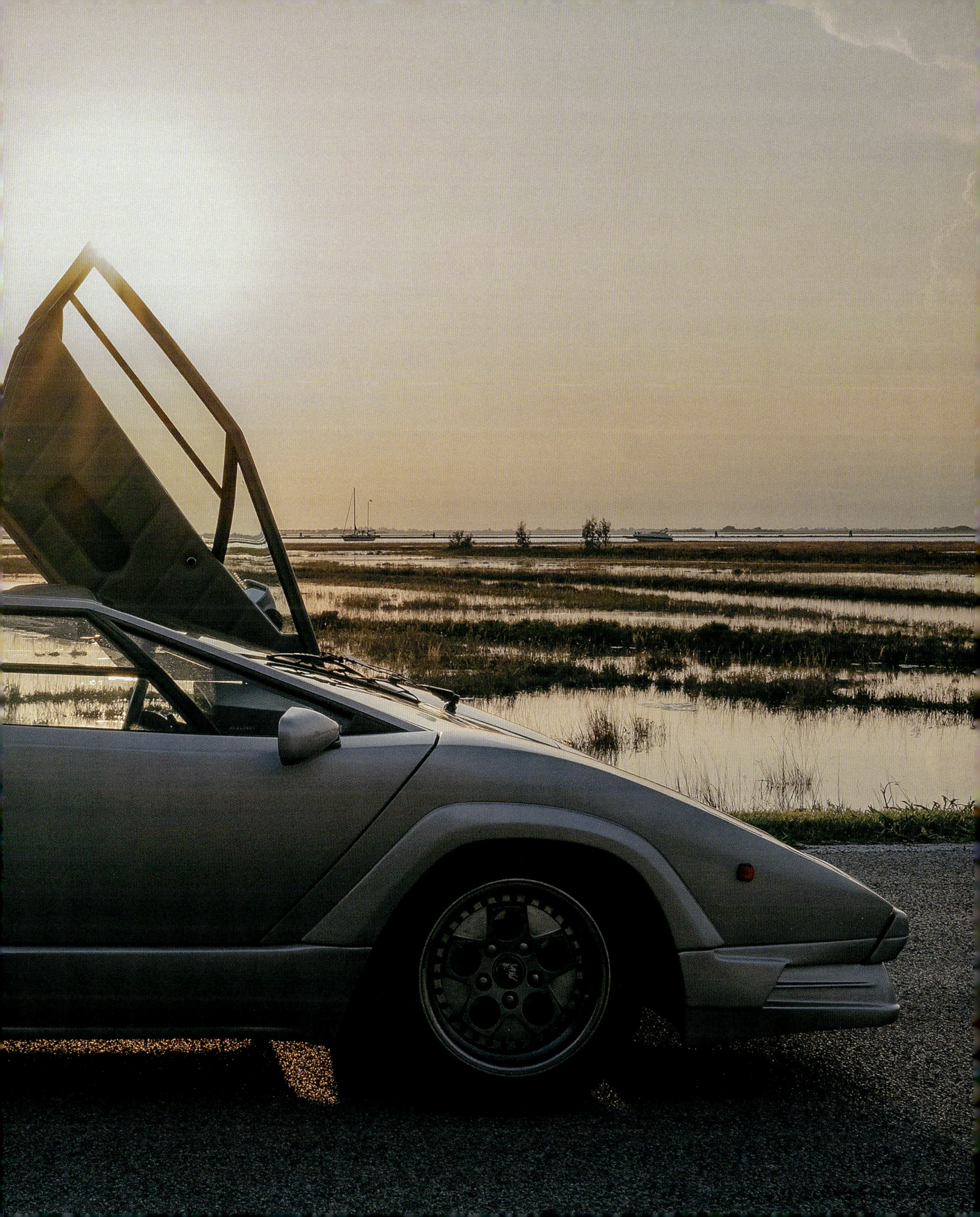

MATTIA BALSAMINI
FRIULI VENEZIA GIULIA

I have portrayed Friuli Venezia Giulia, my home region, in all its mysterious charm. I wanted to bring out its sober elegance, exploring places where the relationship between man and nature is more respectful. From Lake Barcis, in Valcellina, with its unique waters coloured green and sky-blue, to Trieste; from the Sarone quarries to the stone beds of the Magredi, carried down by floodwaters following glaciation. Often hidden, lesser-known sights help to show a different side to this beautiful region, in which the Lamborghini Urus blends in as part of a perfectly harmonious whole. It's as if the car was made to be here.

Lake Barcis dam, Pordenone

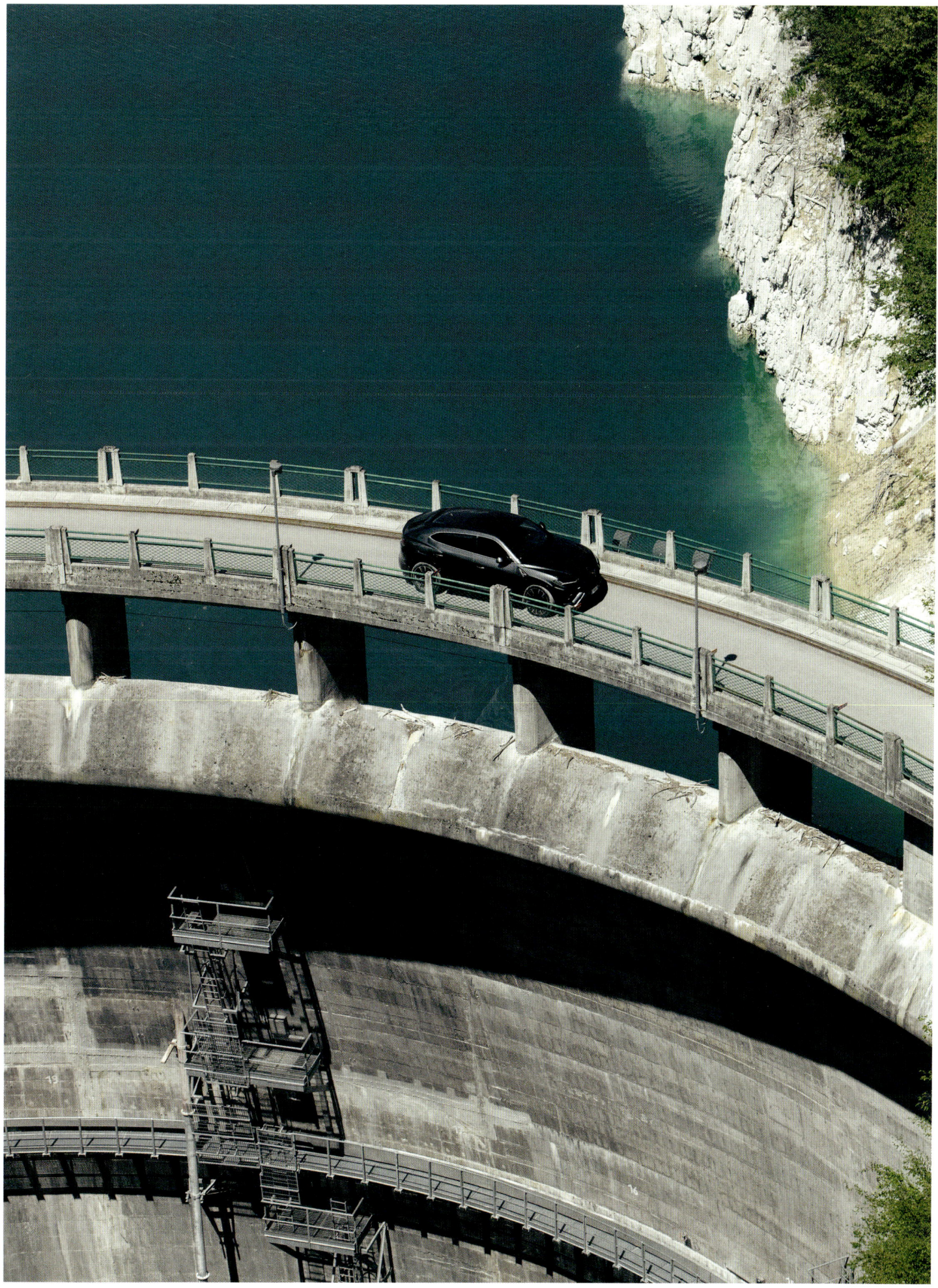

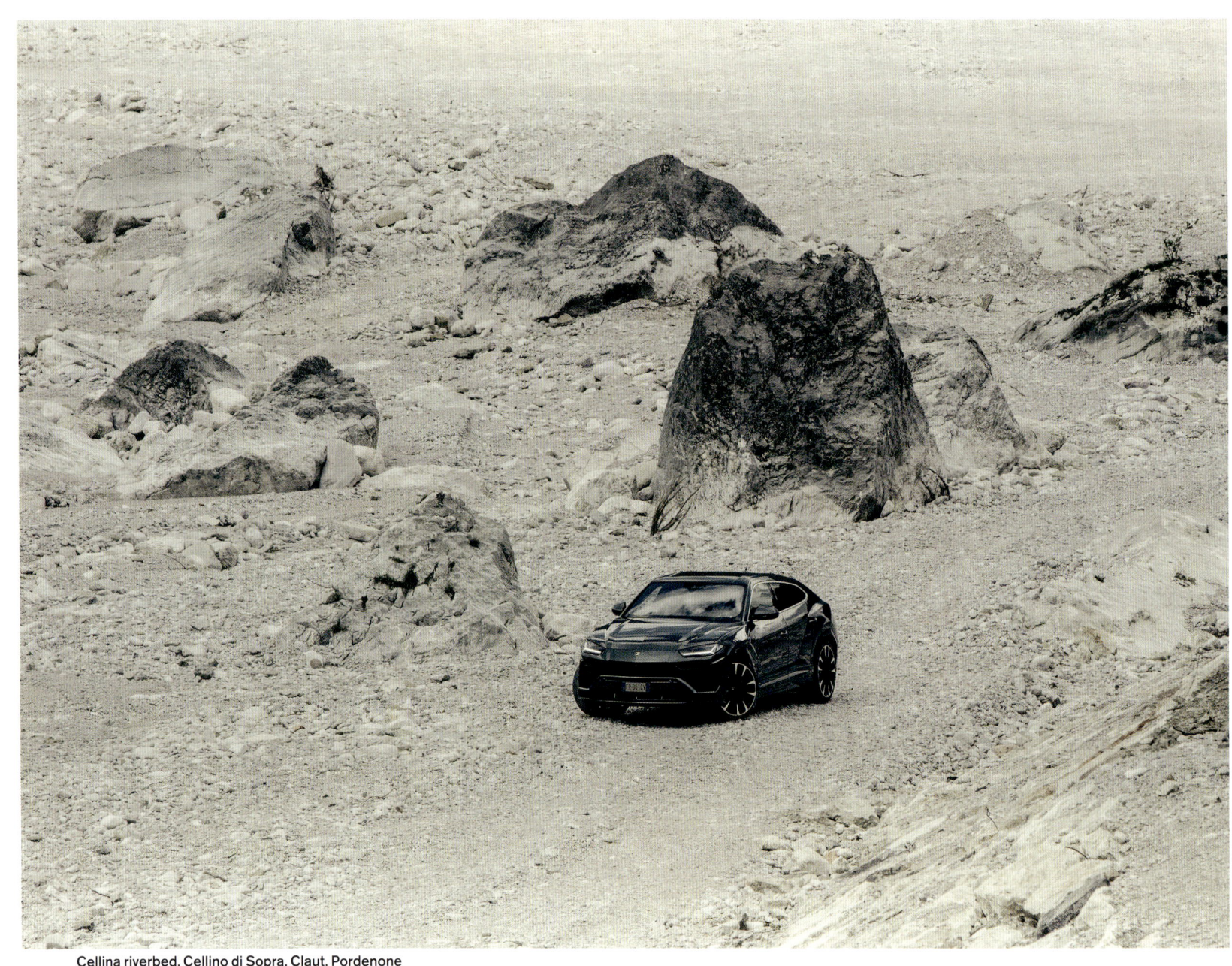

Cellina riverbed, Cellino di Sopra, Claut, Pordenone

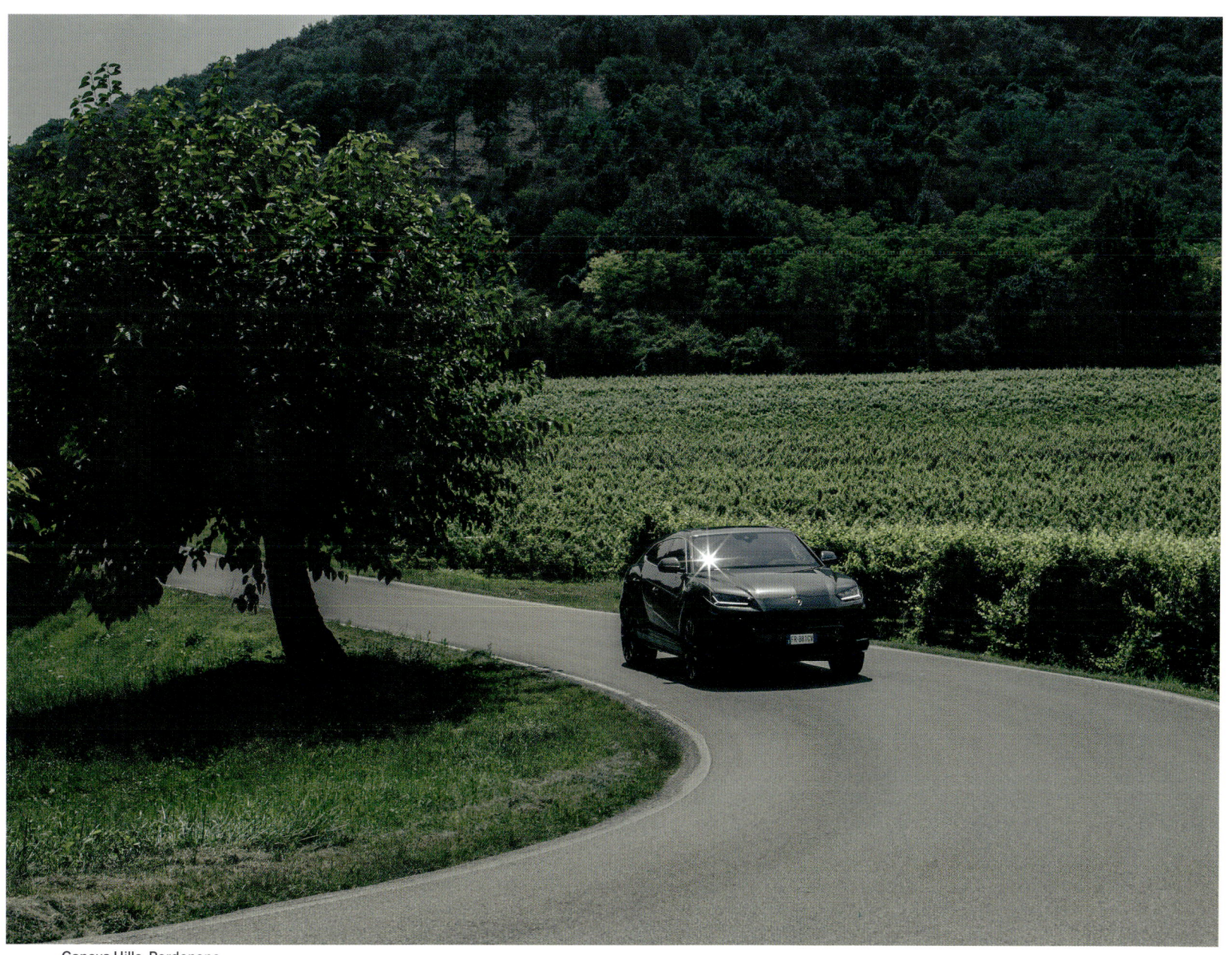
Caneva Hills, Pordenone

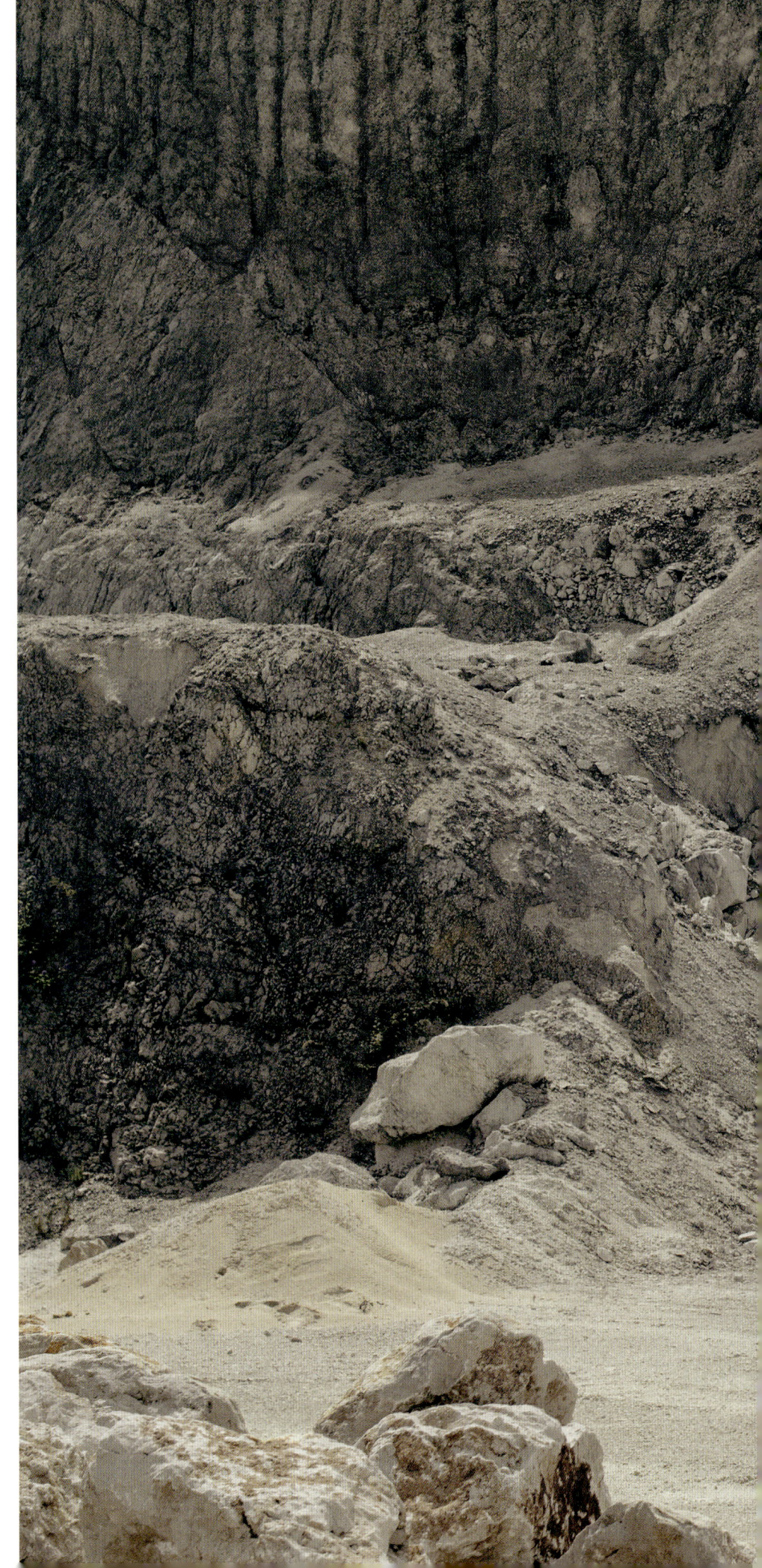

Sarone quarries, Caneva, Pordenone

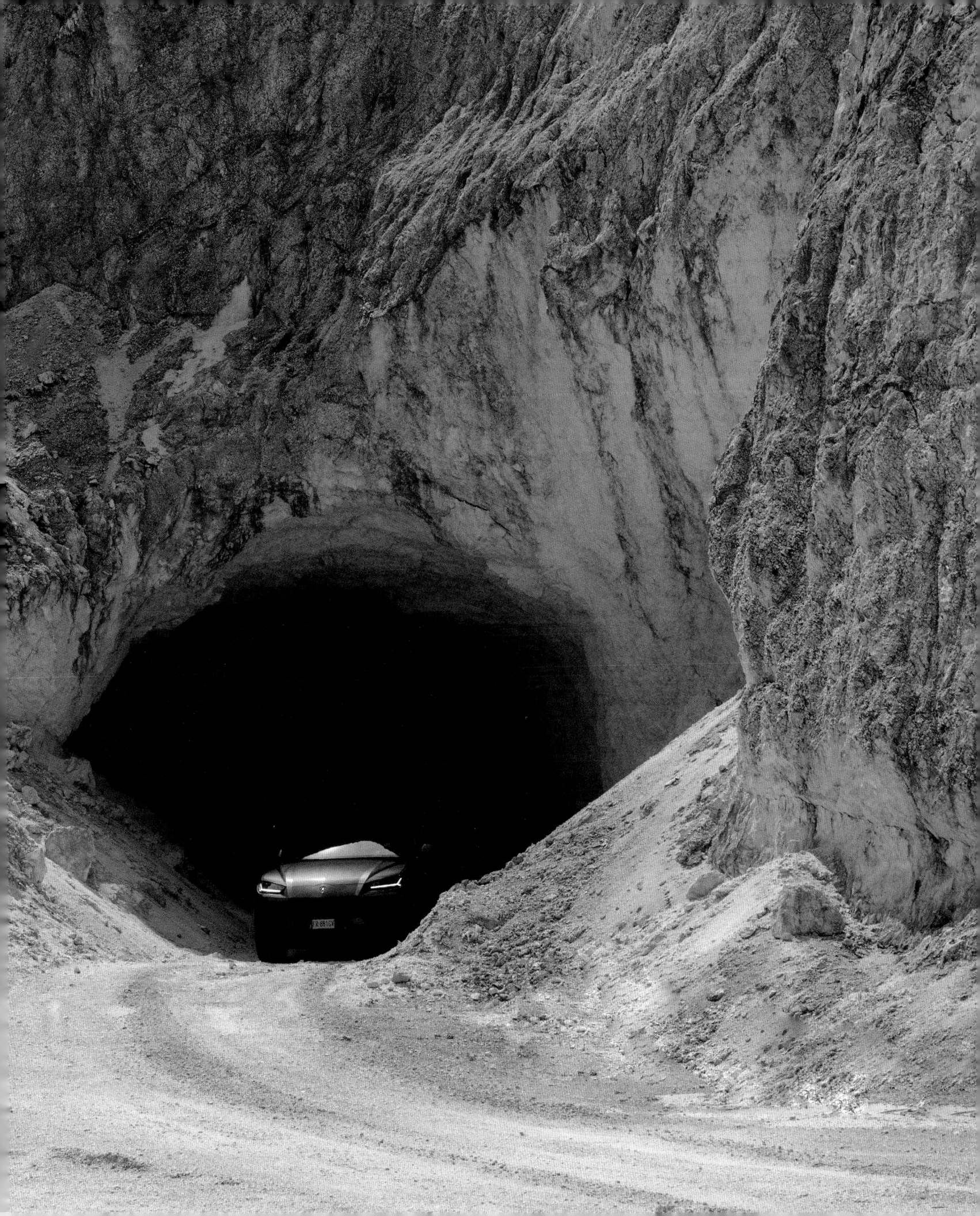

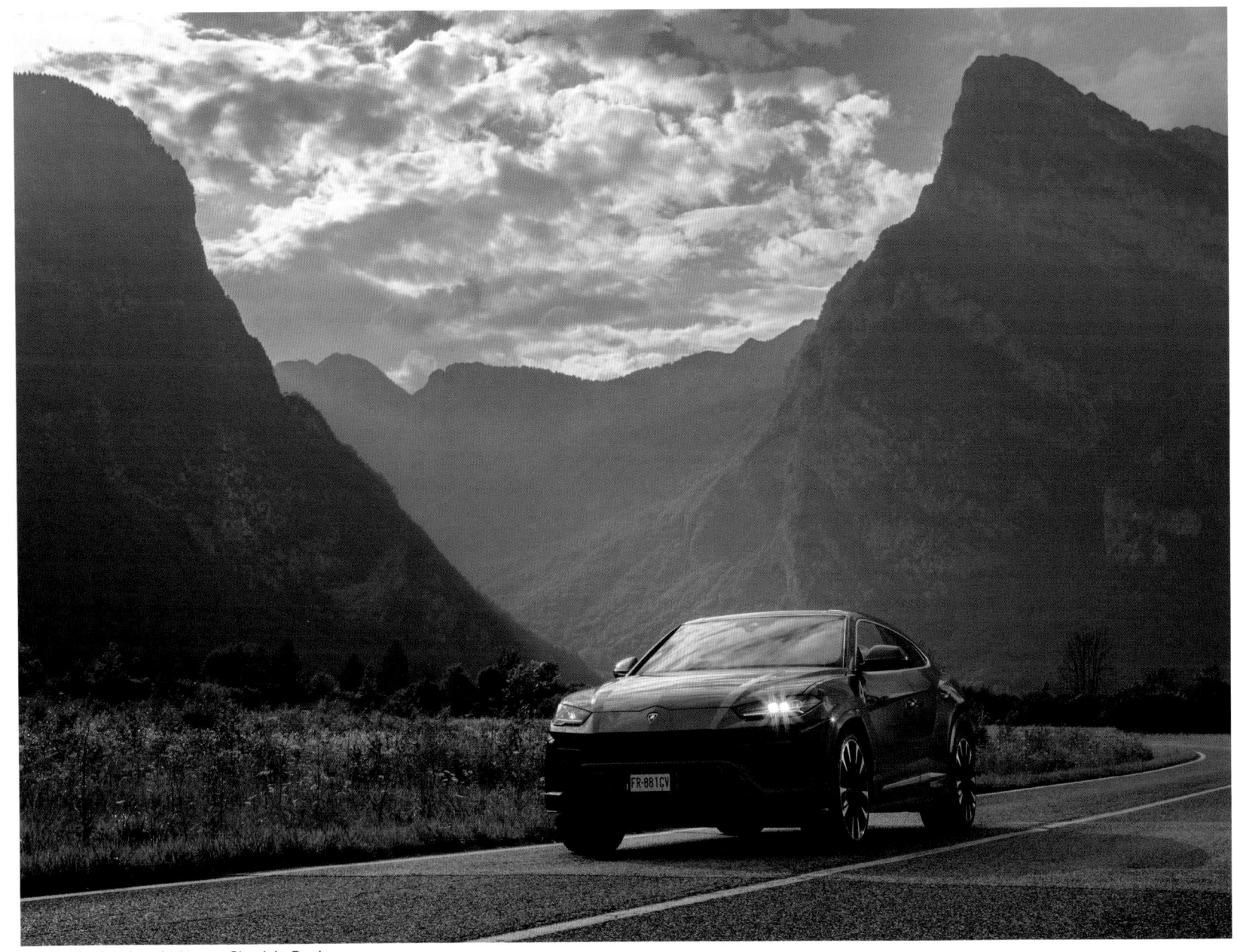

Regional road 251, near Cimolais, Pordenone

Miramare, Trieste

Pages 196–97 Cellina riverbed, Cellino di Sopra, Pordenone

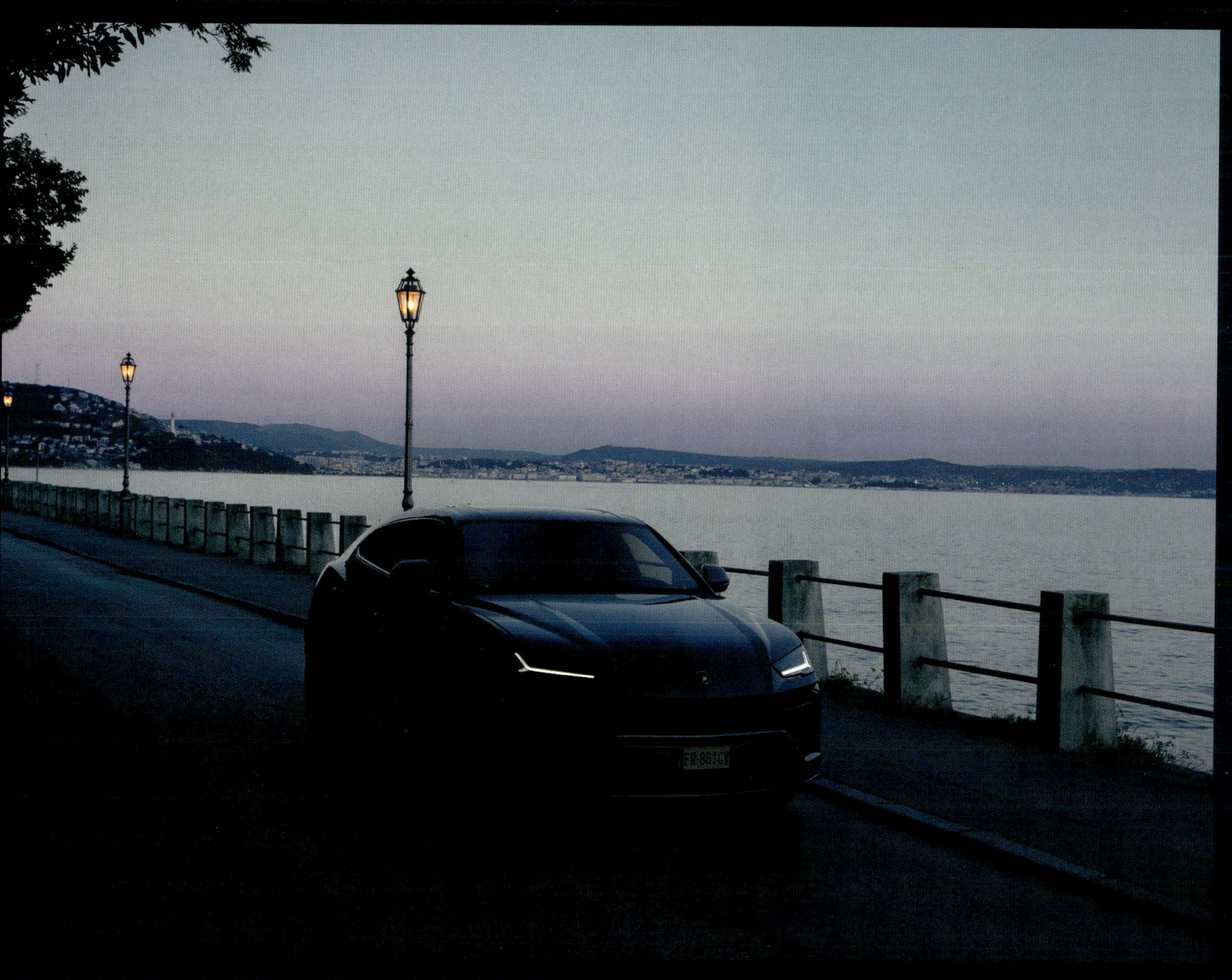

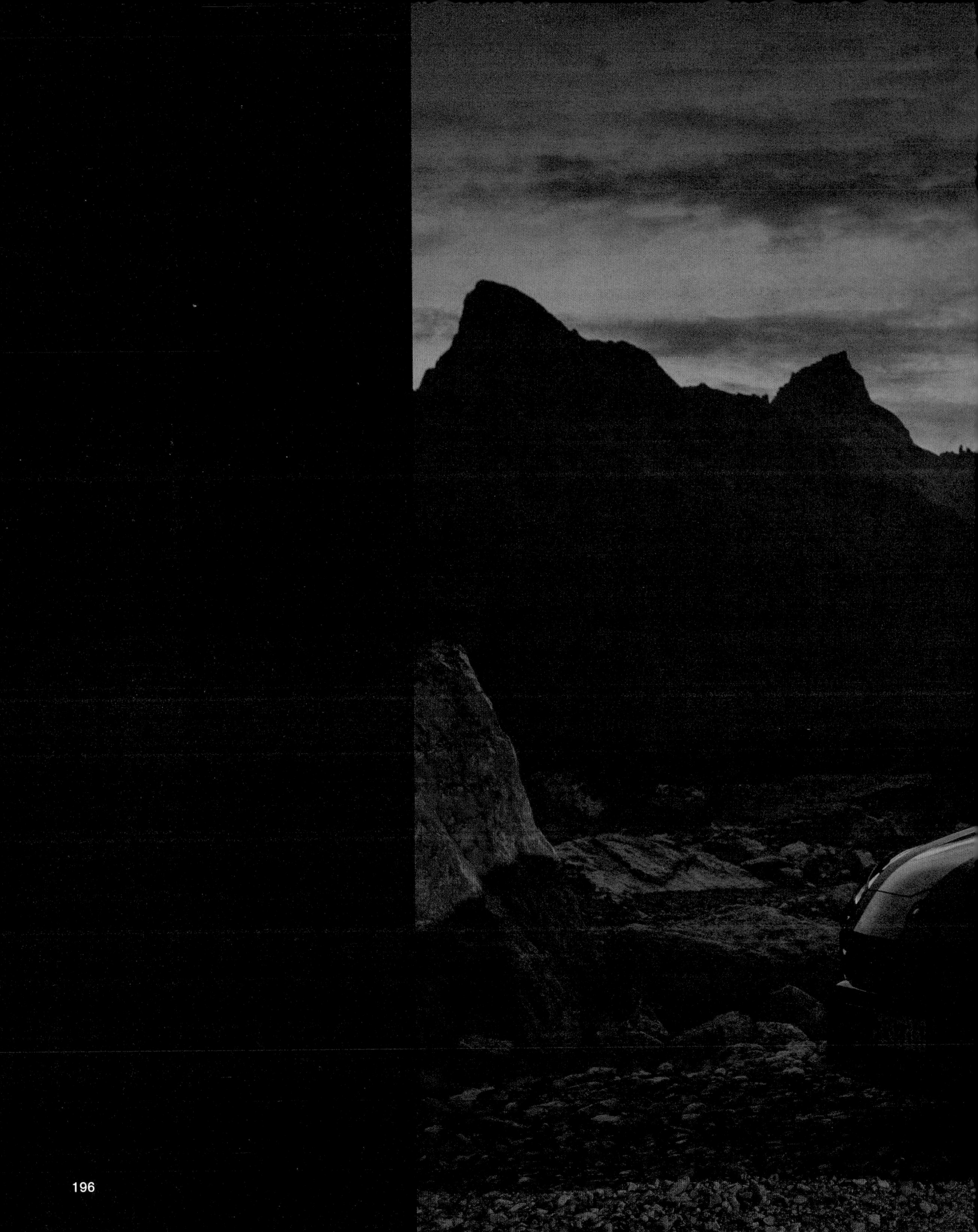

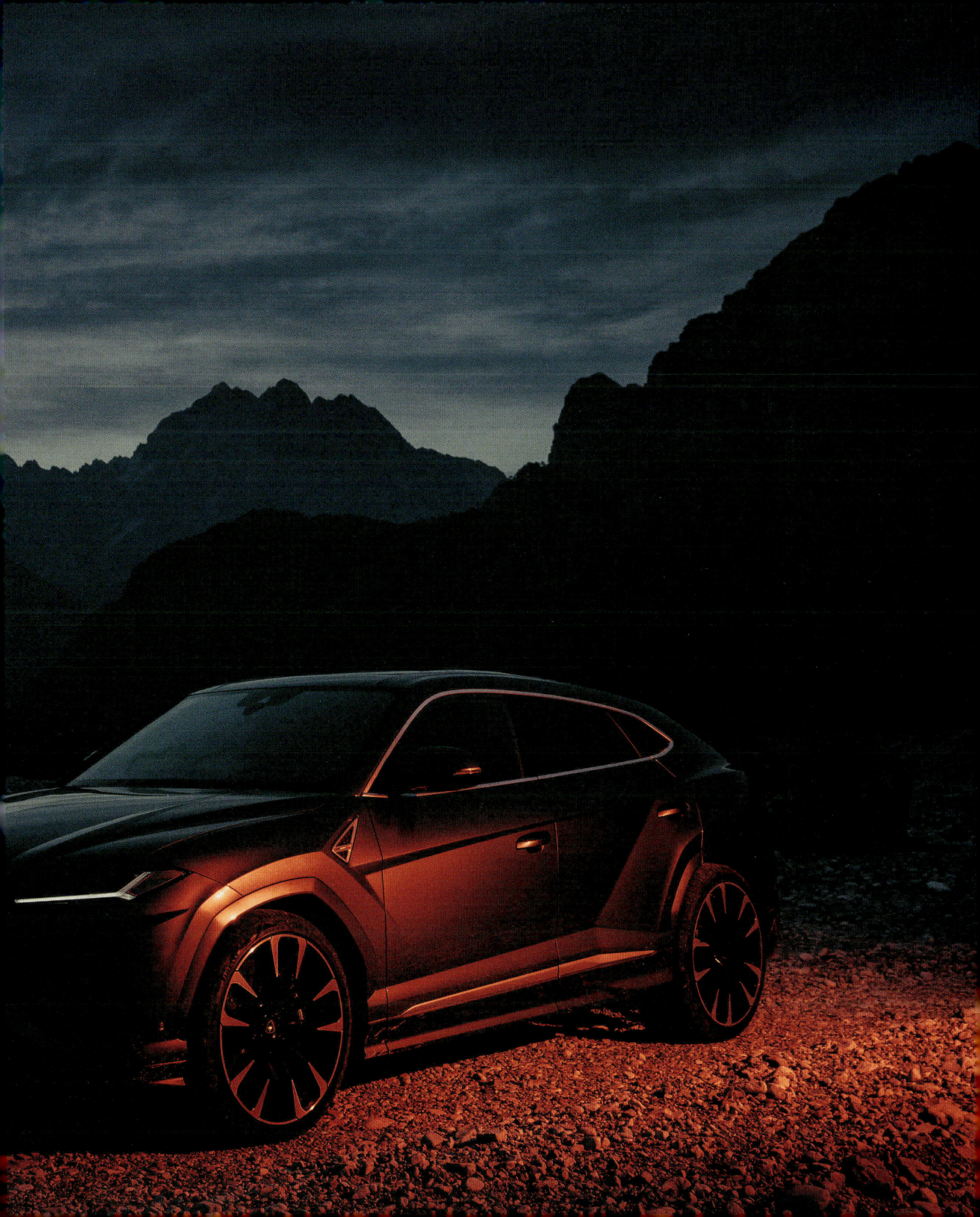

SIMONE BRAMANTE
TRENTINO-ALTO ADIGE

Trentino-Alto Adige is a region with a proud, powerful, high-energy landscape, making its character similar to that of the Lamborghini Huracán EVO RWD. I wanted to strike a narrative coherence through an aesthetic research into shapes and colours, fully integrating car and nature. The sequence of images is designed to highlight the beauty of this Alpine territory with precise, chromatic associations and the elements that make it up, including matching and contrasting lines: vertical, outstretched to the sky in nature; low and horizontal like the Huracán.

Series of shots taken in Val San Nicolò and Val Venegia

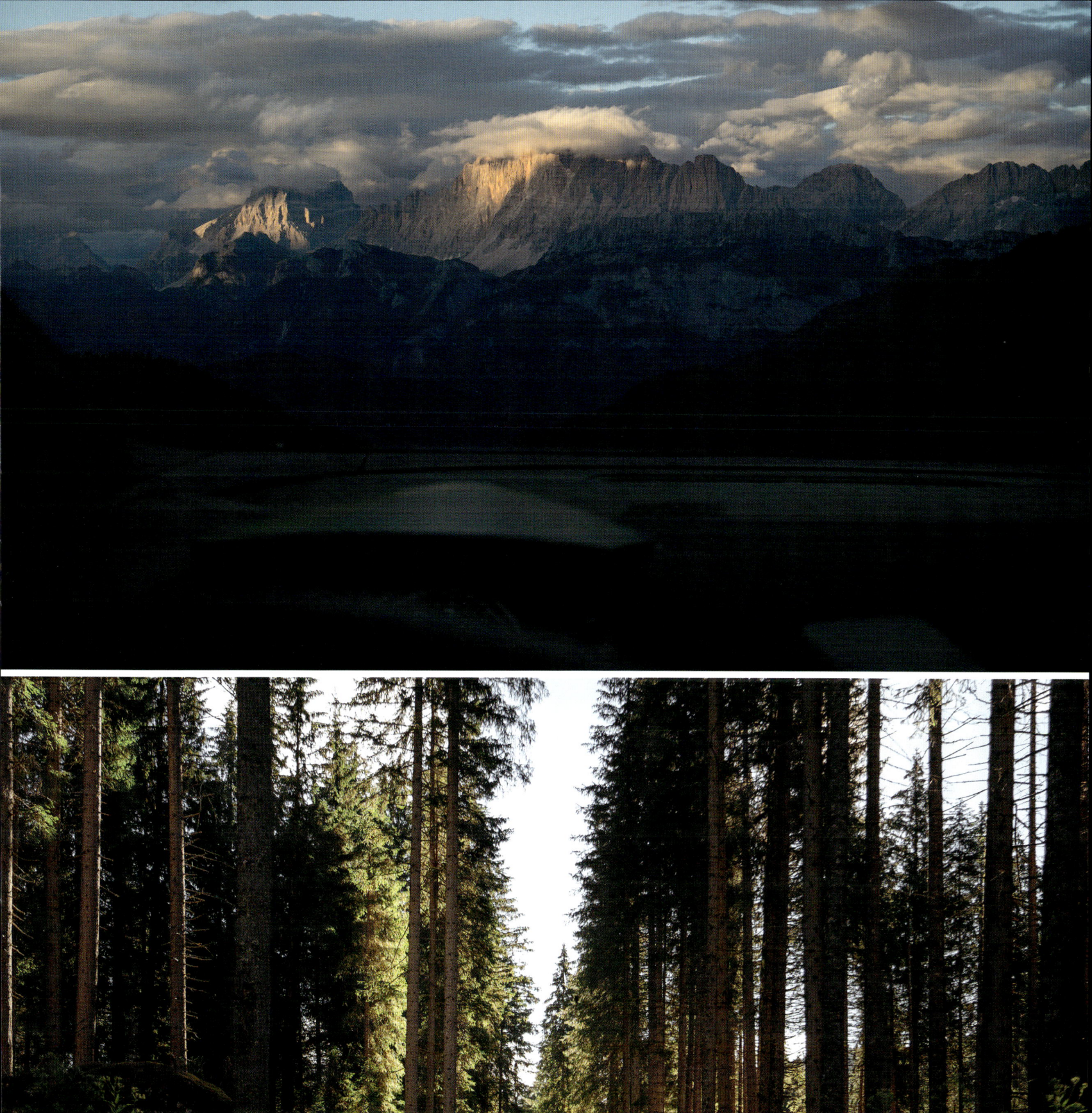

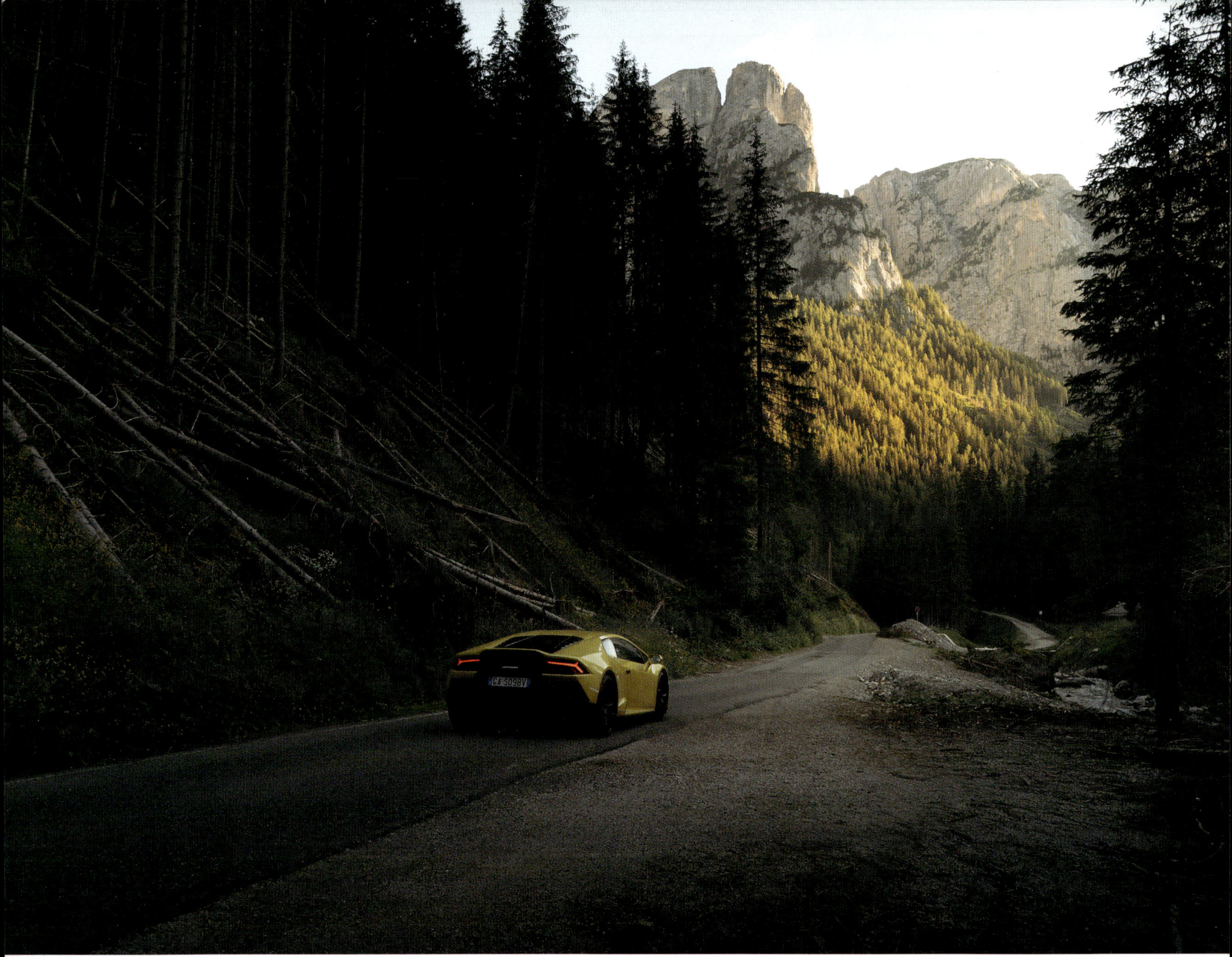

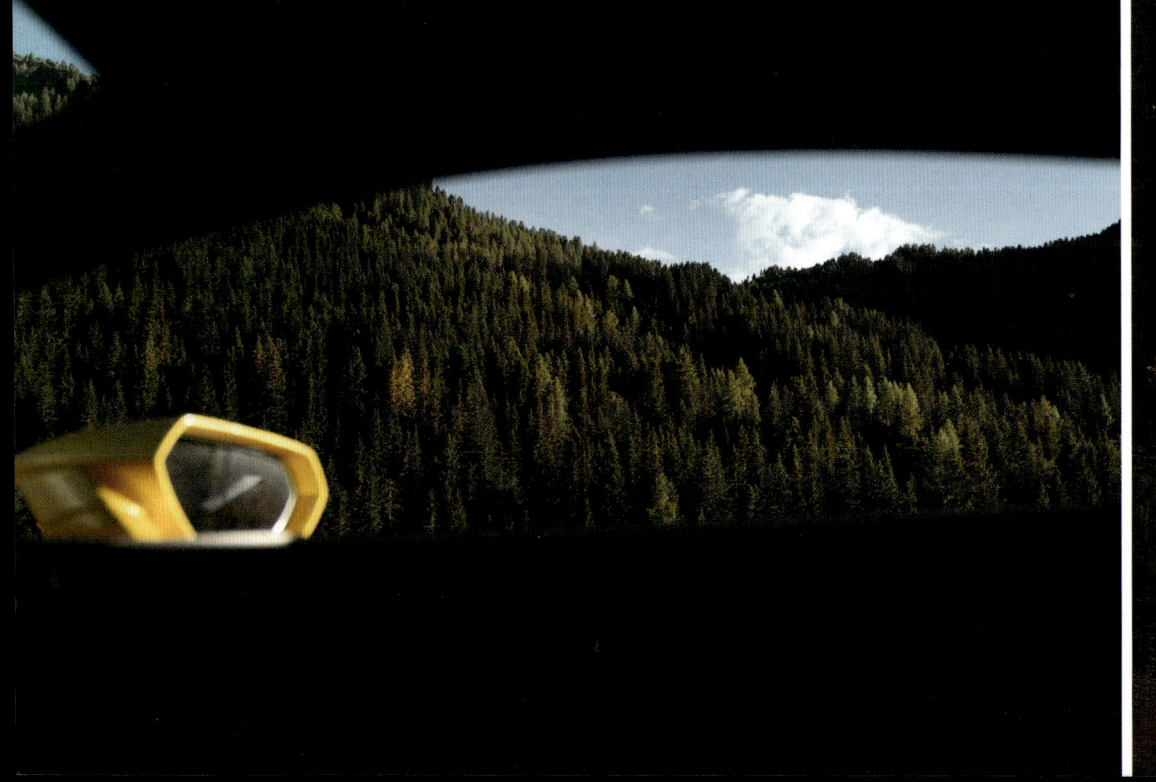
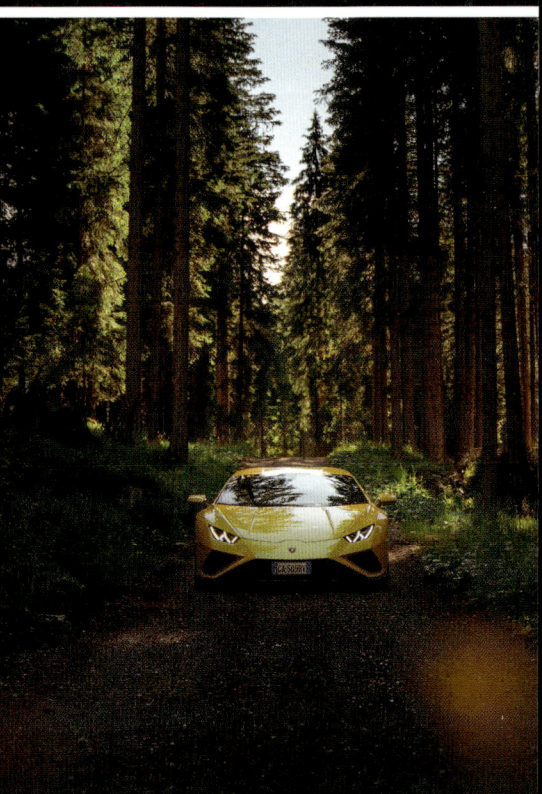

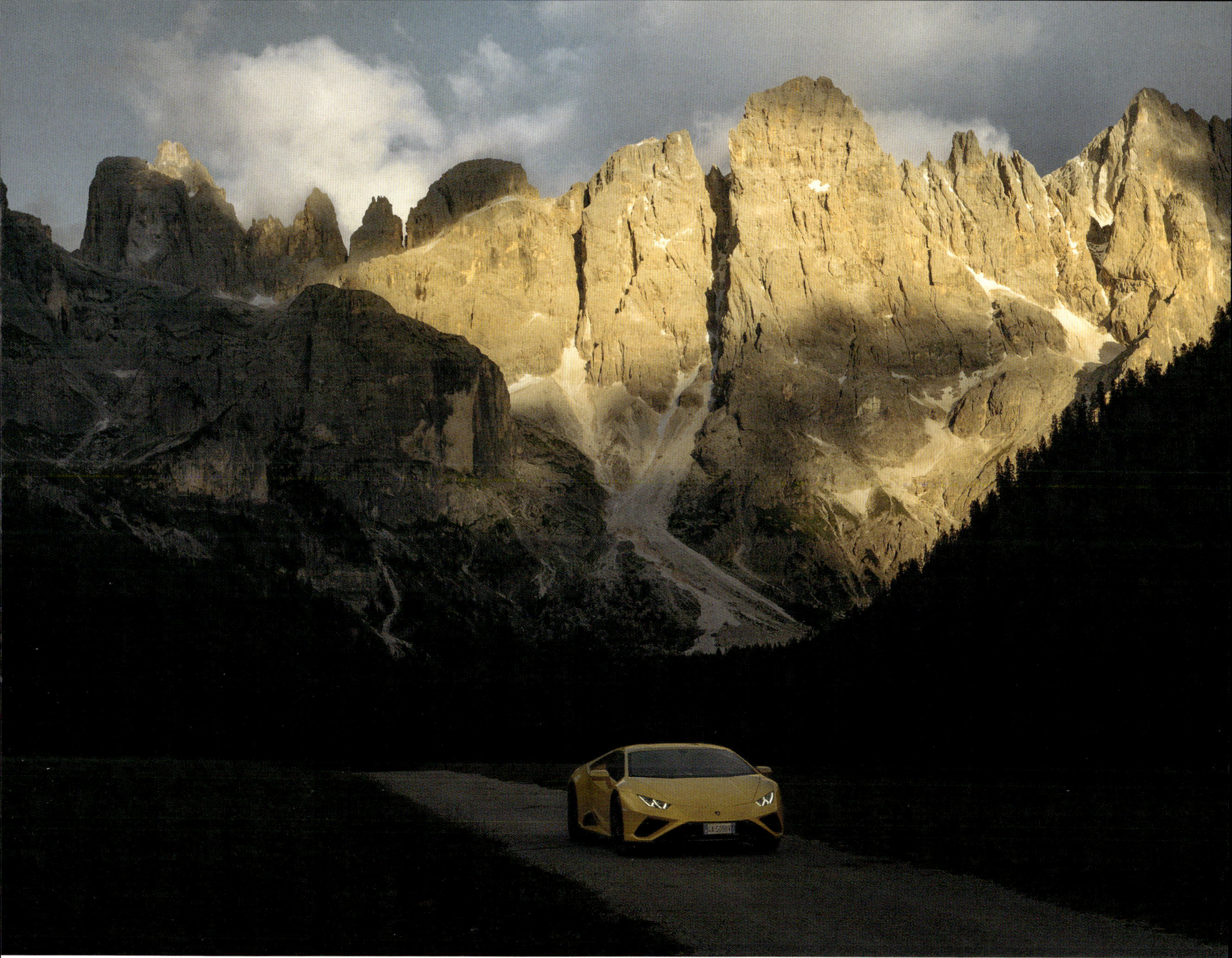

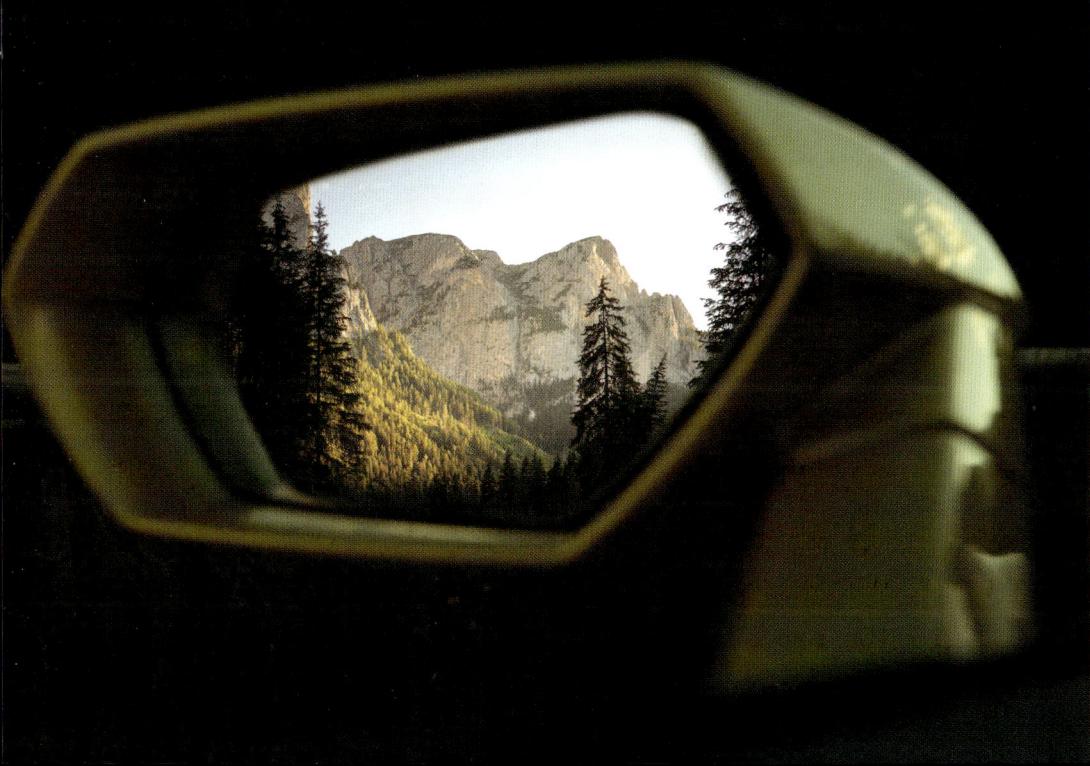
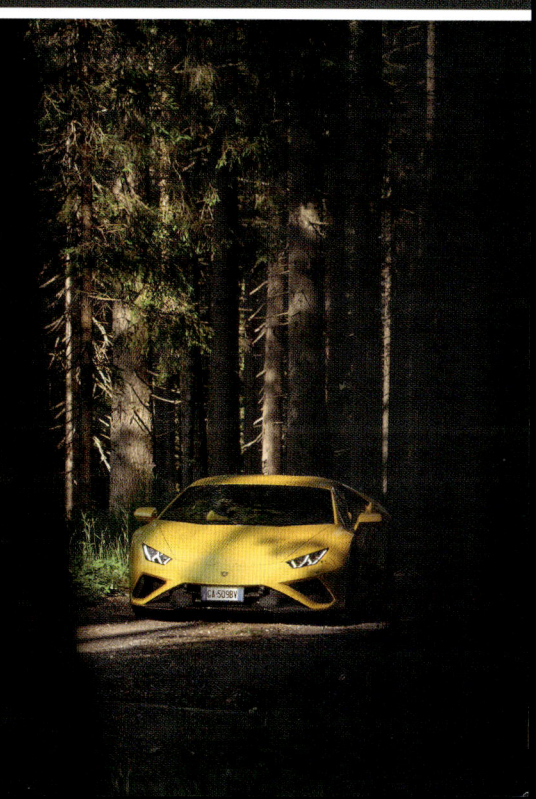

VINCENZO GRILLO
LOMBARDIA

A journey back in time across Lombardy, where a Huracán EVO drives through millennia of history in a single region. An undisputed protagonist of street style shots, Lamborghini is a modern interpretation of the medieval charm of cities like Mantua and Cremona, as well as the Pavia Charterhouse, the views from Bergamo Alta or the elegance of Lake Como, immersed in the most authentic nature and the dynamism of a city like Milan, capital of fashion and design. A true journey to rediscover an infinite love for our Italy.

Milan Cathedral

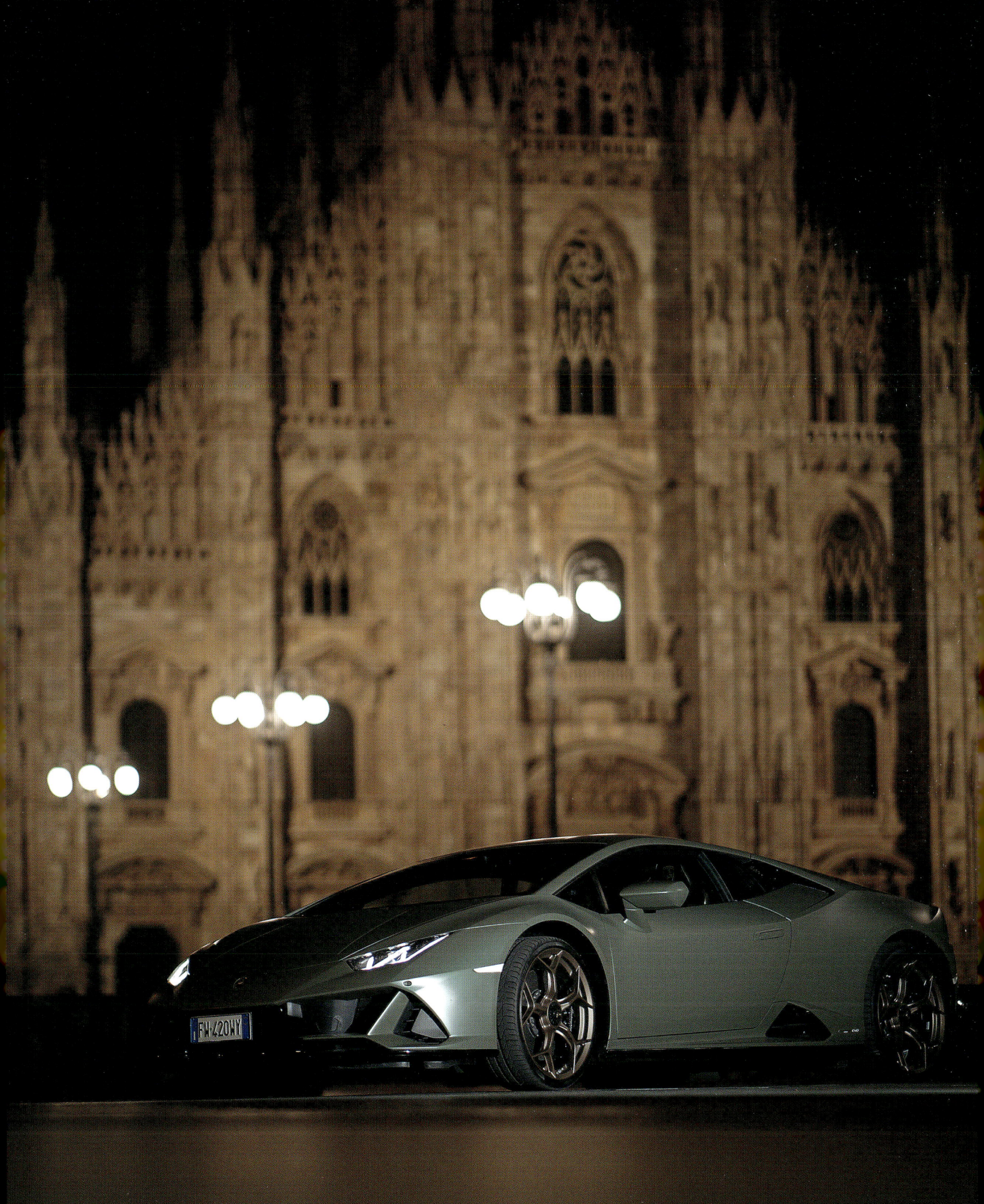

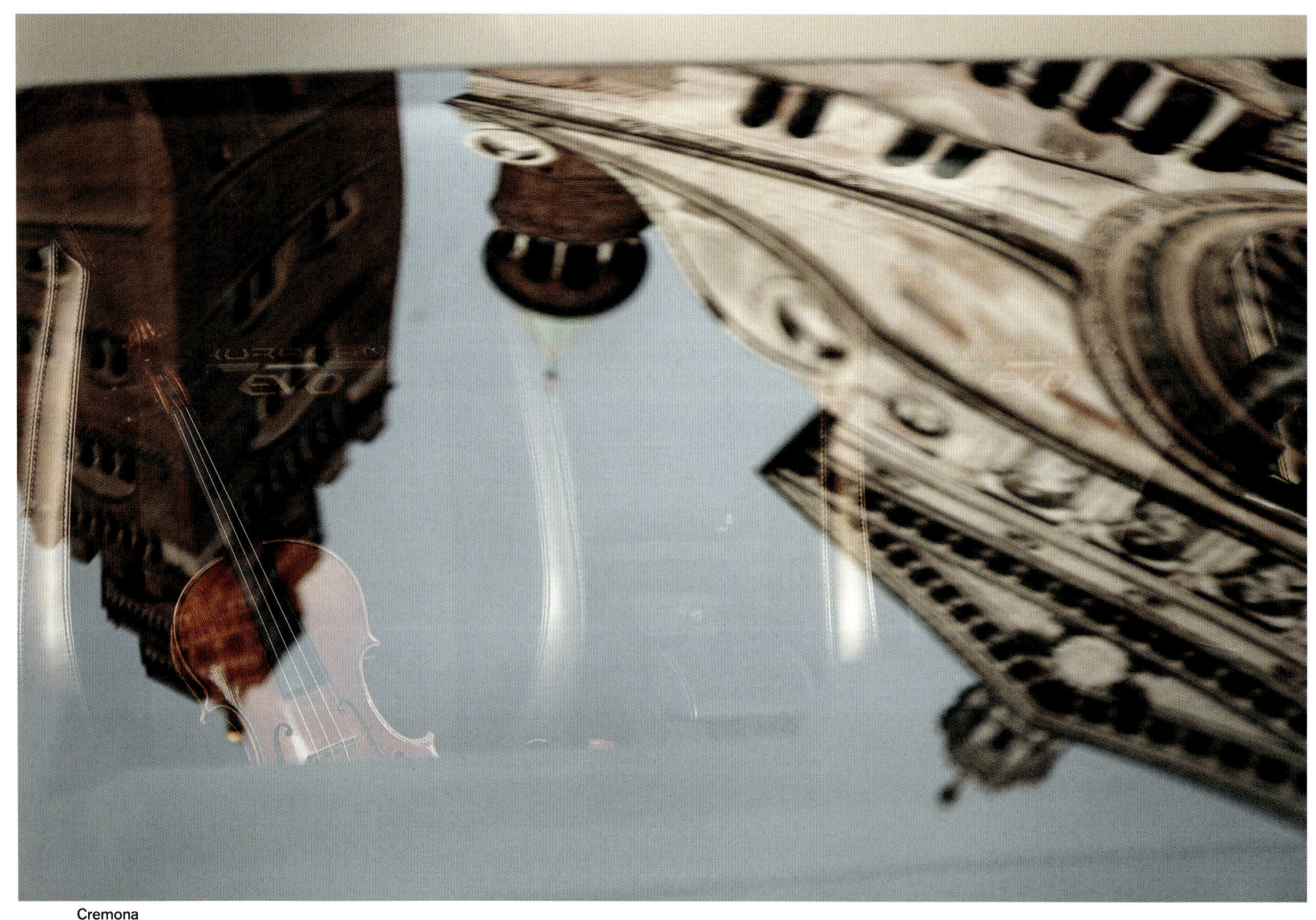

Cremona

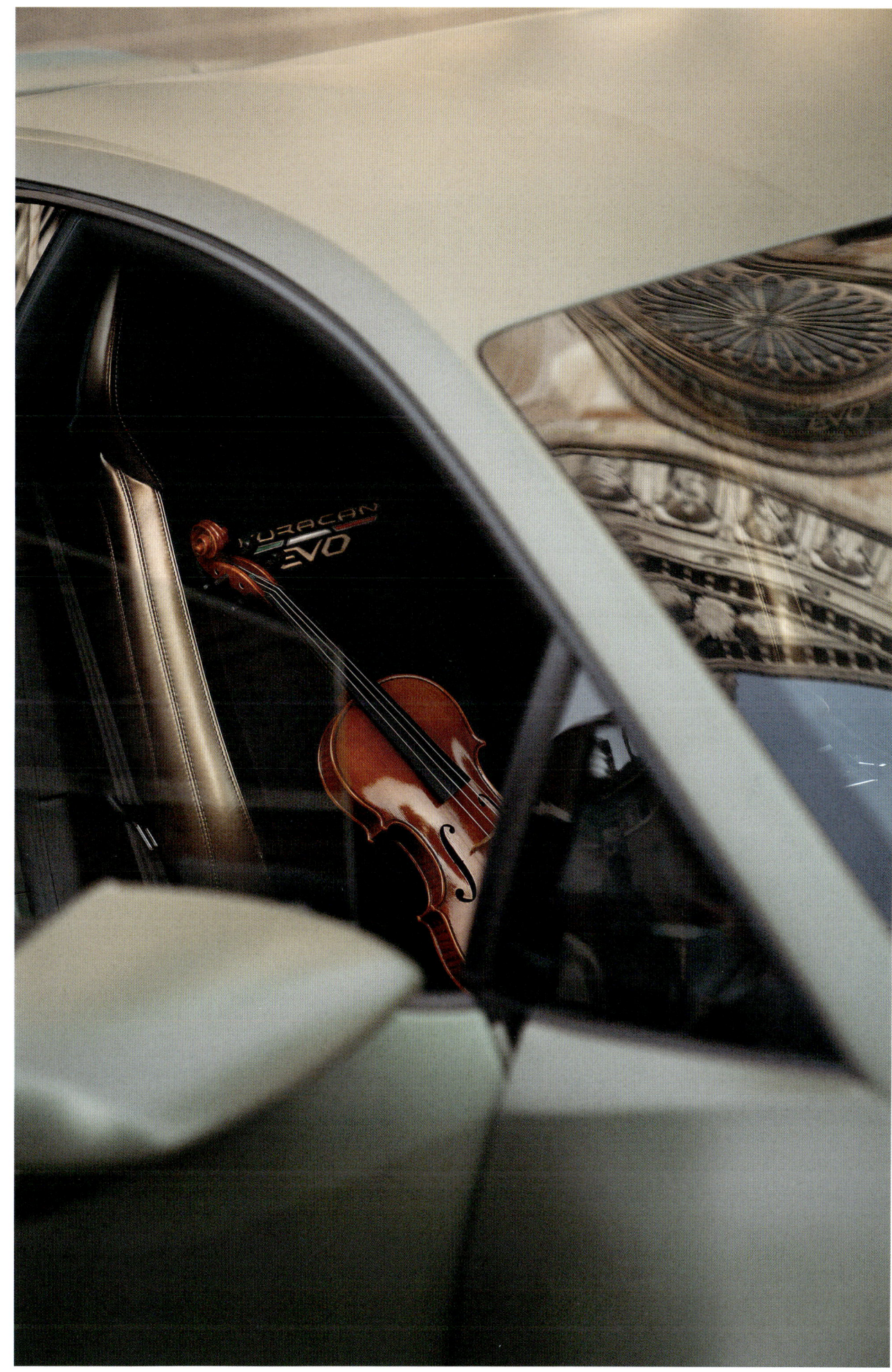

Cremona

Pages 206–07
Villa Orlando, Bellagio,
Lake Como

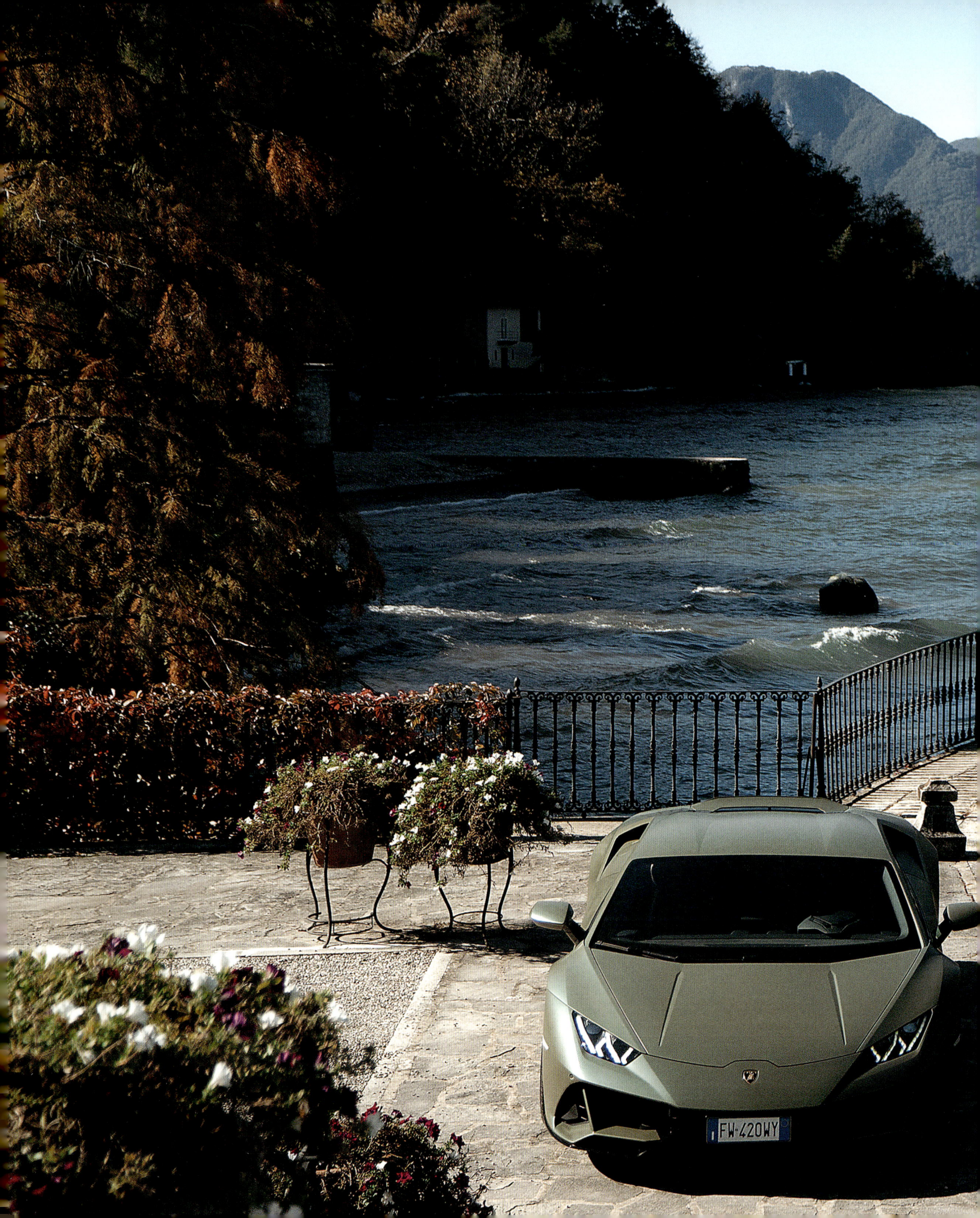

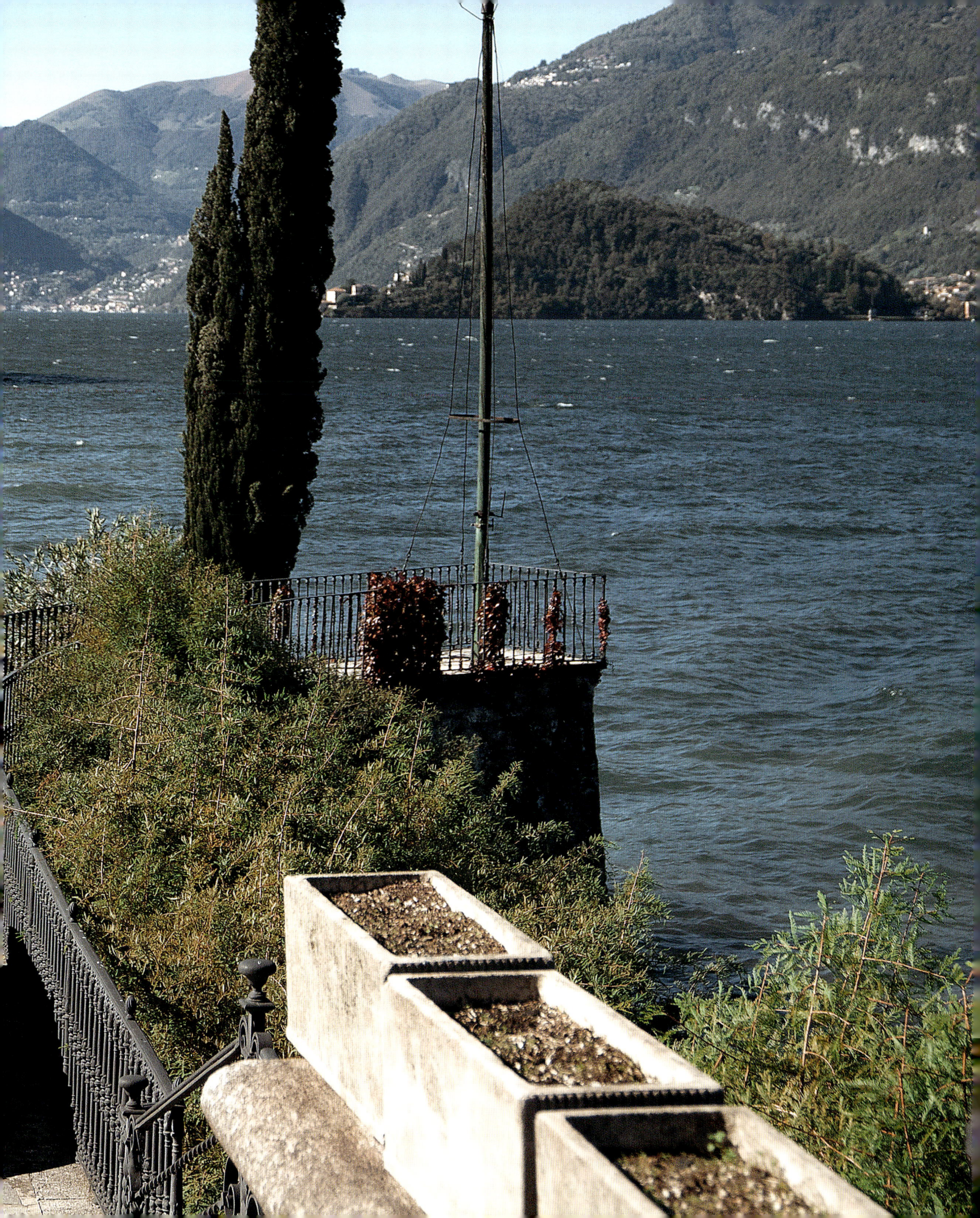

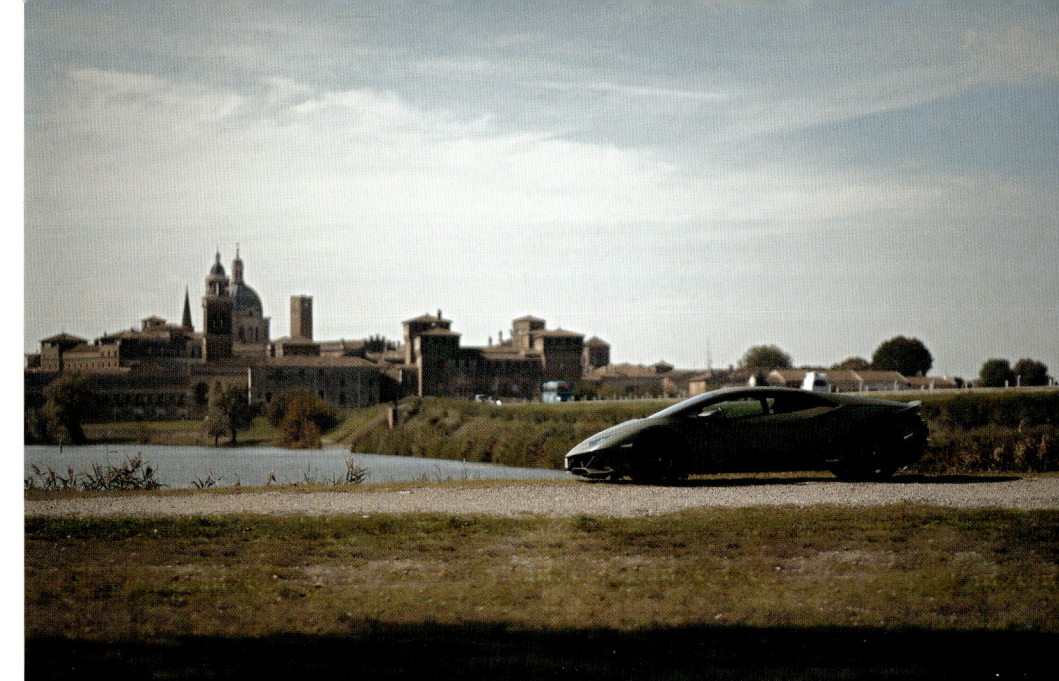

Mantua

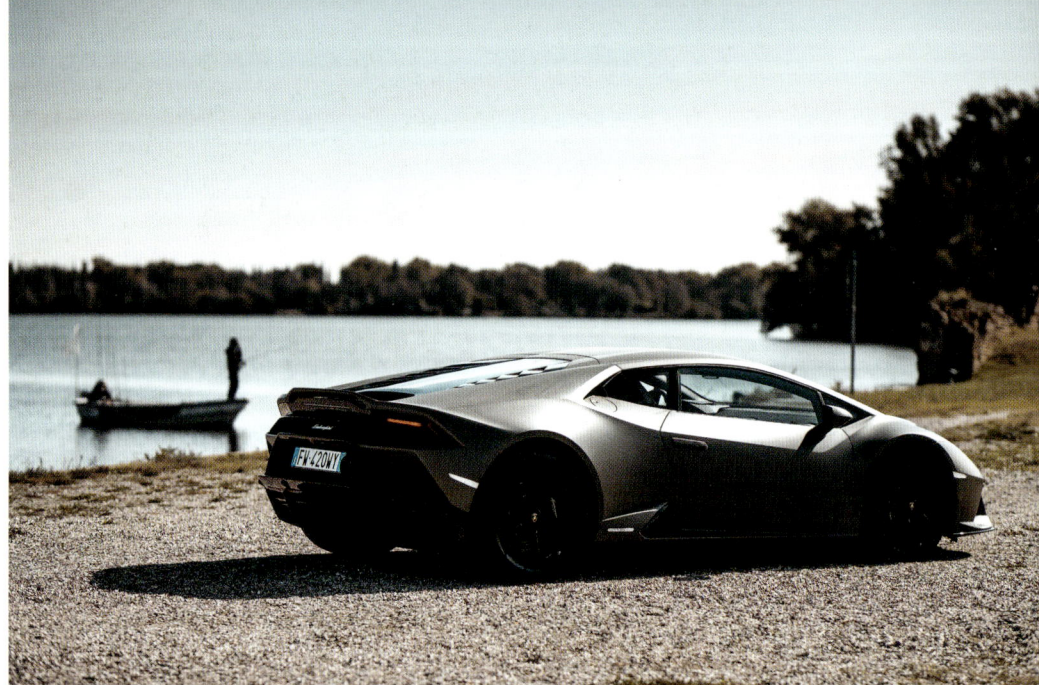

Mincio, Mantua

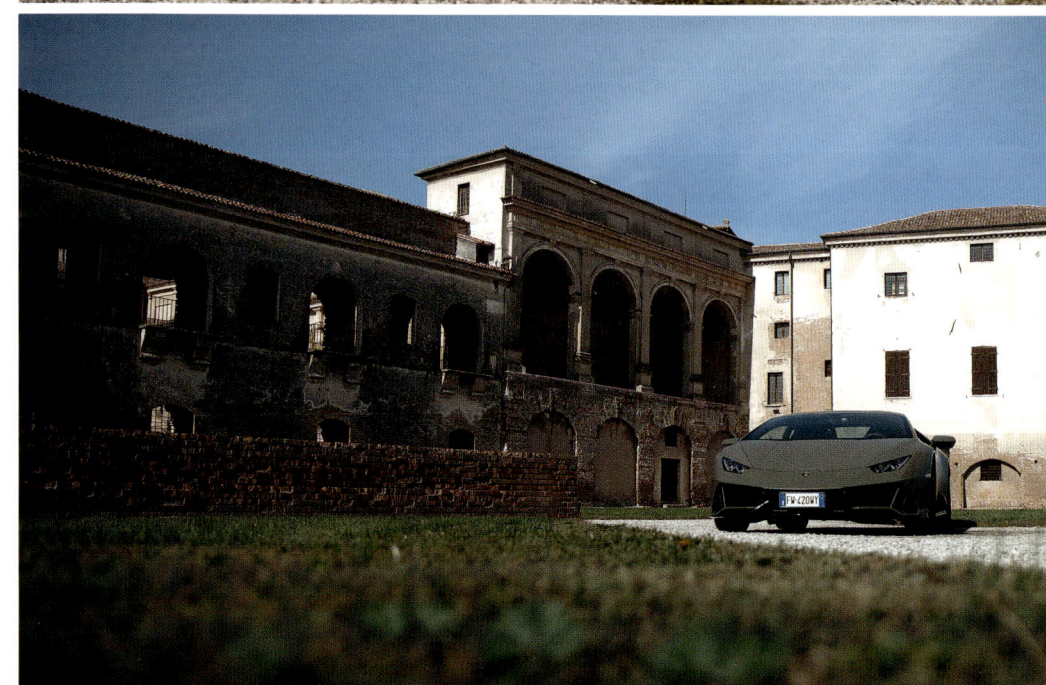

Ducal Palace, Mantua

Pavia Charterhouse

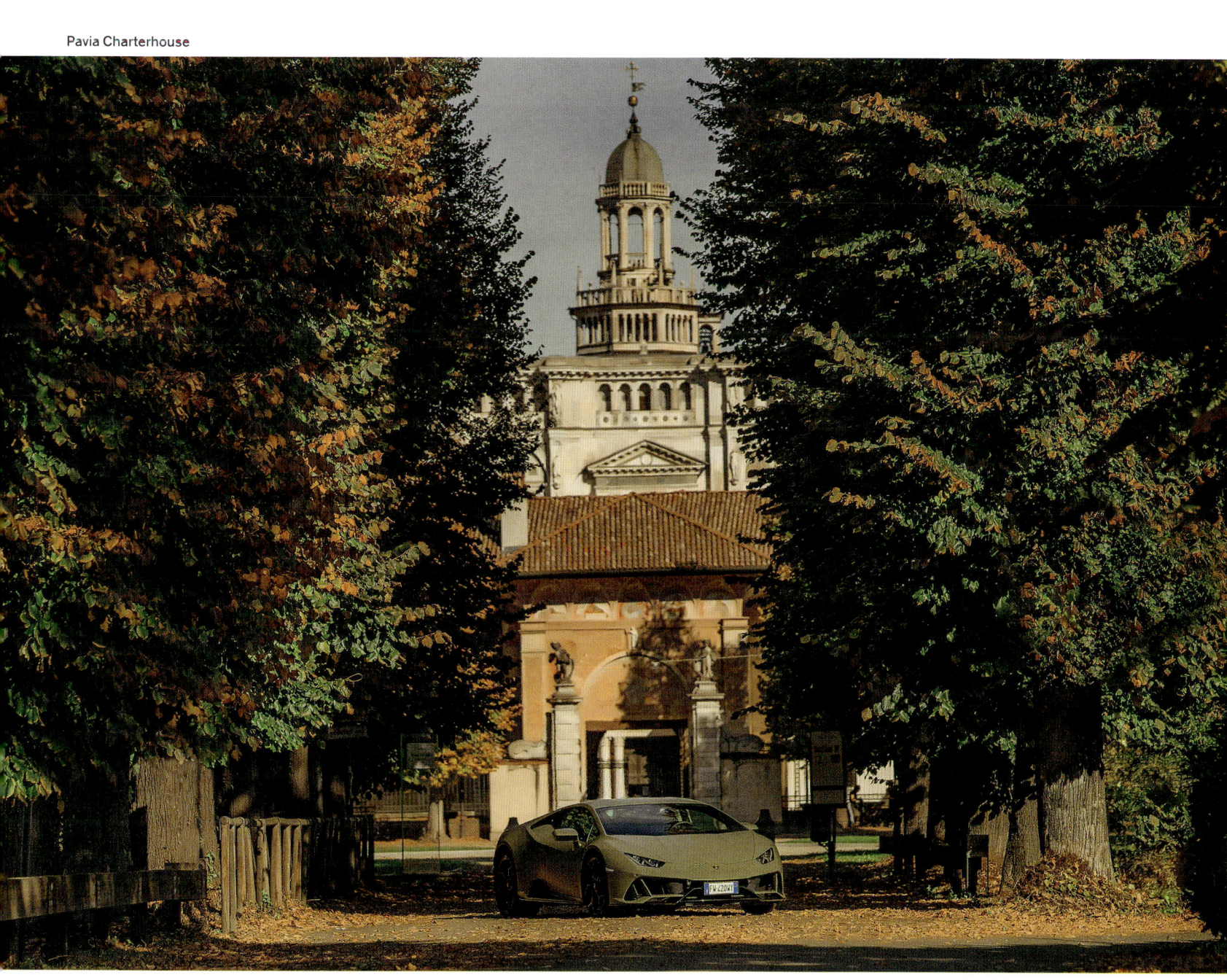

CHIARA MIRELLI
PIEMONTE

The Huracán EVO Spyder whooshed me on a sightseeing tour of Turin and the Piedmont region. I told the story of this part of the country by shedding light on its underground and musical culture, which is my photographic bailiwick. Fragments of life, emotions, cultural and musical pulses interweave with historical and urban backgrounds. With a common thread: the pulsating energy of a creative region, ever-changing and in continuous ferment. Last but not least, an open-air gig by the Magliano Alfieri band in the middle of the Roero vineyards, paying homage to the beauty of a region that is proud of its uniqueness and cultural identity.

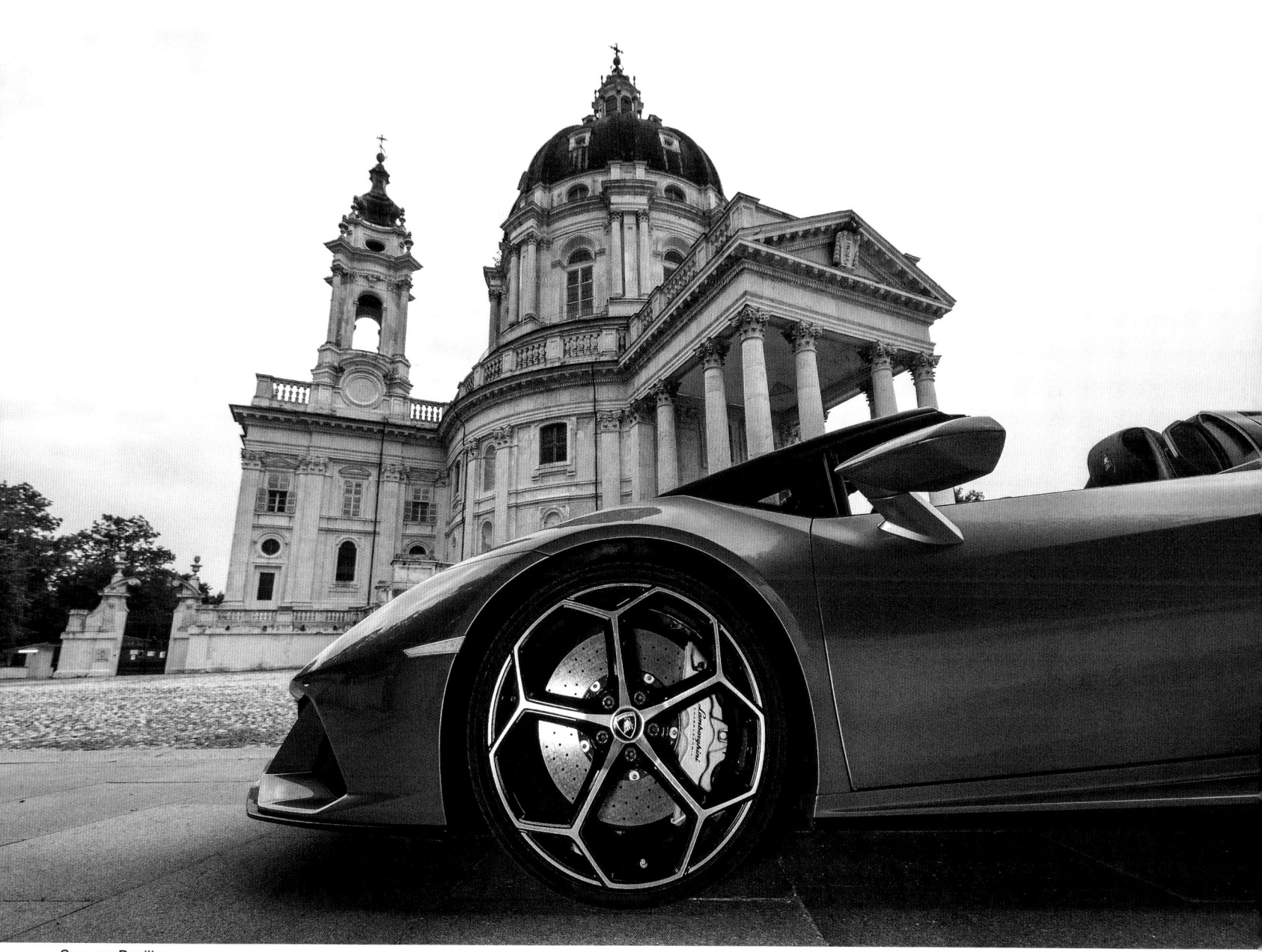
Superga Basilica

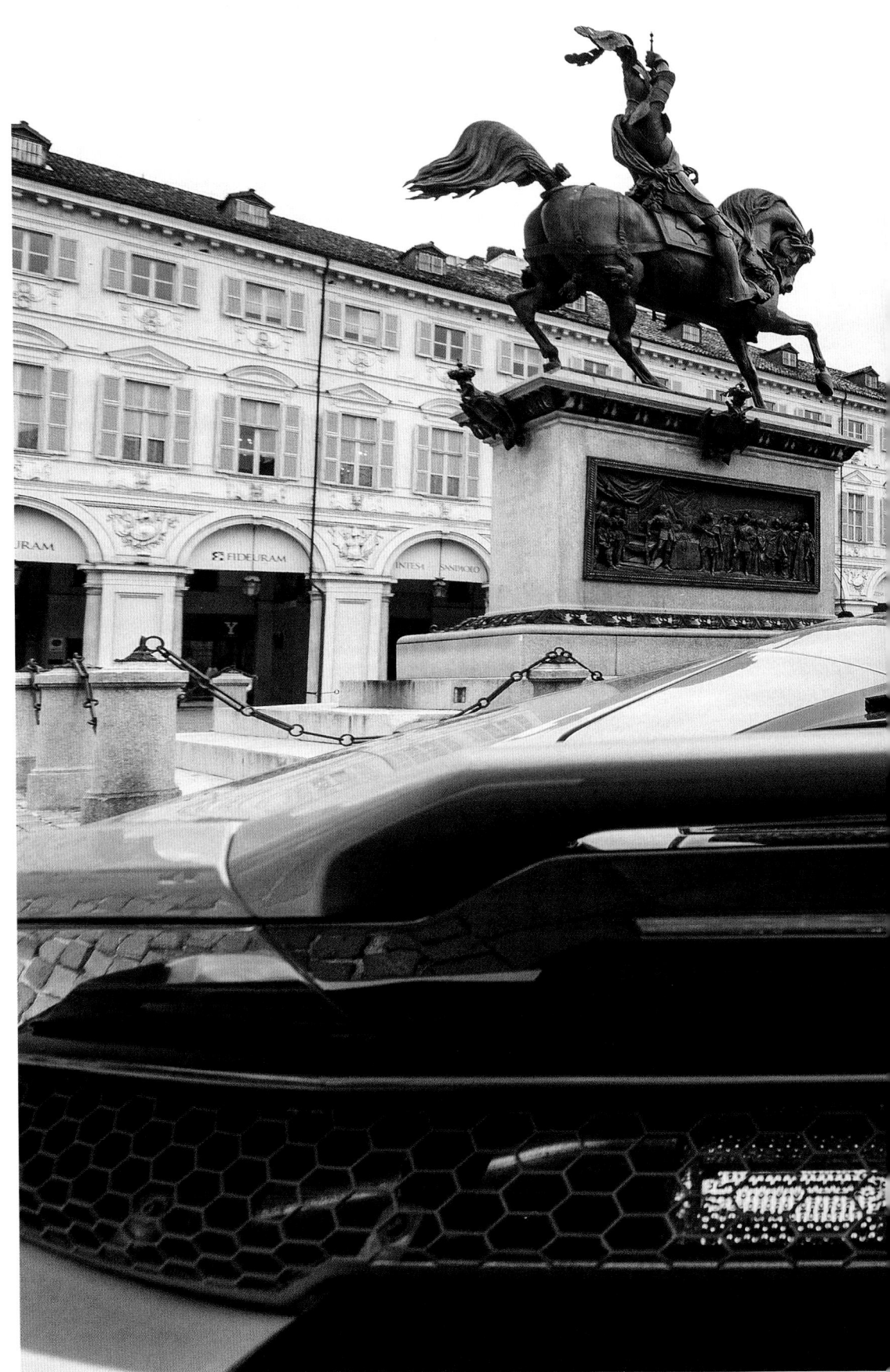

Piazza San Carlo, Turin

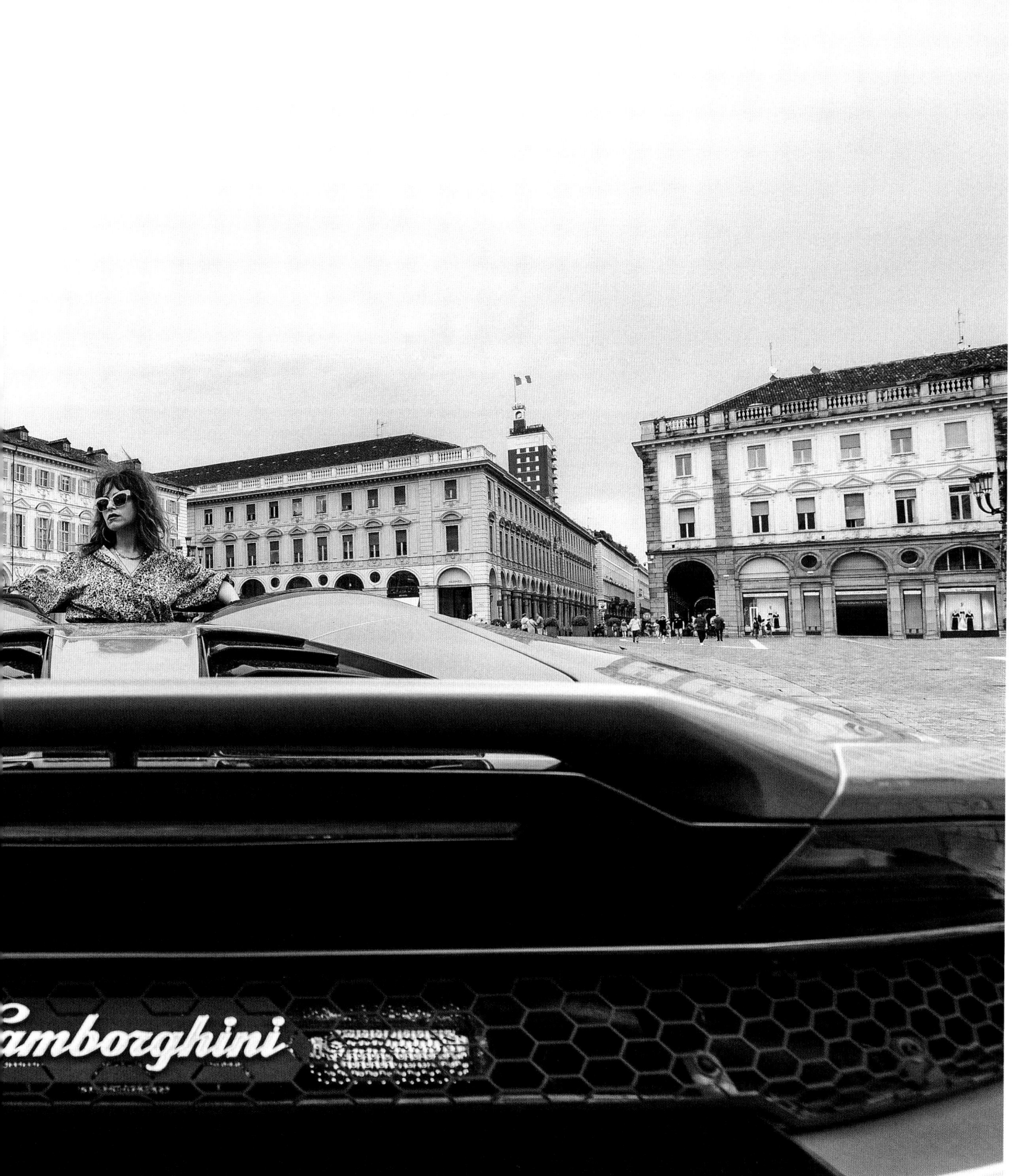

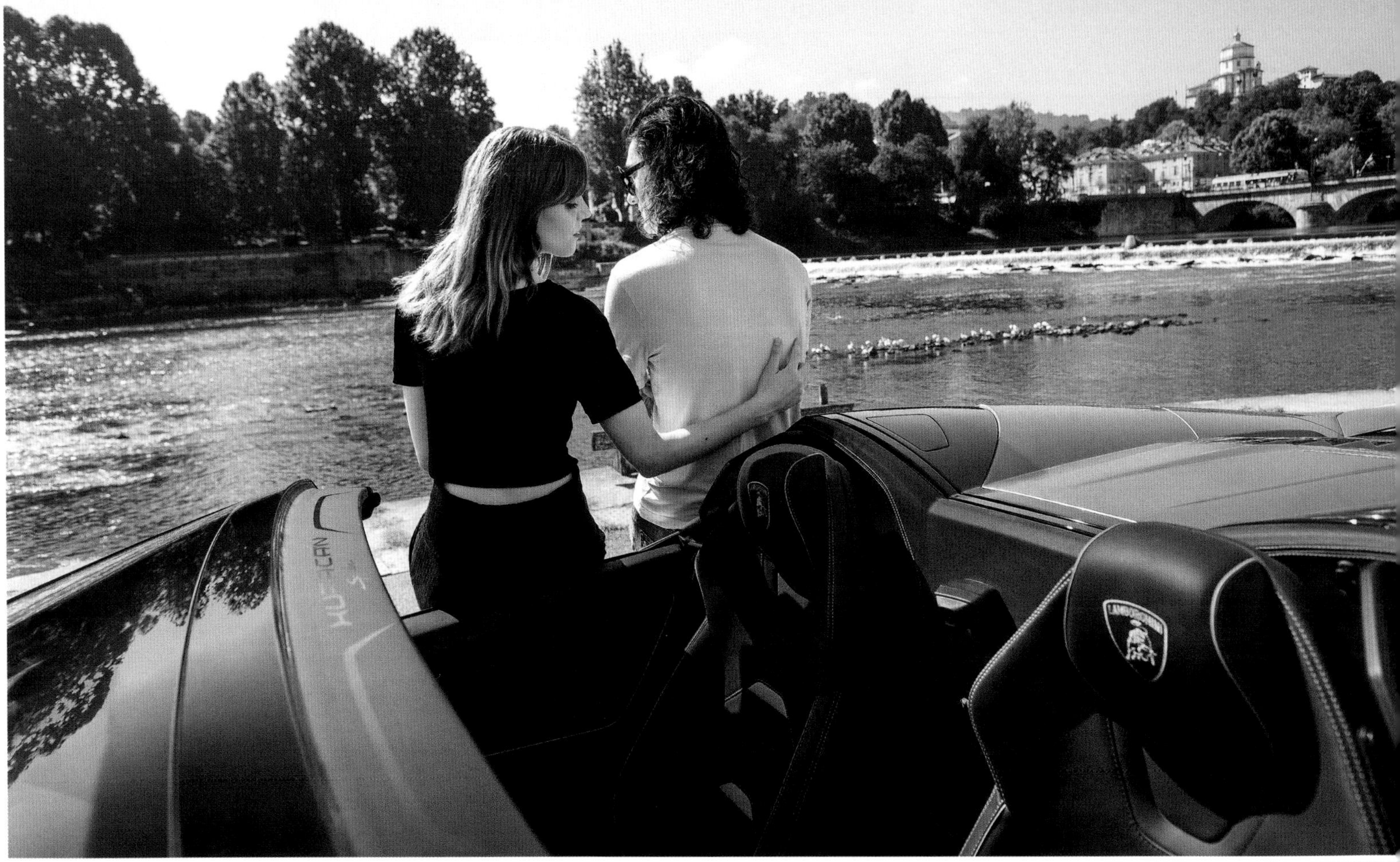

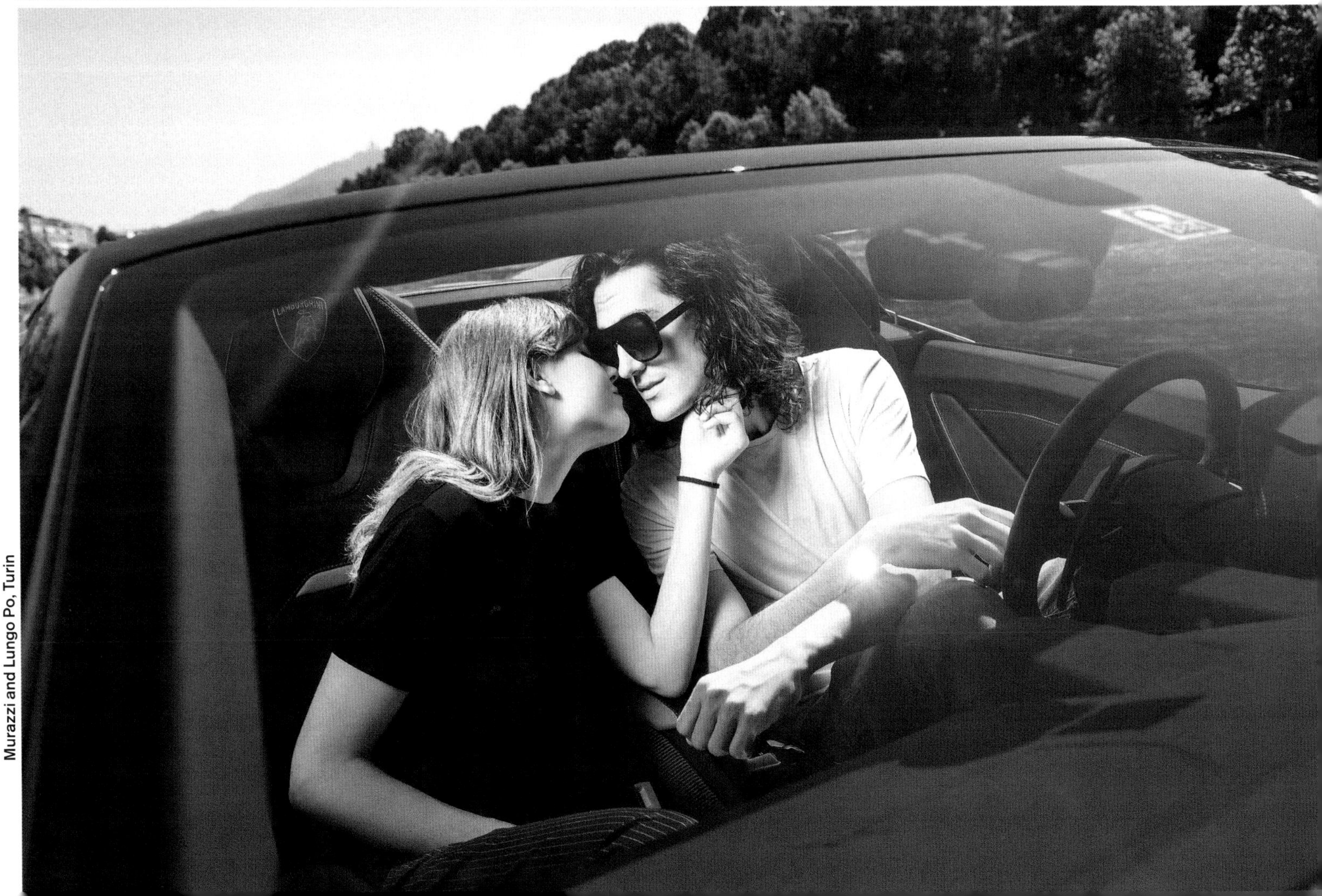

Murazzi and Lungo Po, Turin

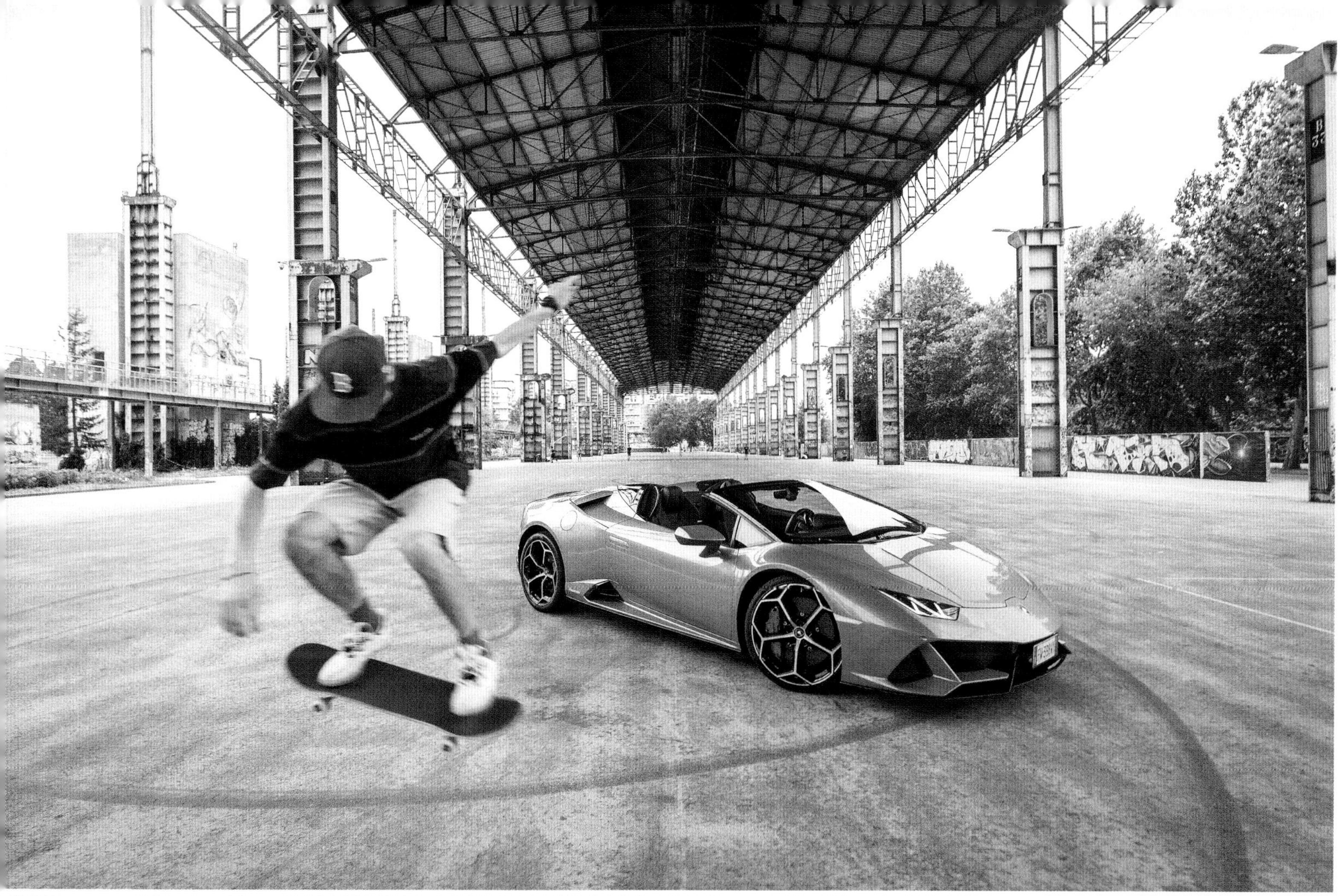

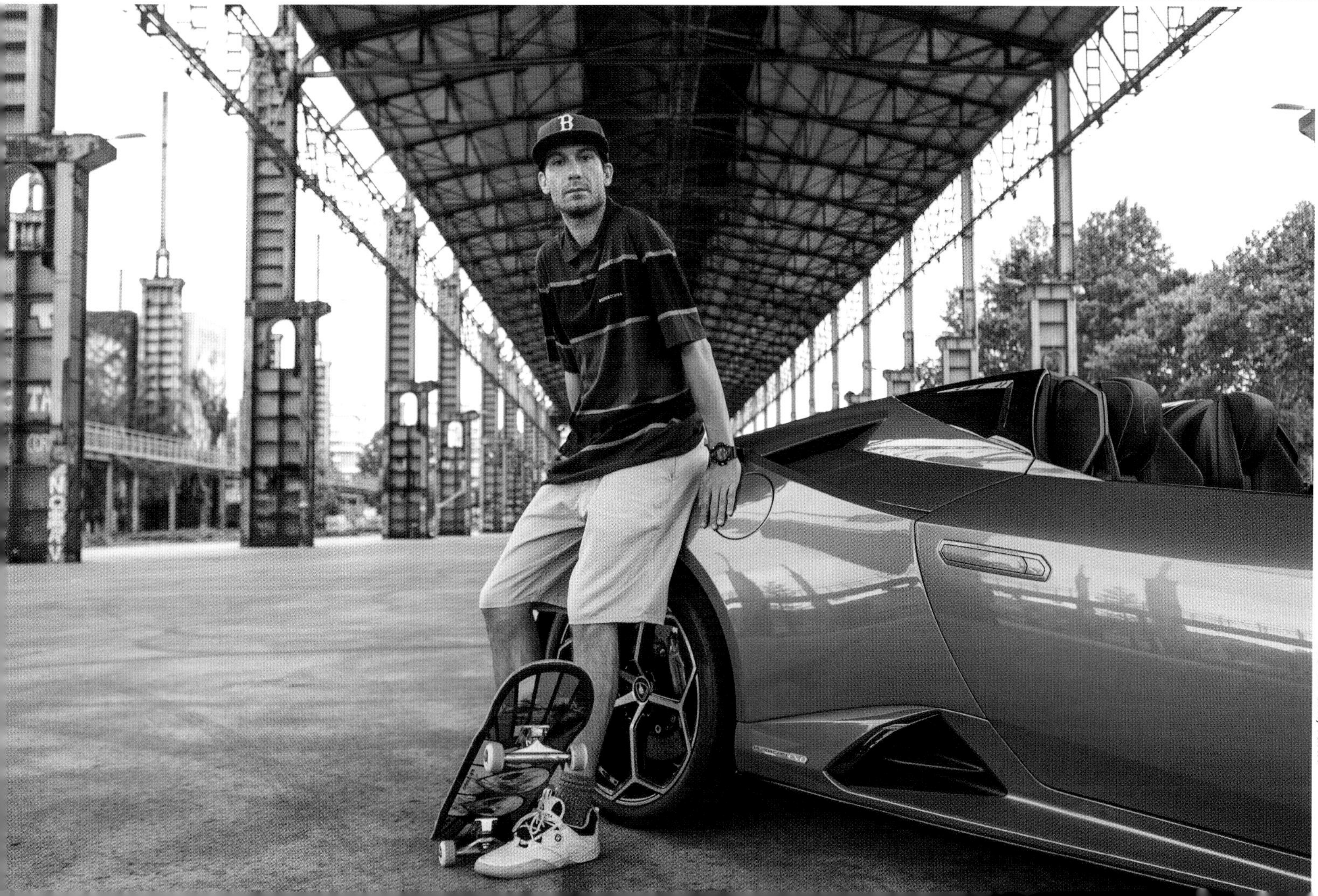

Dora Park, Turin

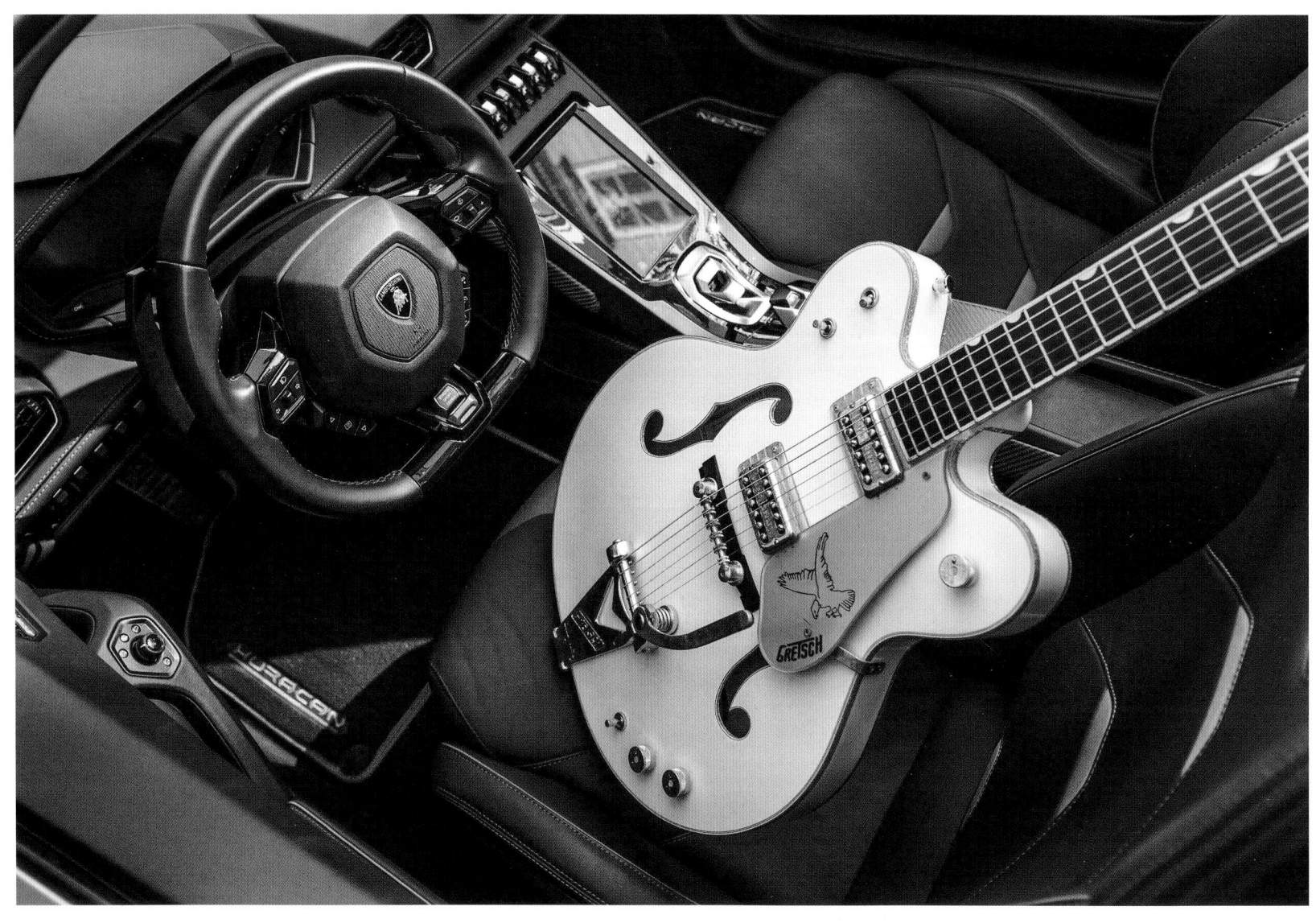

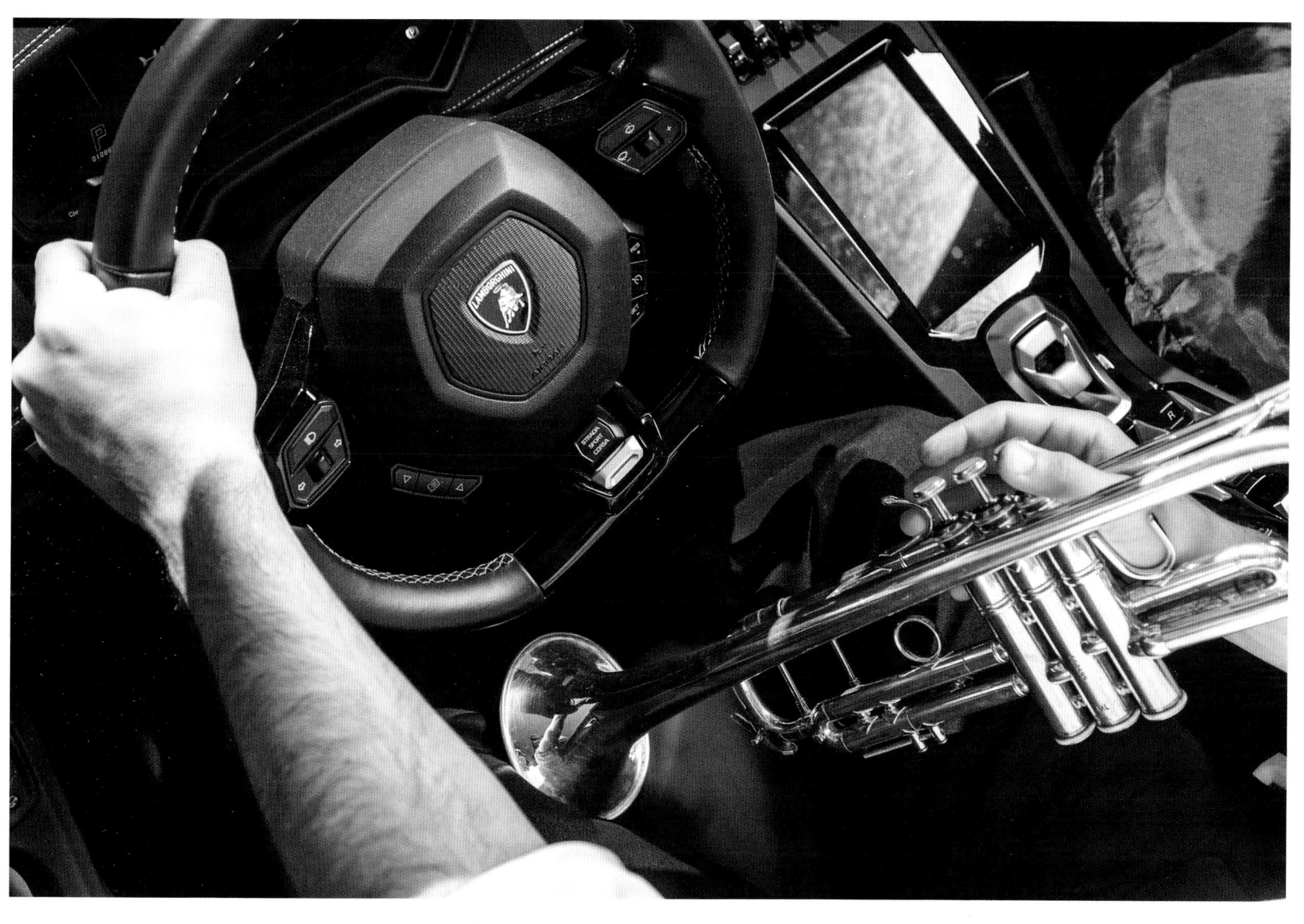

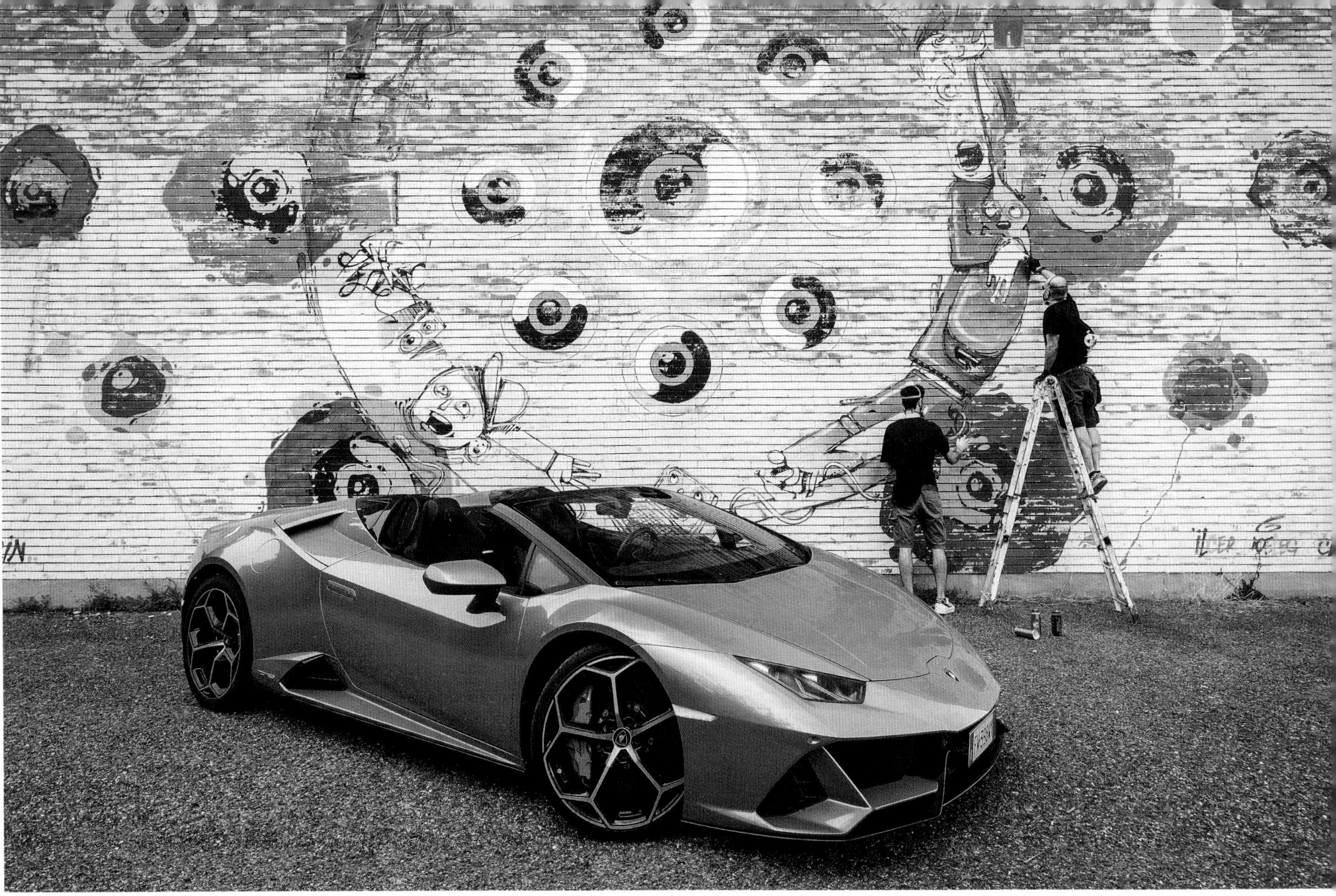

"Hiroshima mon amour" mural, Turin

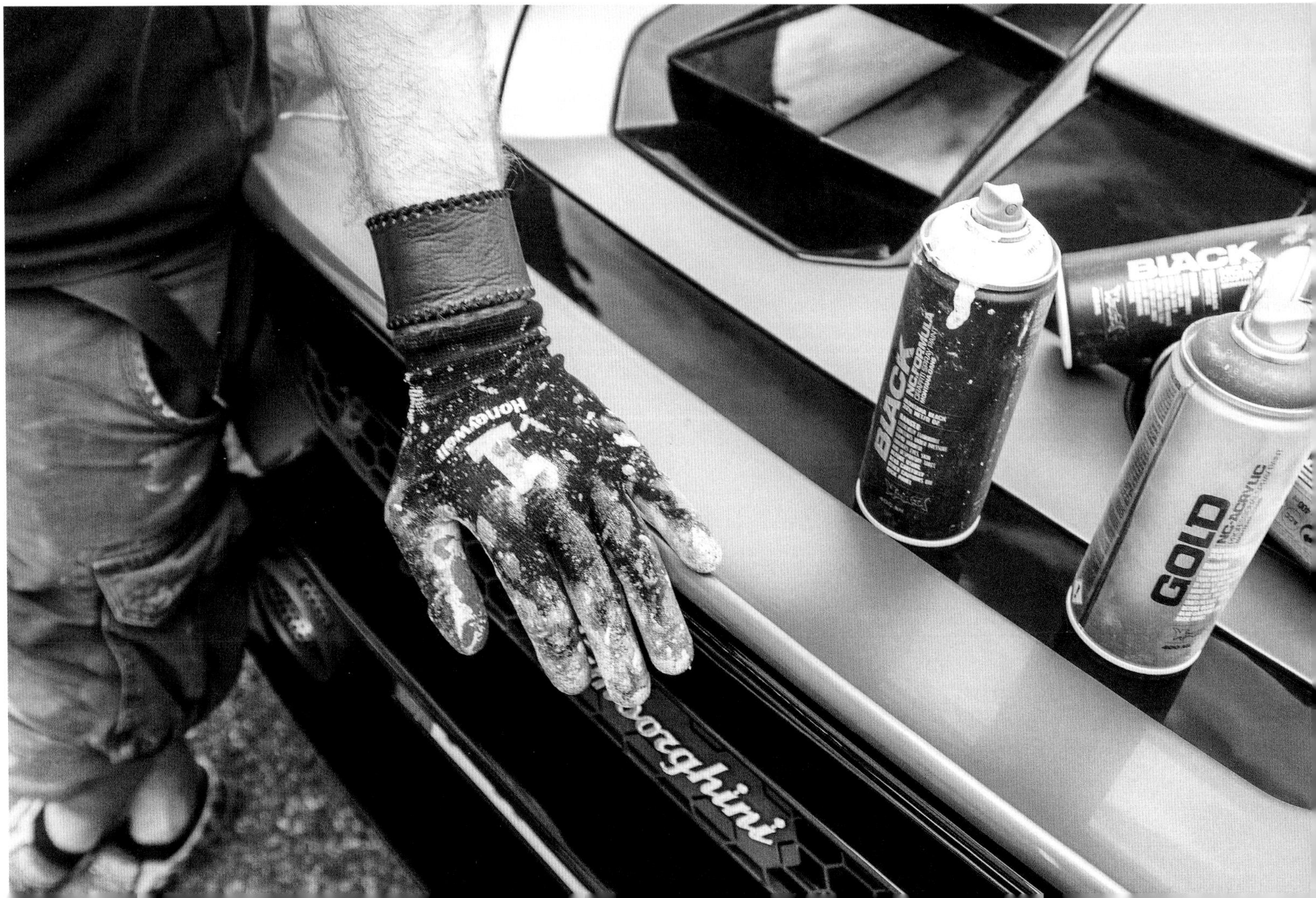

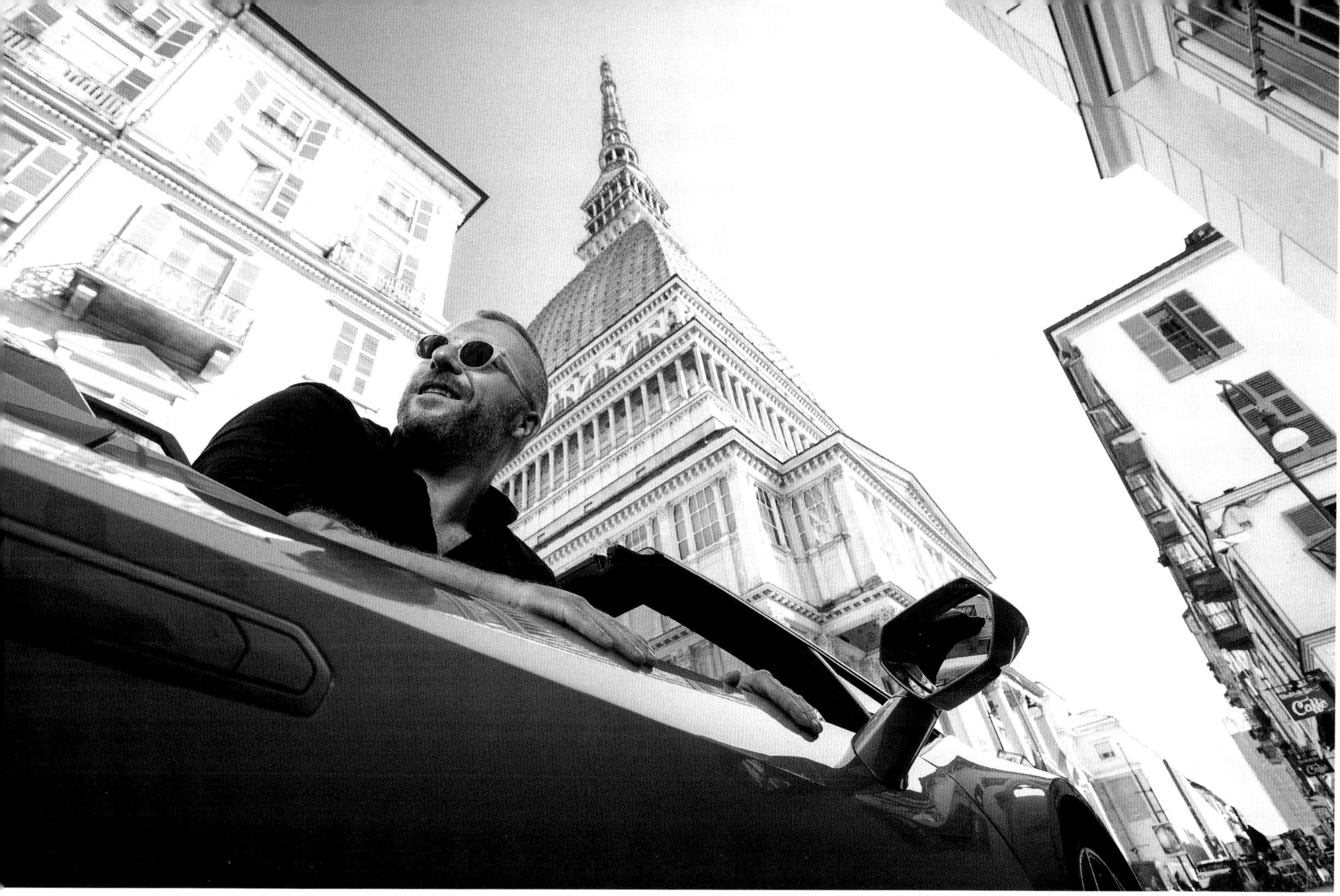

Mole Antonelliana, Turin

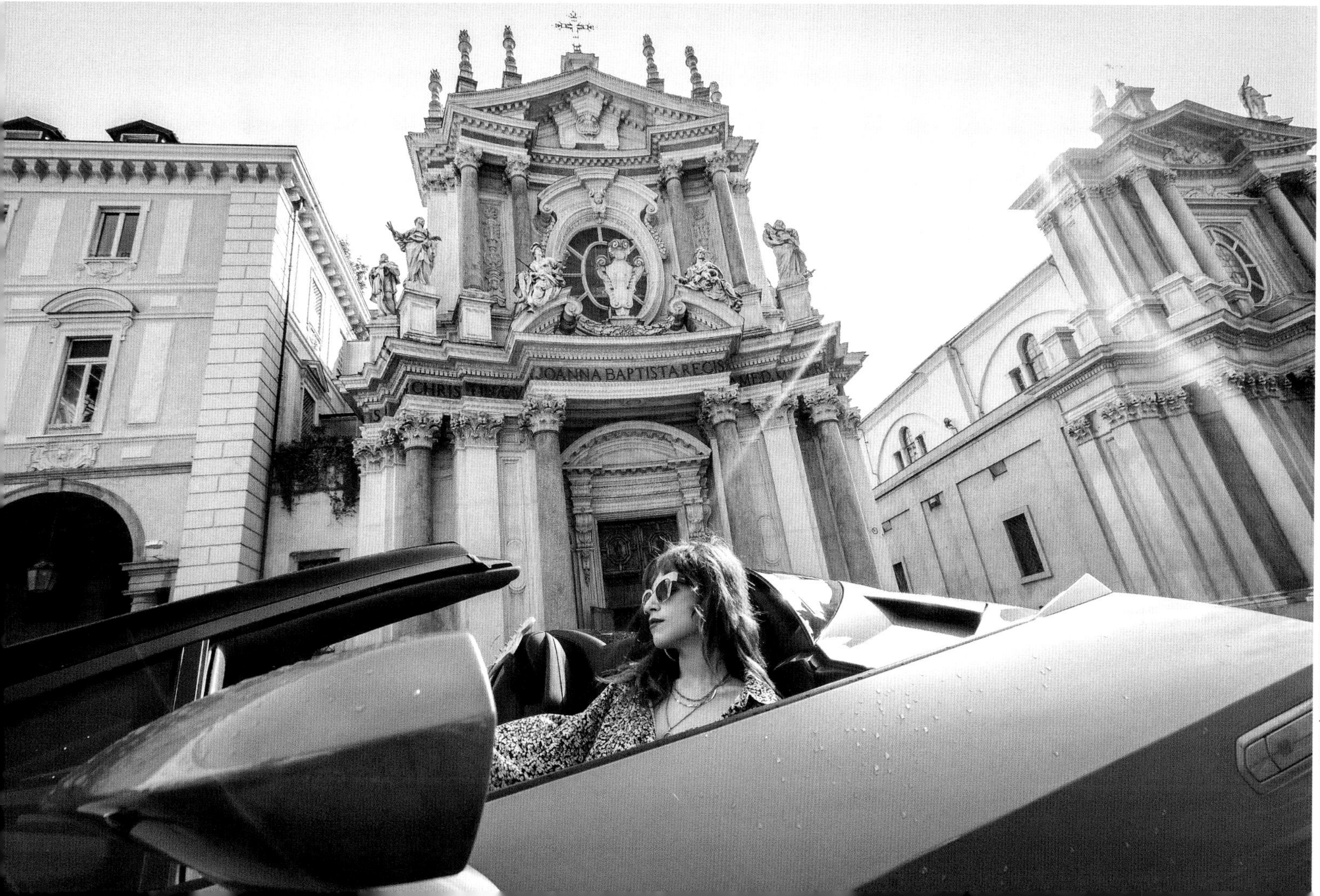

Piazza San Carlo, Turin

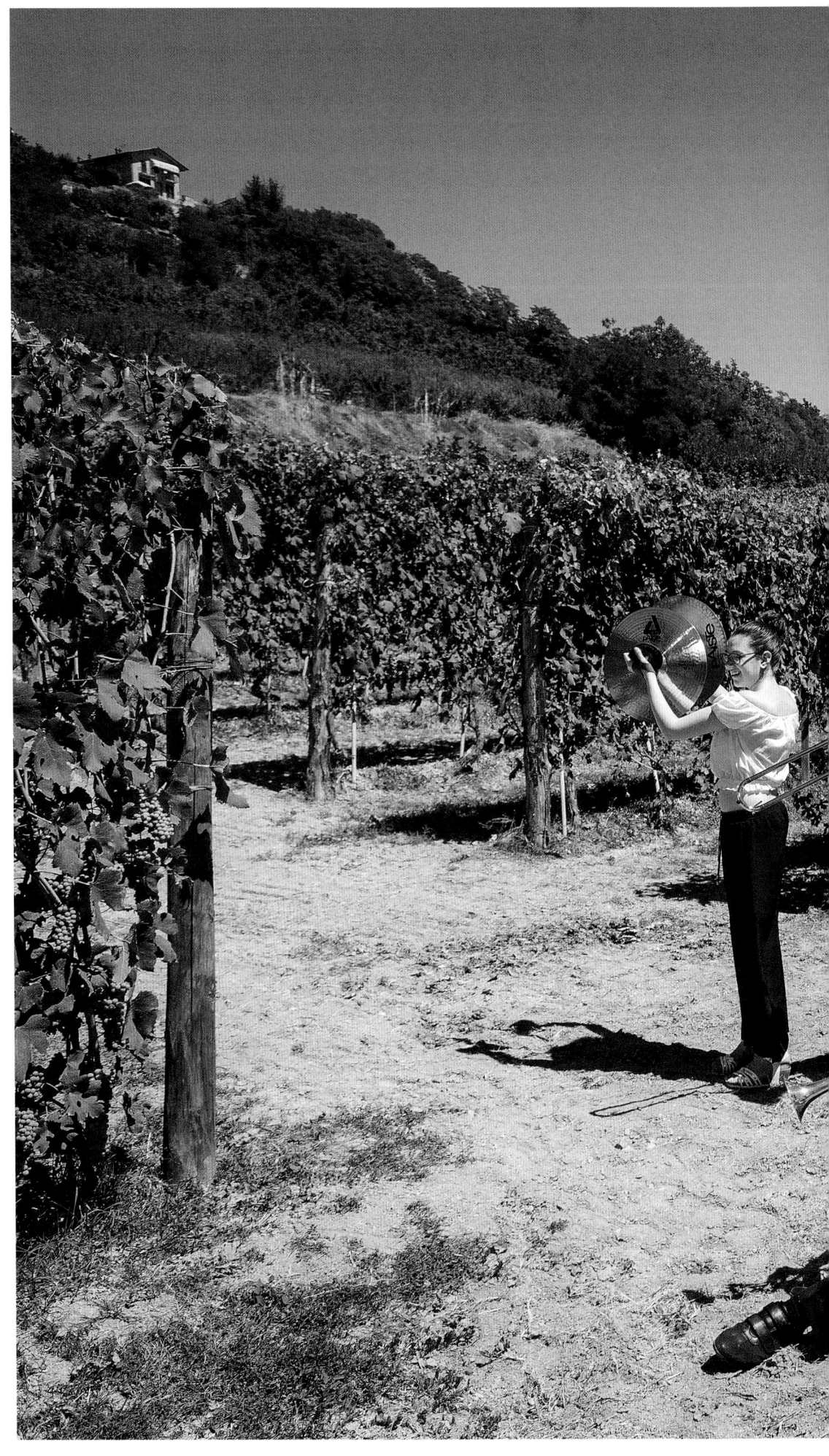

Roero vineyards, Langhe

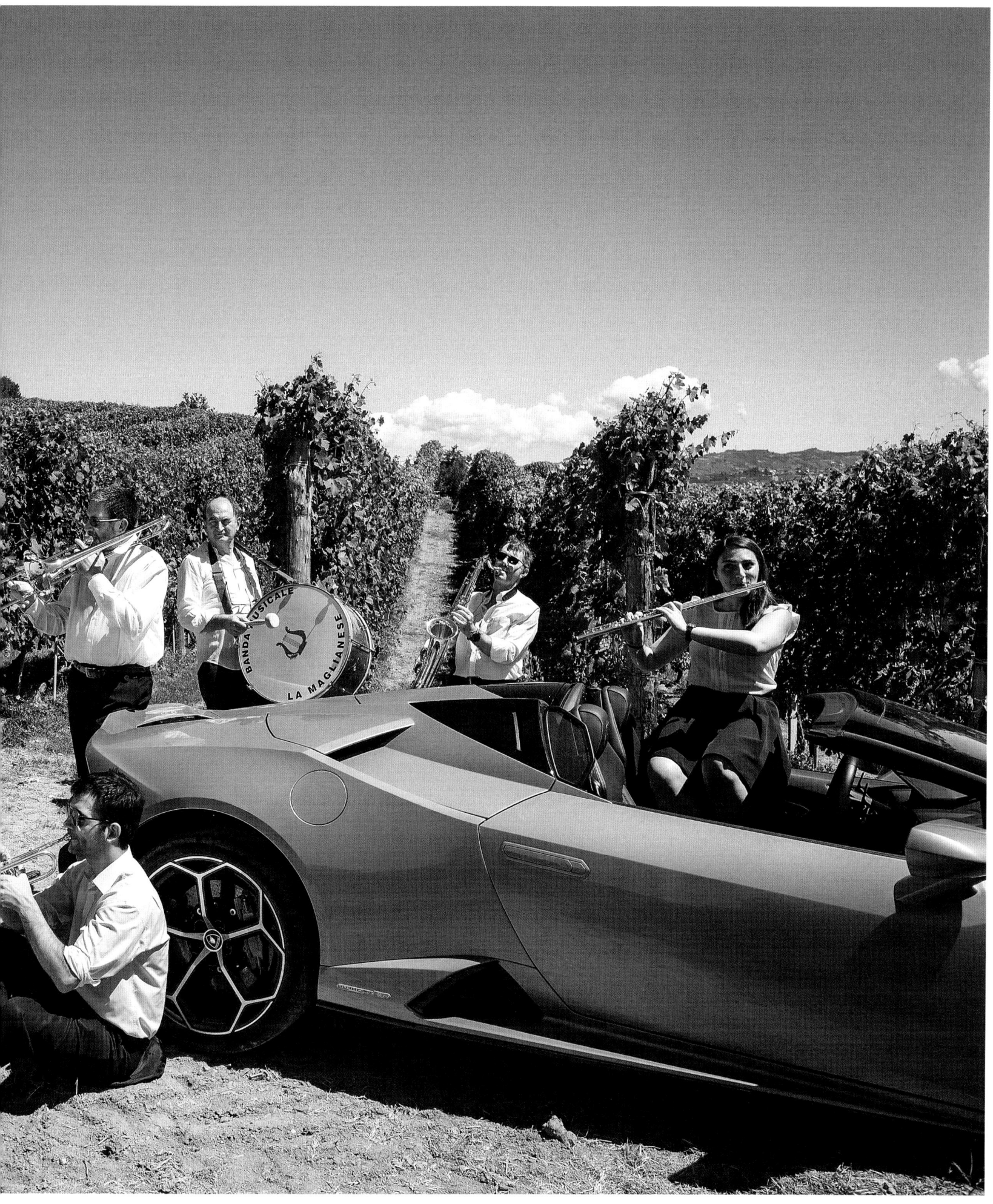

ALBERTO SELVESTREL
LIGURIA

This series of photographs represent a metaphorical path to redemption: from a dark period, one arrives at a new, more aware new beginning. The narrative begins at night, where the moonlight and the Huracán EVO's headlights create an interplay among light sources. Then it's morning time and our journey continues through the magnificent Liguria, its mountains overlooking the sea. The Lamborghini seems to have been sculpted by the same wind that, for millennia, has shaped the rocks along the roads. In the end, the Huracán enters into symbiosis with the surrounding landscape and, as we reach the coast, welcomes us at the dawn of a new day. Rebirth.

Sestri Levante, Genoa

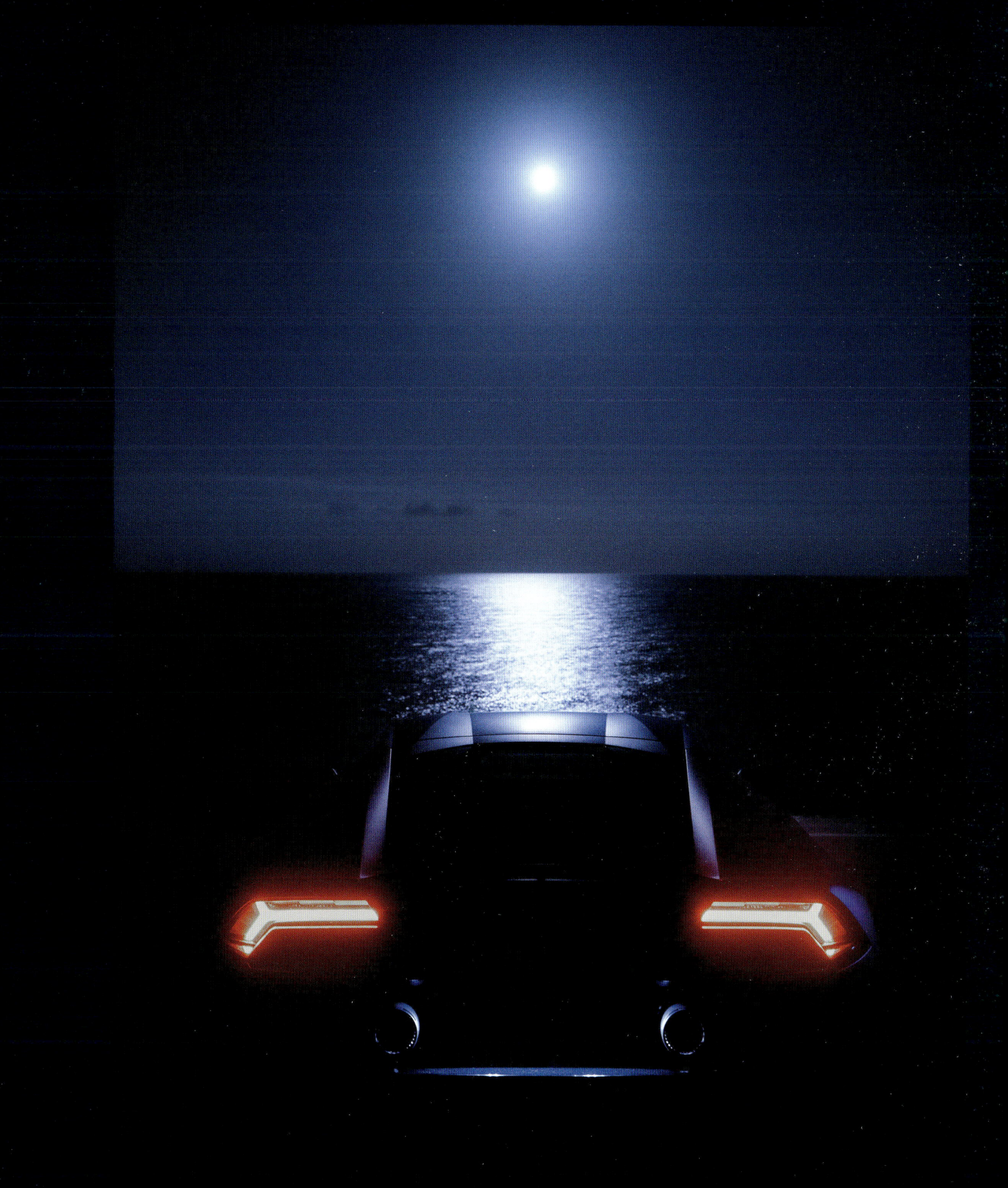

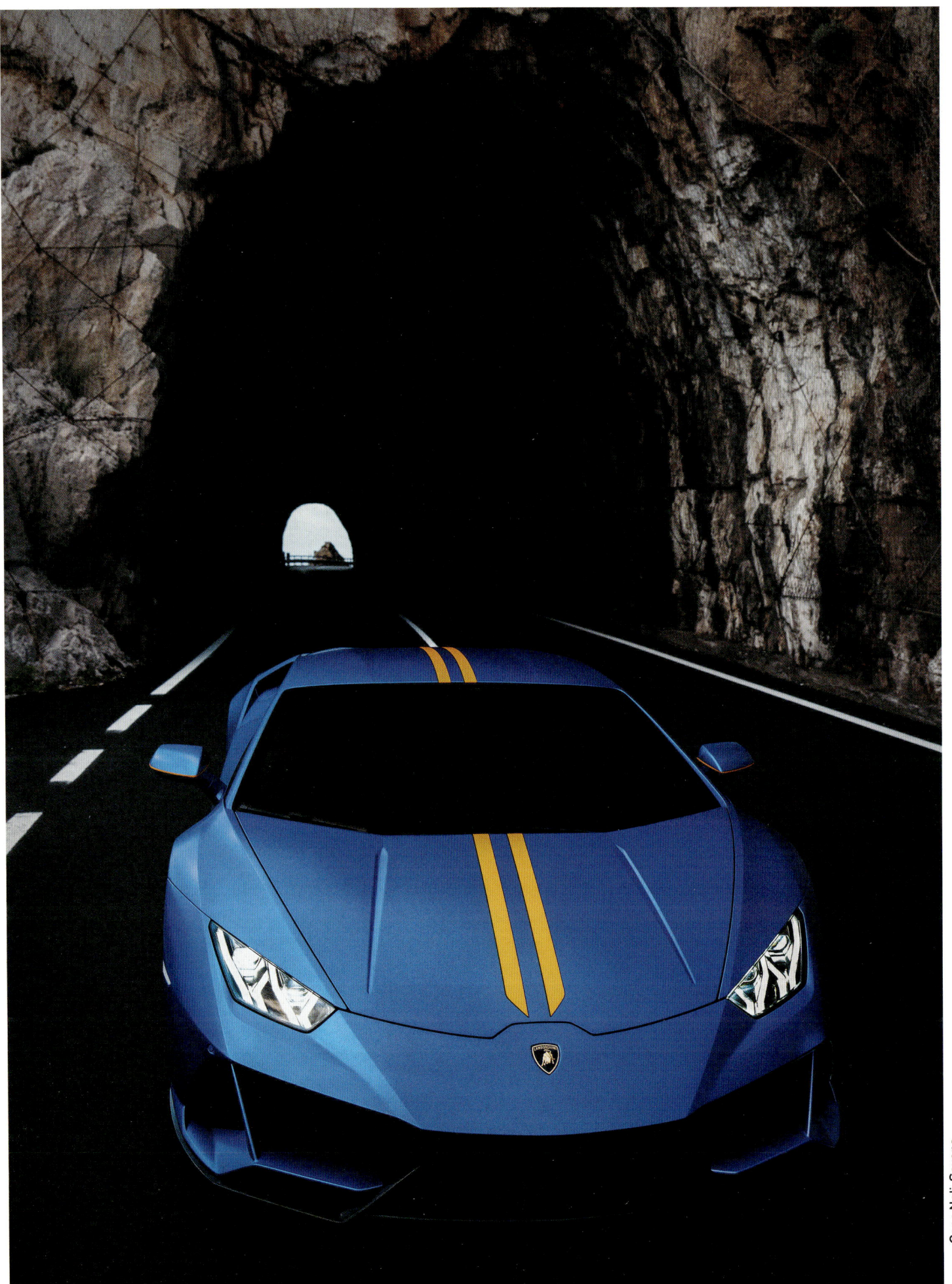

Cape Noli, Savona

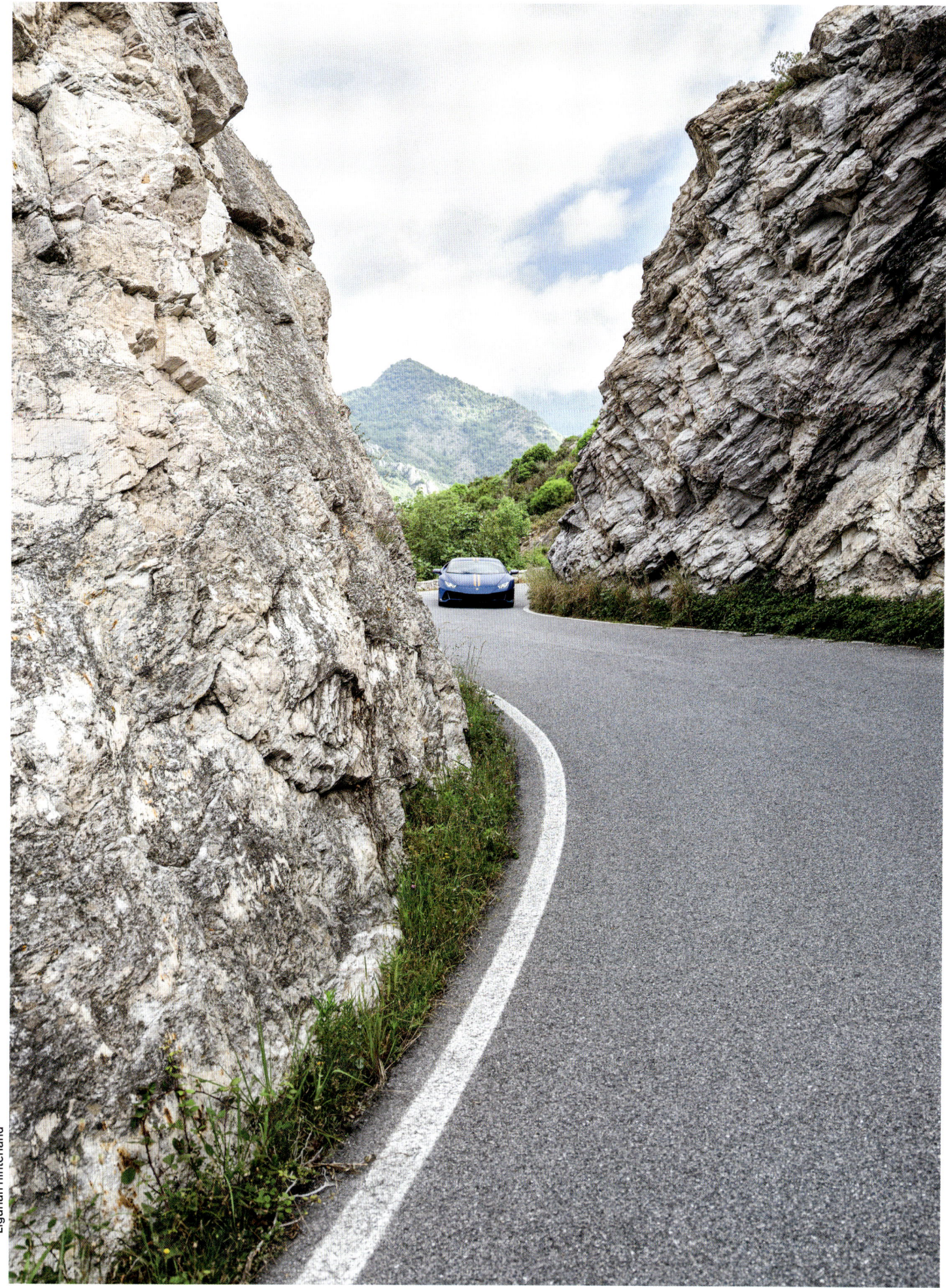

Ligurian hinterland

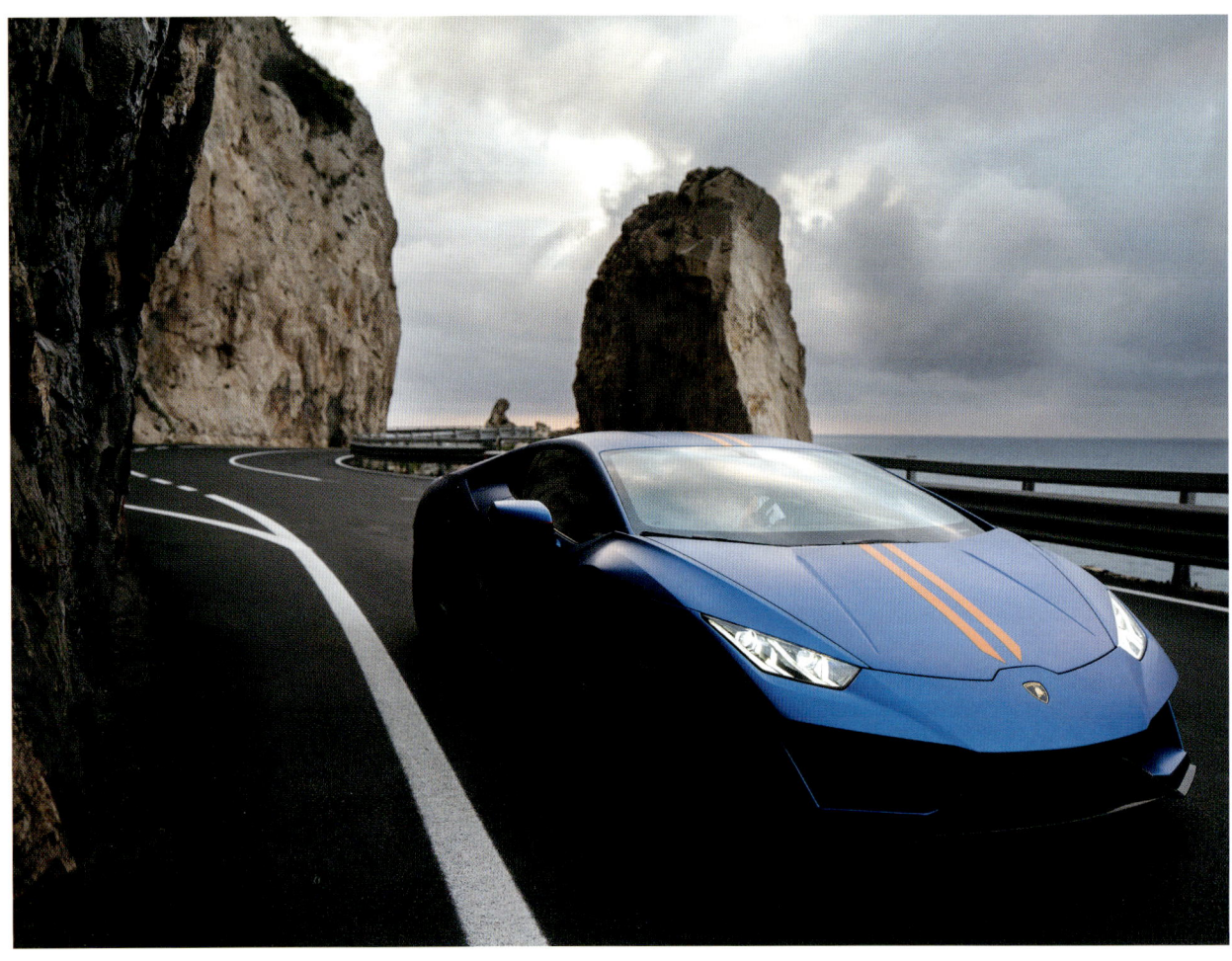

Cape Noli, Savona

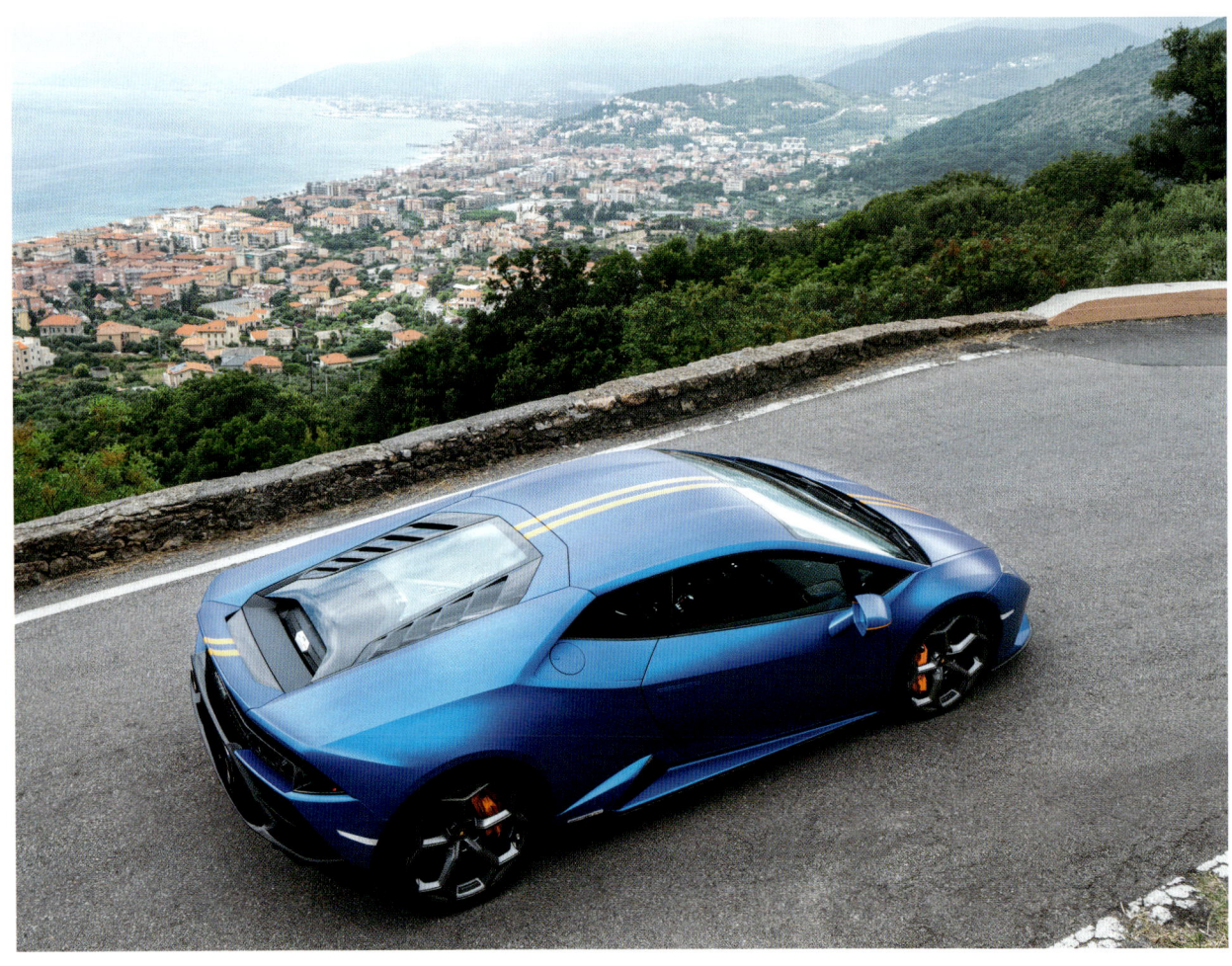

Verezzi, Savona

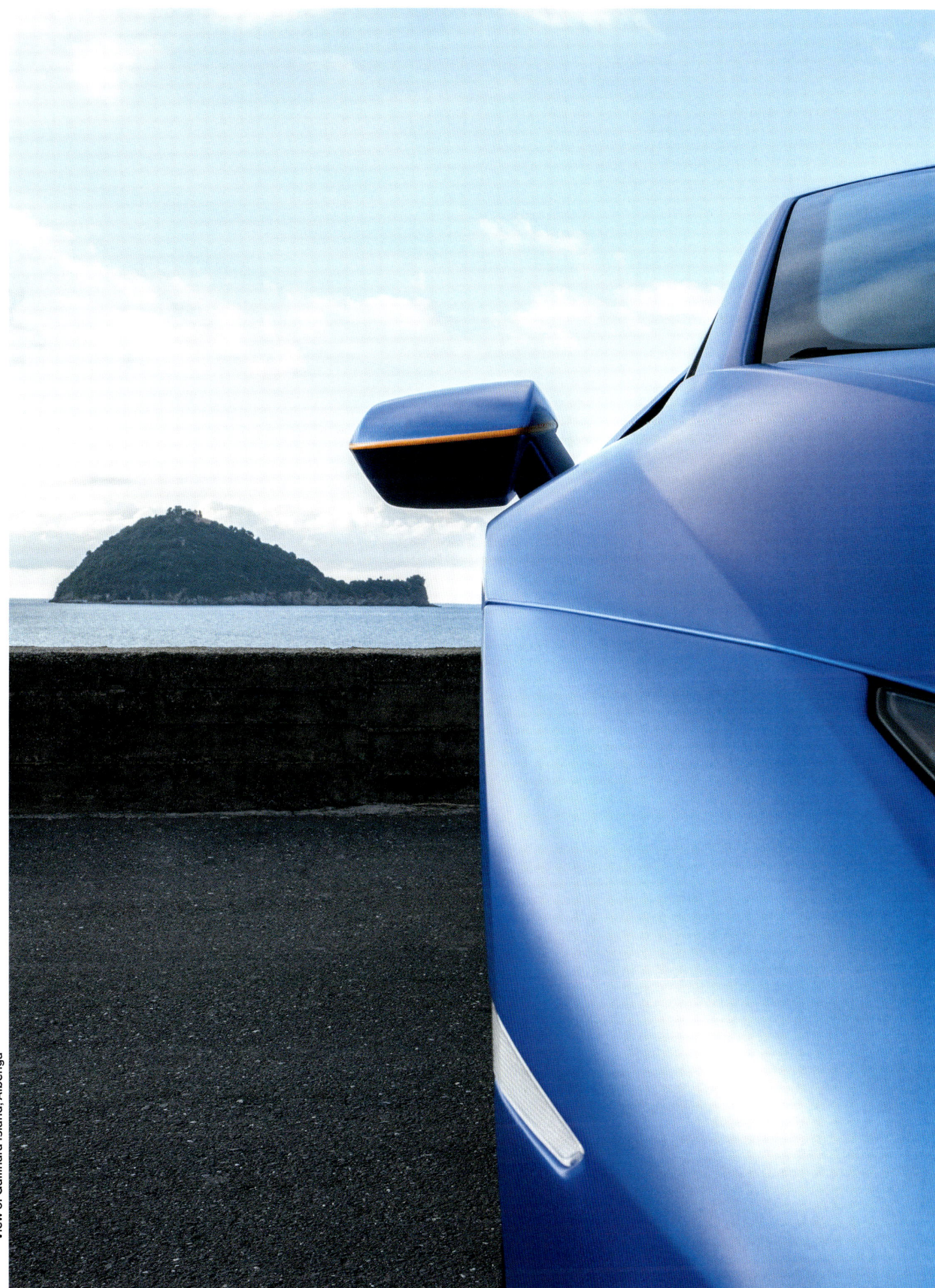

View of Gallinara Island, Albenga

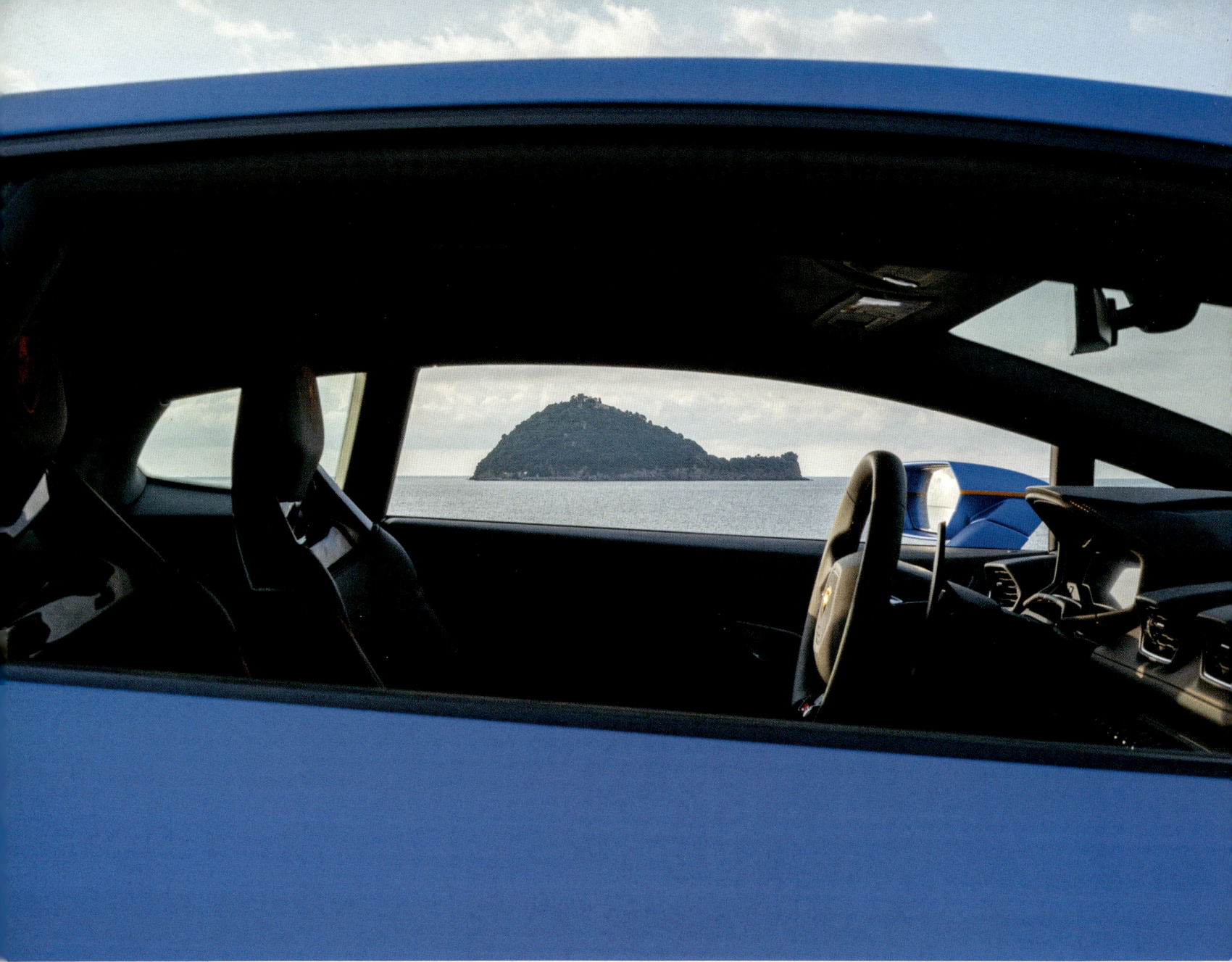
Gallinara Island, Albenga

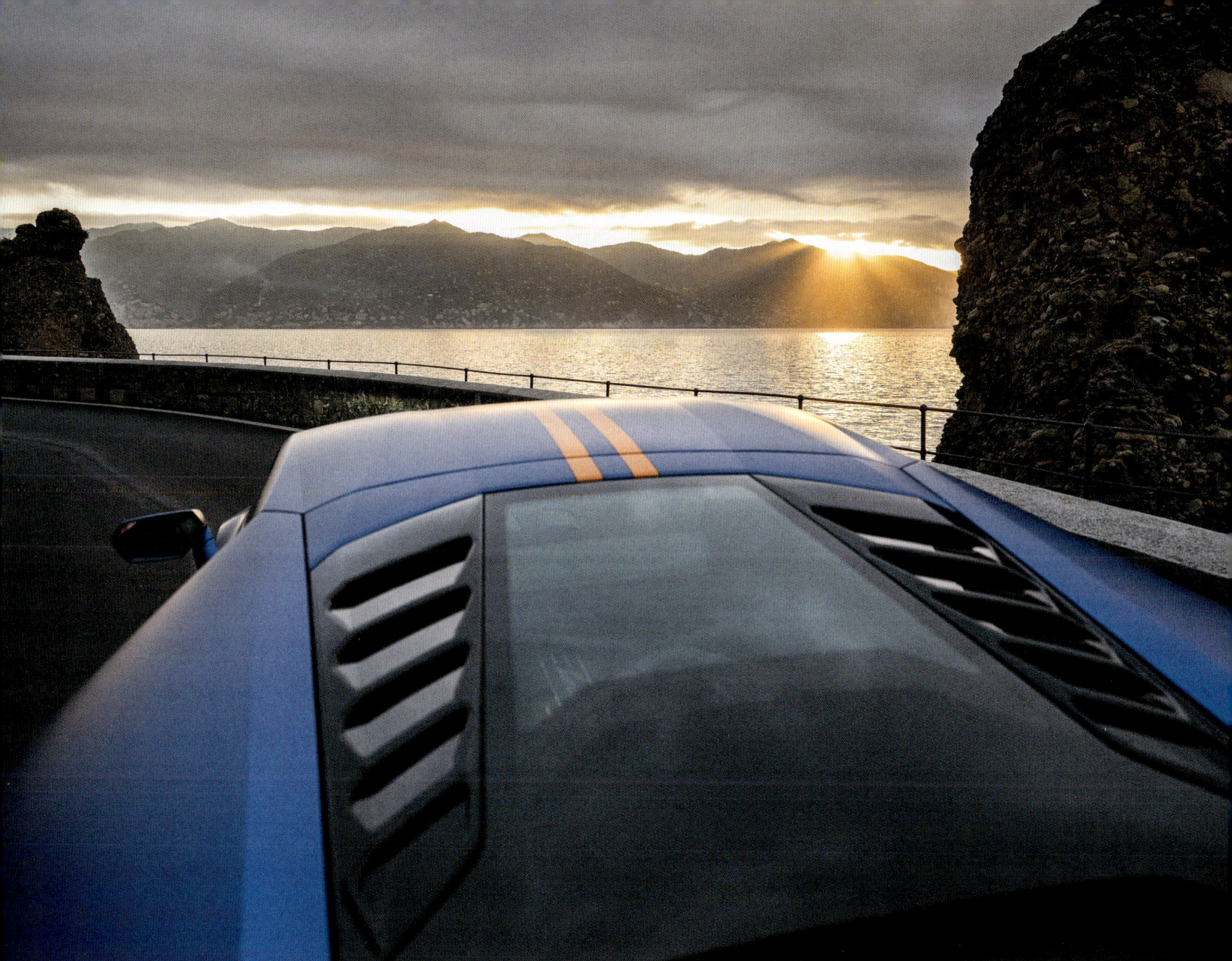

Paraggi, Santa Margherita Ligure

"The Lamborghini seems to have been sculpted by the same wind that, for millennia, has shaped the rocks along the roads"

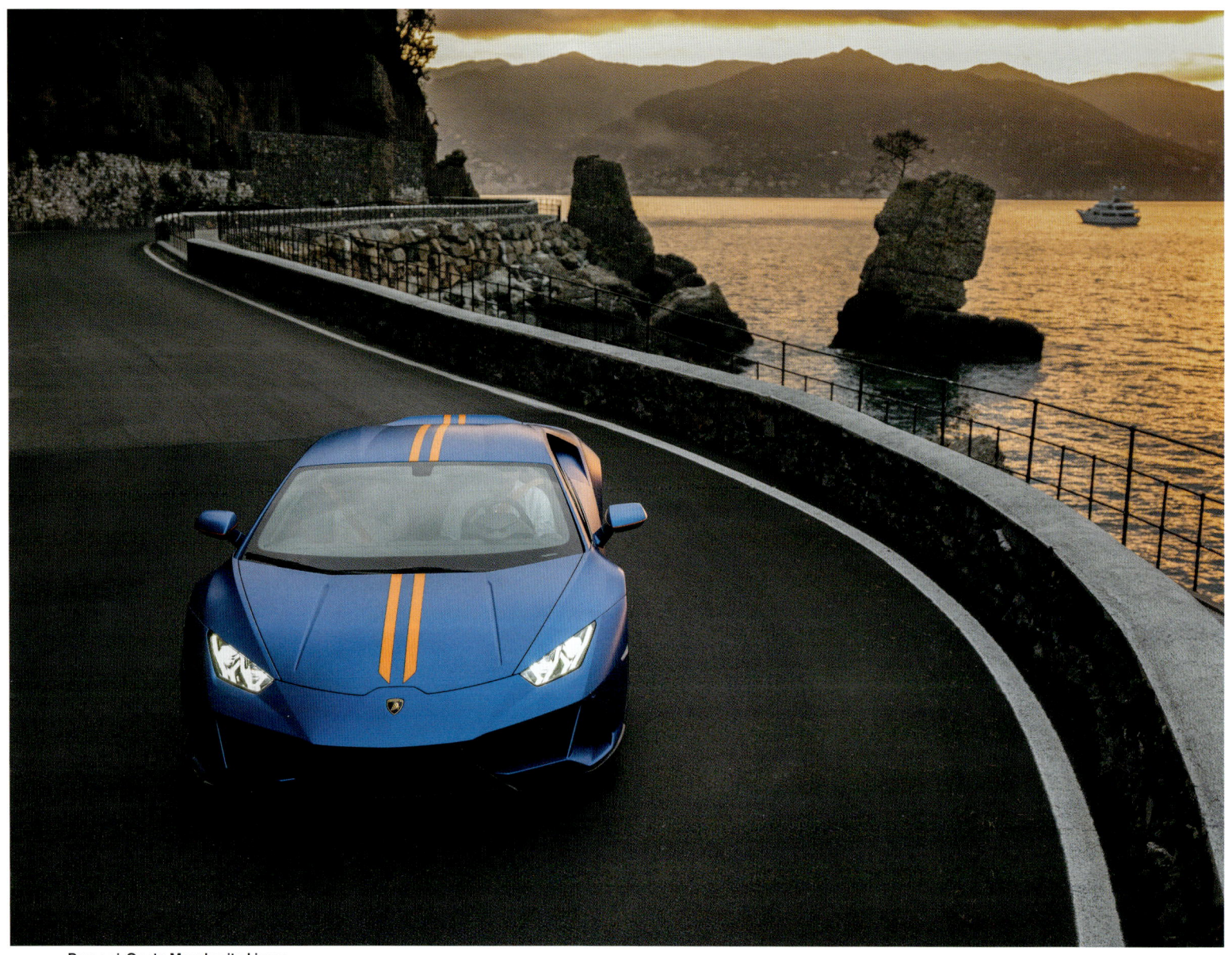

Paraggi, Santa Margherita Ligure

Ceriale, Savona

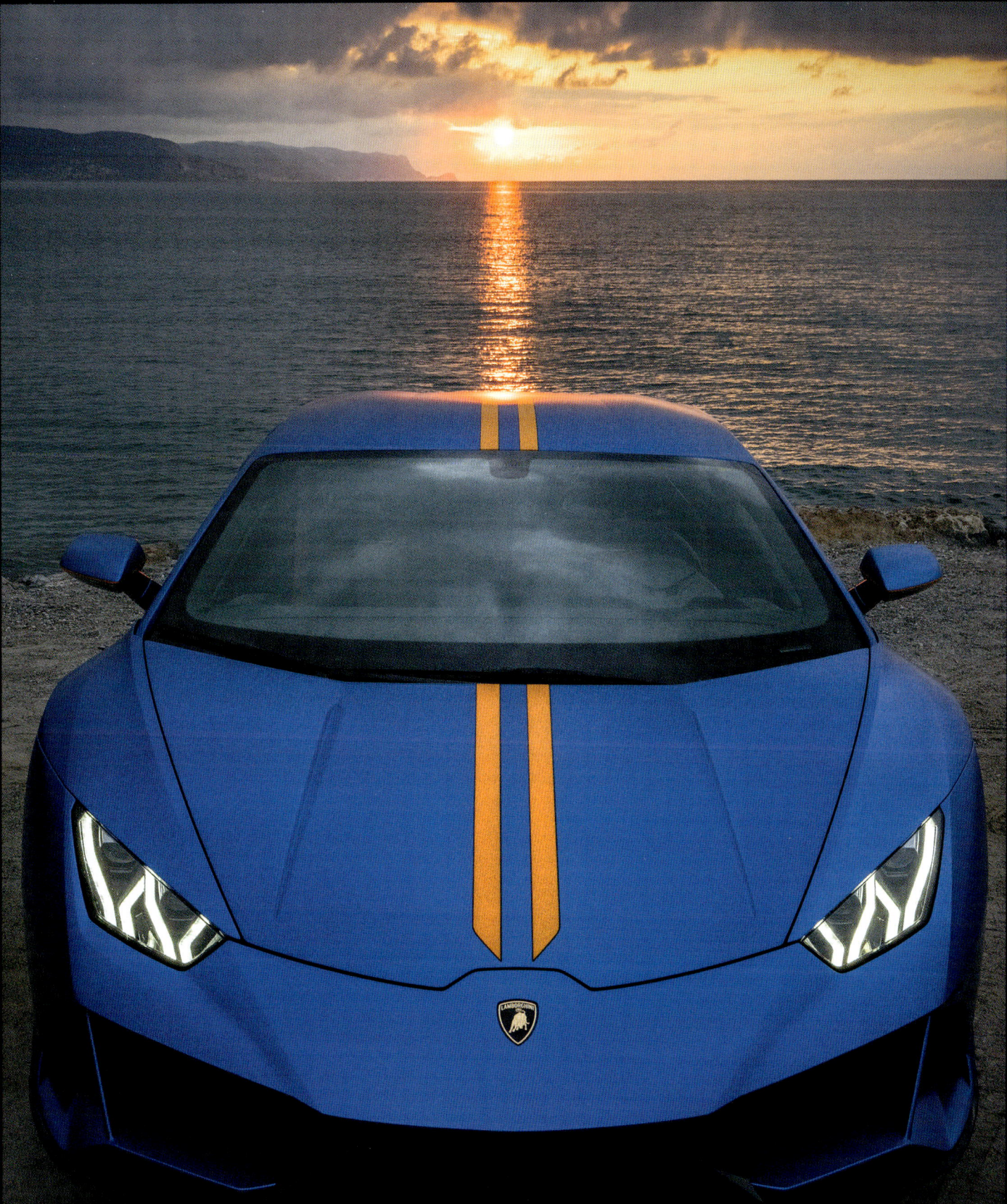

FULVIO BUGANI
VALLE D'AOSTA

I wanted to twin the essence of the Sián Roadster, the Lamborghini super sports car with hybrid technology, and the Aosta Valley, a territory that is deeply-rooted in a virtuous respect for nature. This region has such a rich and glorious past, as we see in the countless fortifications and the splendid Arch of Augustus in Aosta. The area is dominated by wild nature and majestic peaks, the grandeur and verticality of the Alps' highest mountains forming a backdrop to the geometric shapes of a masterpiece of automobile technology and design. I combined the Sián's Blu Uranus with the night sky photographed from the Saint-Barthélemy Valley's astronomical observatory, acknowledged as one of the finest stargazing spots in the world. Against this backdrop of imposing massifs and magnificent valleys, bursting with rivers and lakes, nestled between the peaks is Europe's most spectacular mountain, the Matterhorn, a top destination for extreme sports enthusiasts and a symbol of challenging our limits, of seeking freedom… The very same values symbolized by this rare and extraordinary Lamborghini, of which just nineteen have been made.

Mount Cervino, Breuil-Cervinia, Aosta

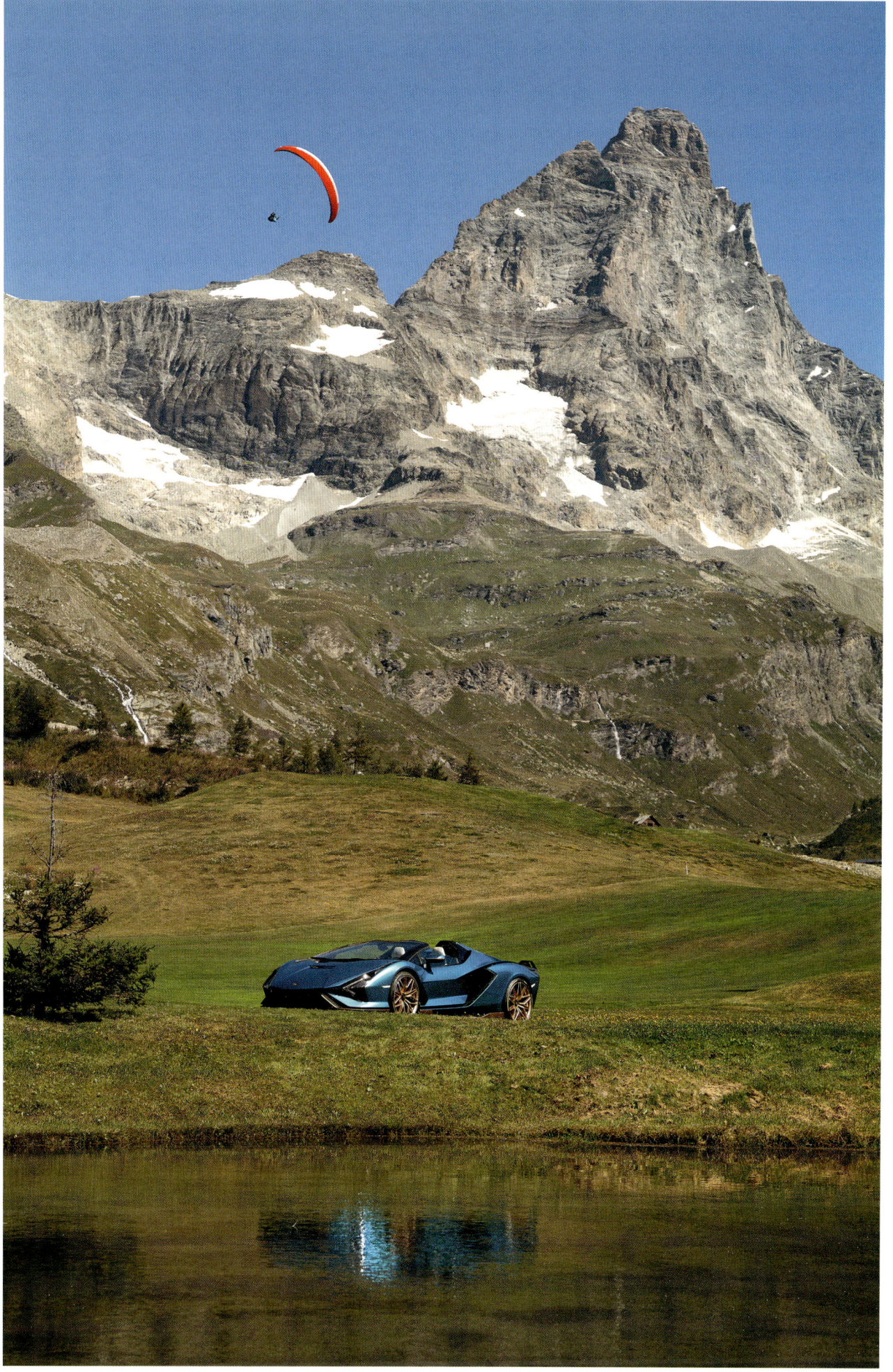

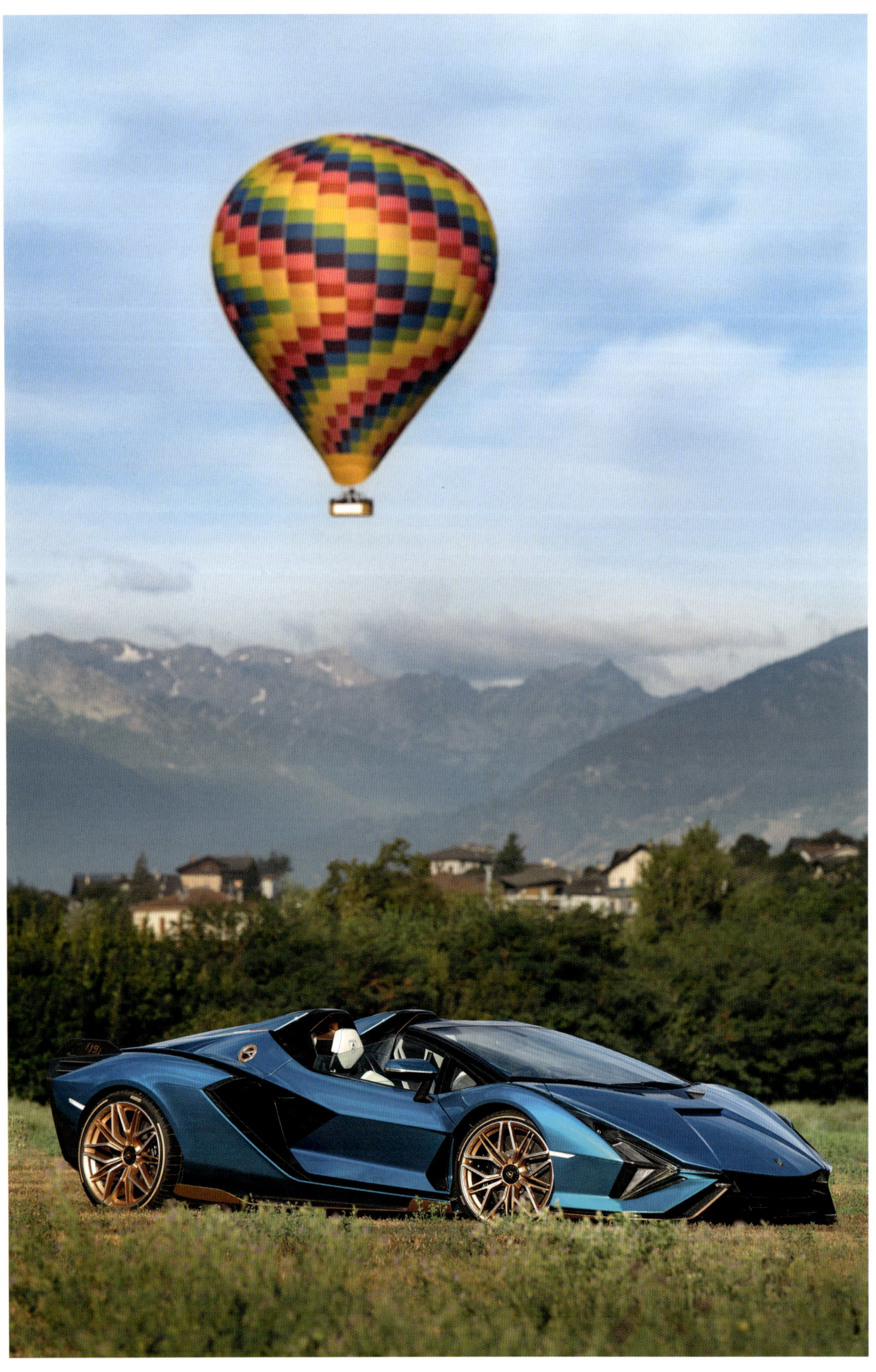

Italy's largest hot air balloon,
Fénis, Aosta

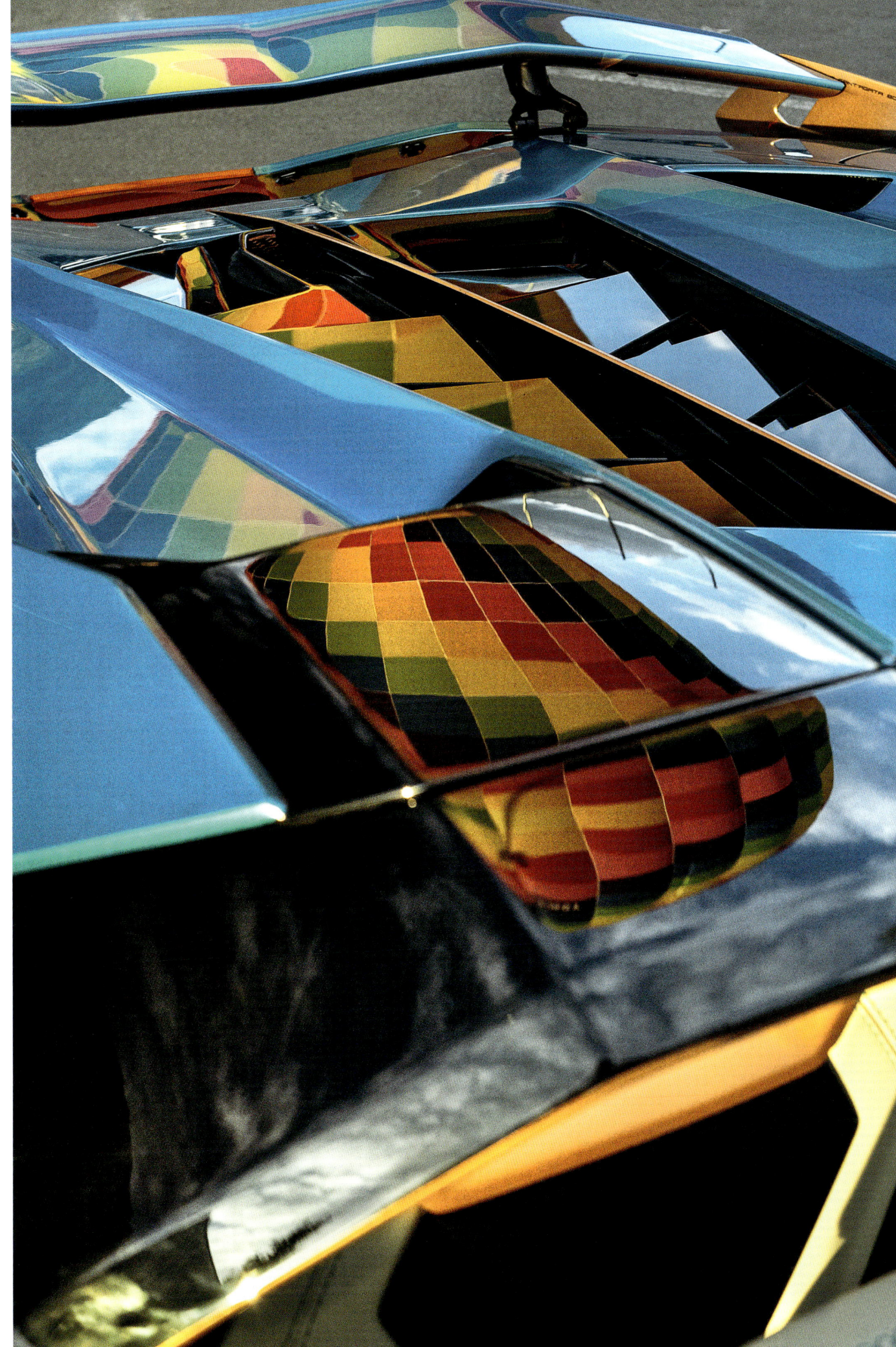

Reflections

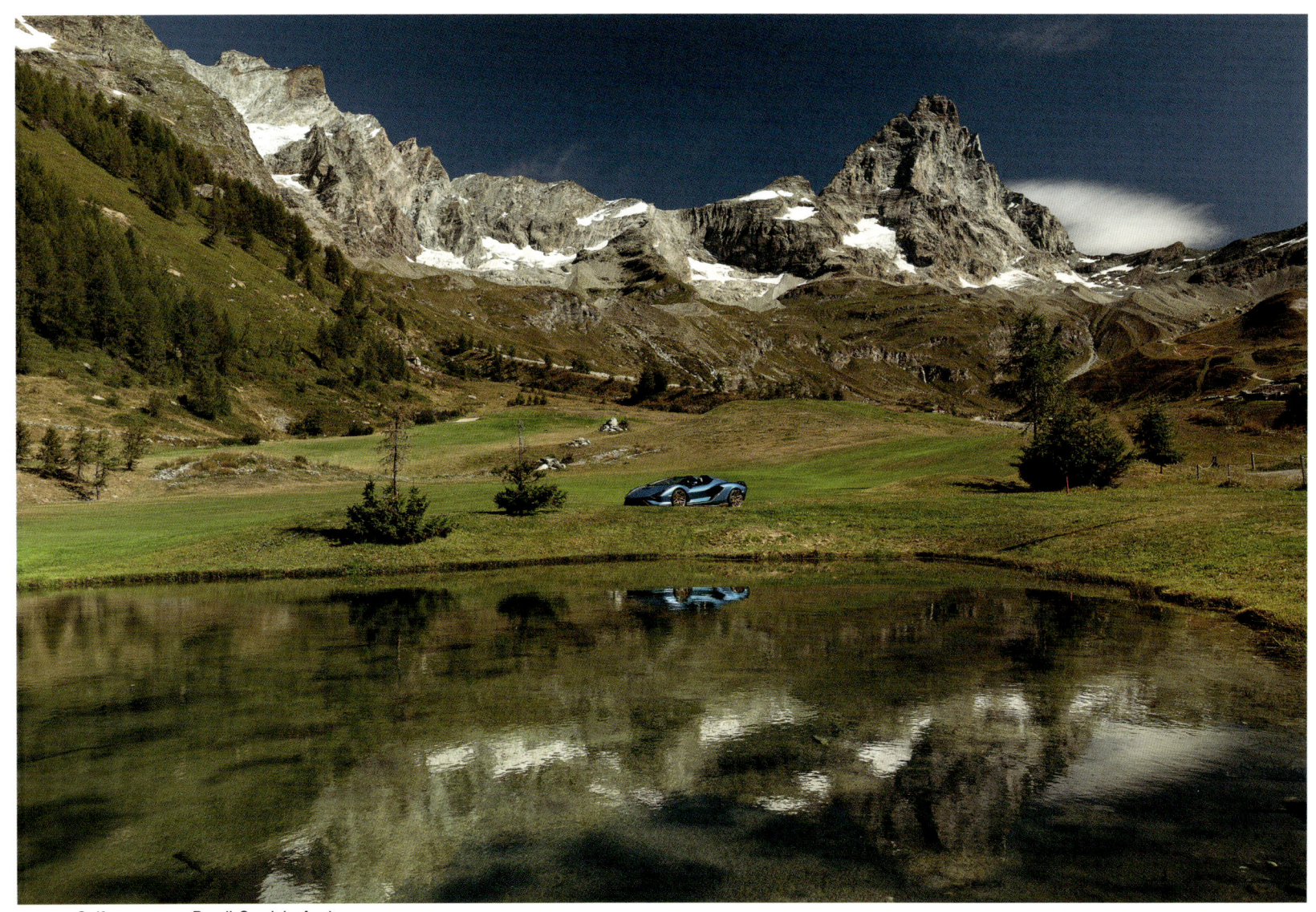

Golf course near Breuil-Cervinia, Aosta

Reflection of Mount Cervino on Lake Layet

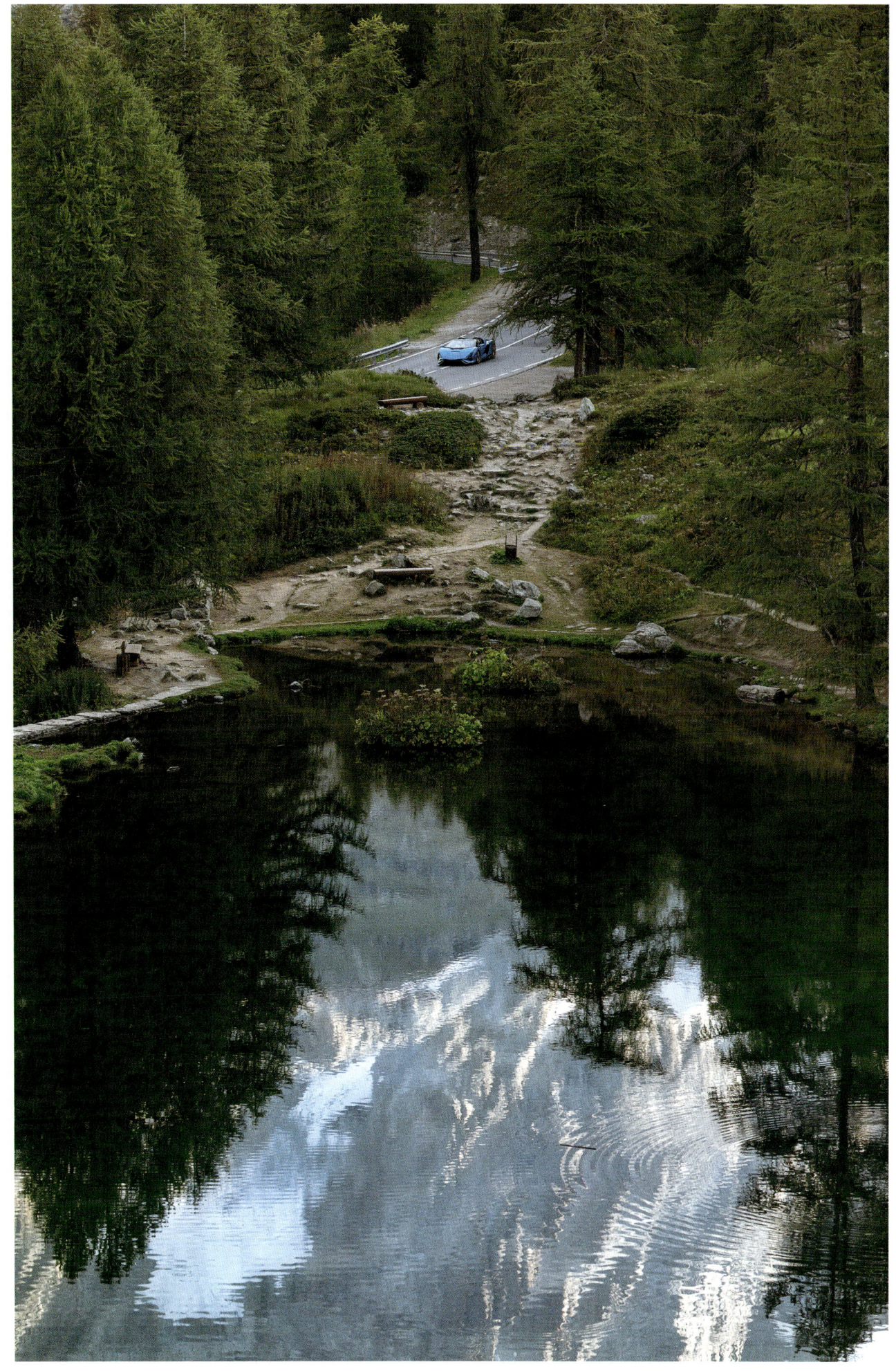

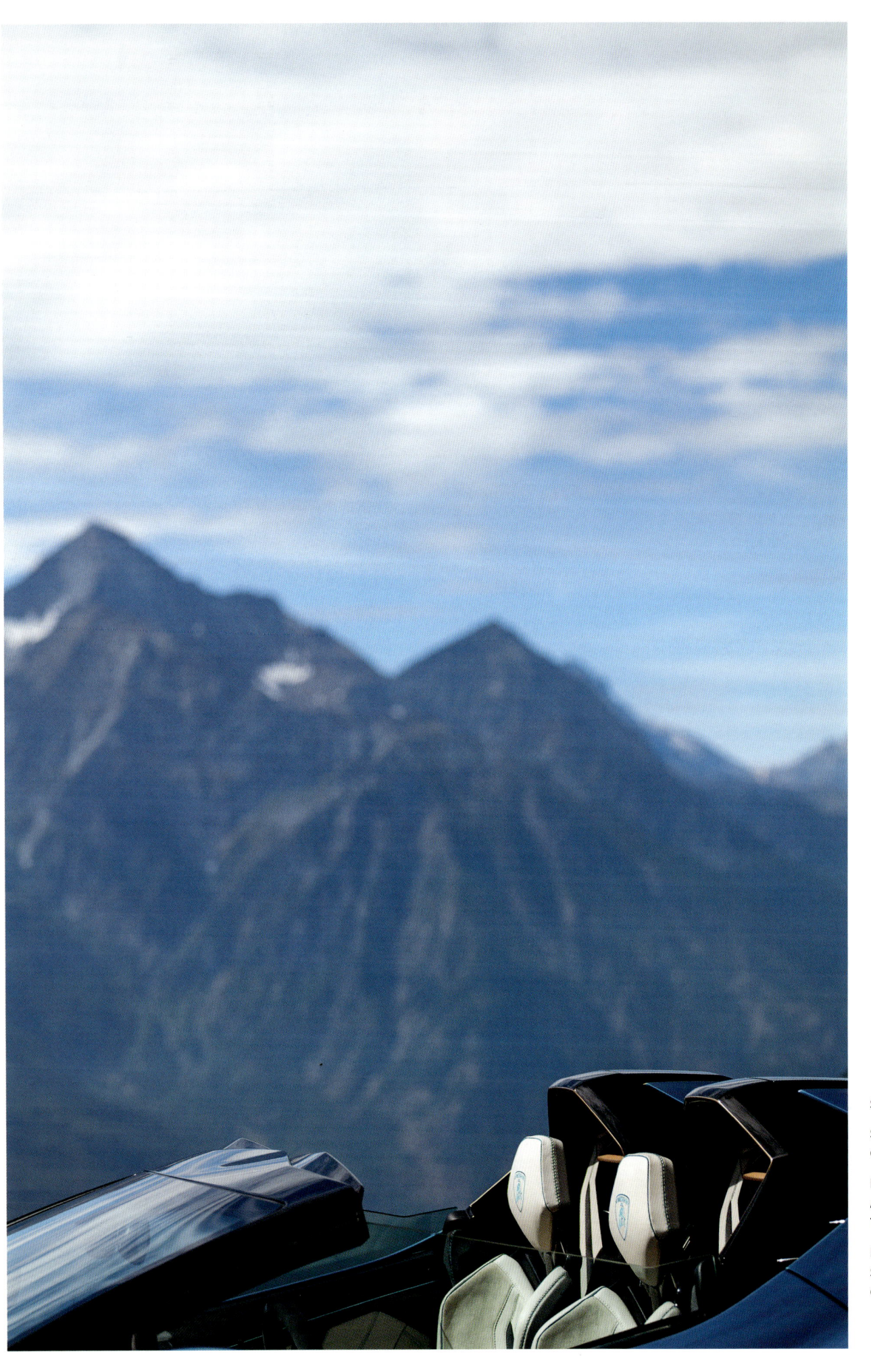

Saint-Barthélemy Valley

Saint-Barthélemy's Astronomical Observatory, Lignan

Pages 240 and 241
Aosta–Valtournenche–Cervinia Tunnel, Aosta

Pages 242–43
Saint Barthélemy's Astronomical Observatory, Lignan

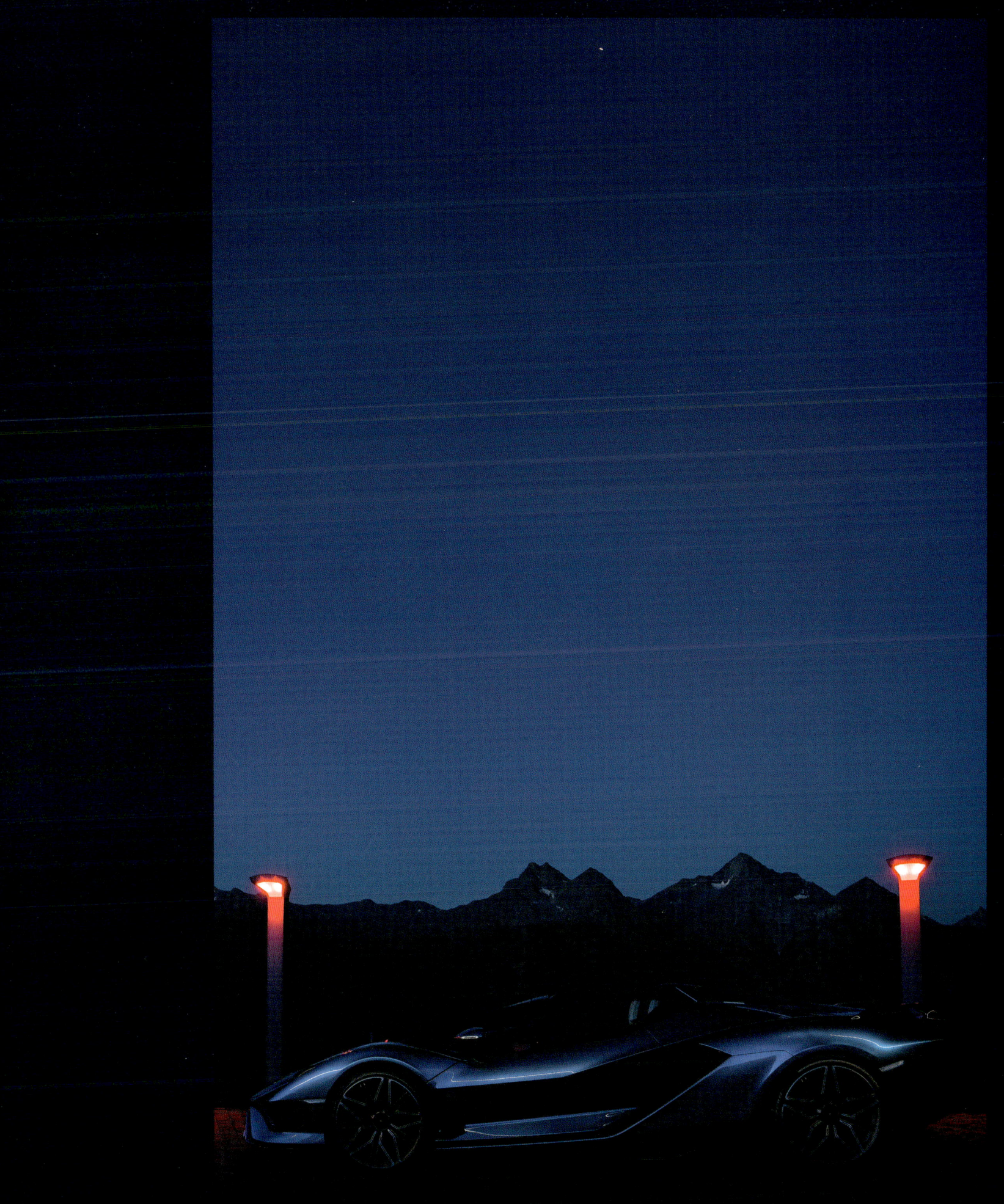

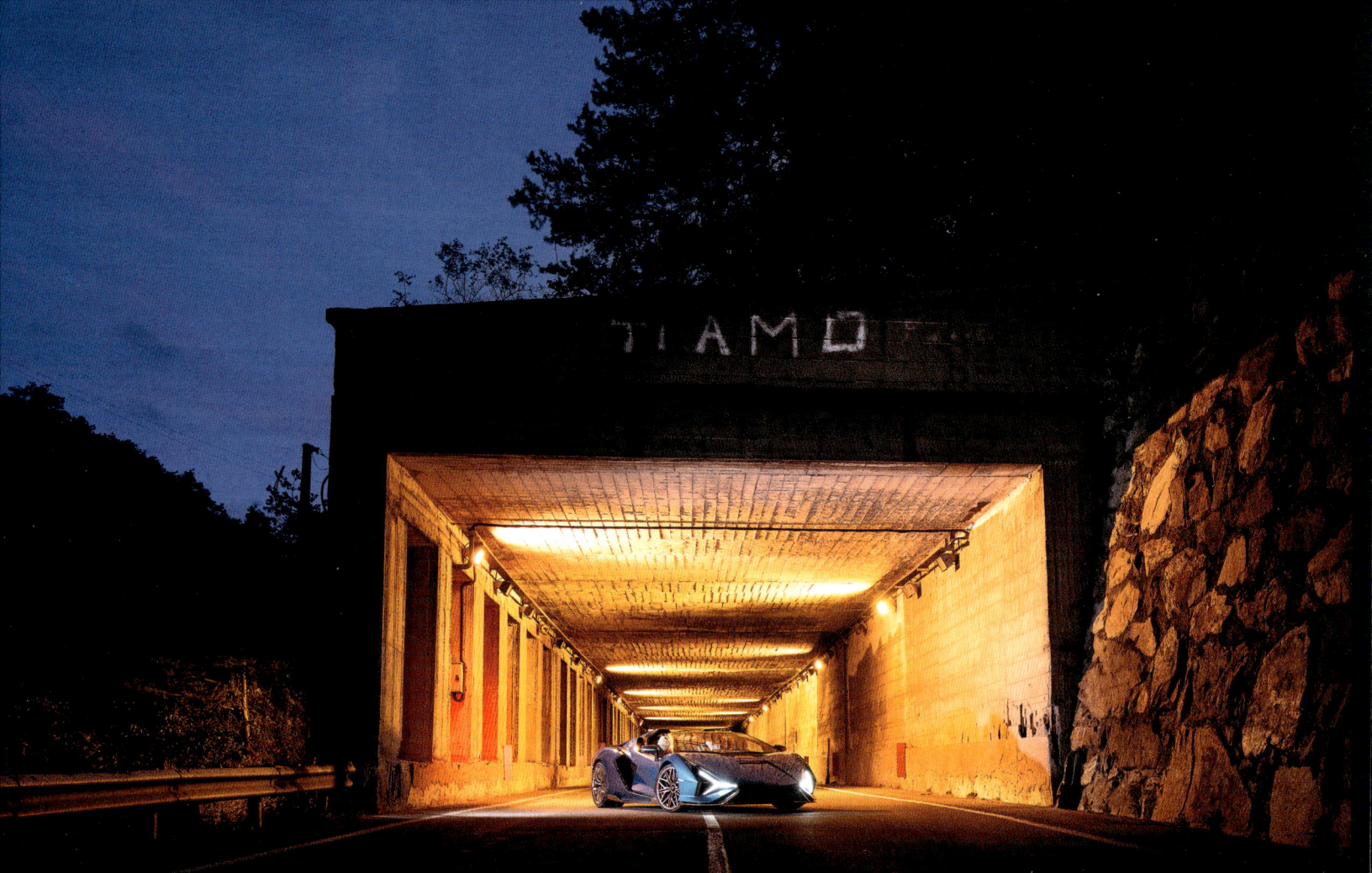

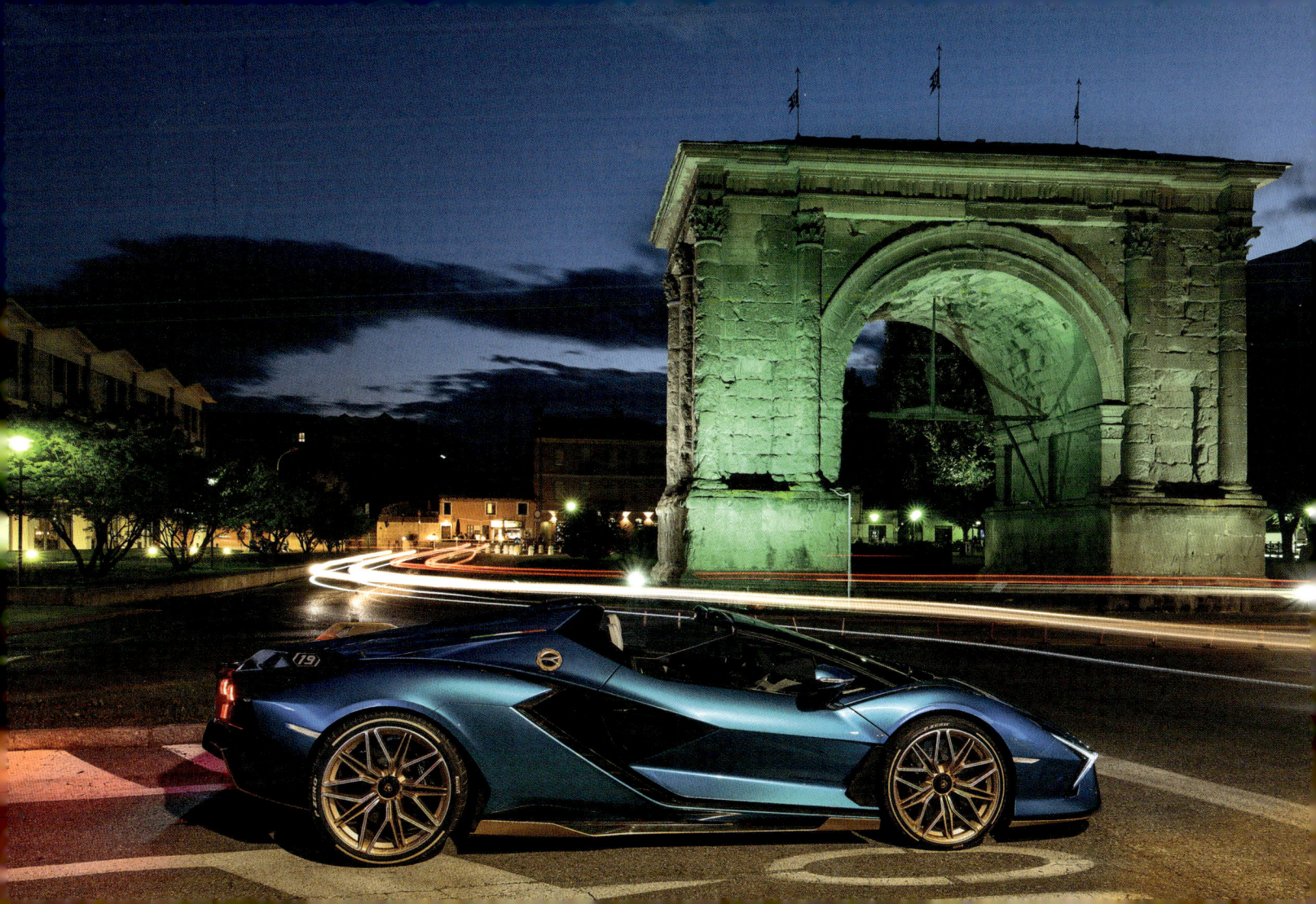

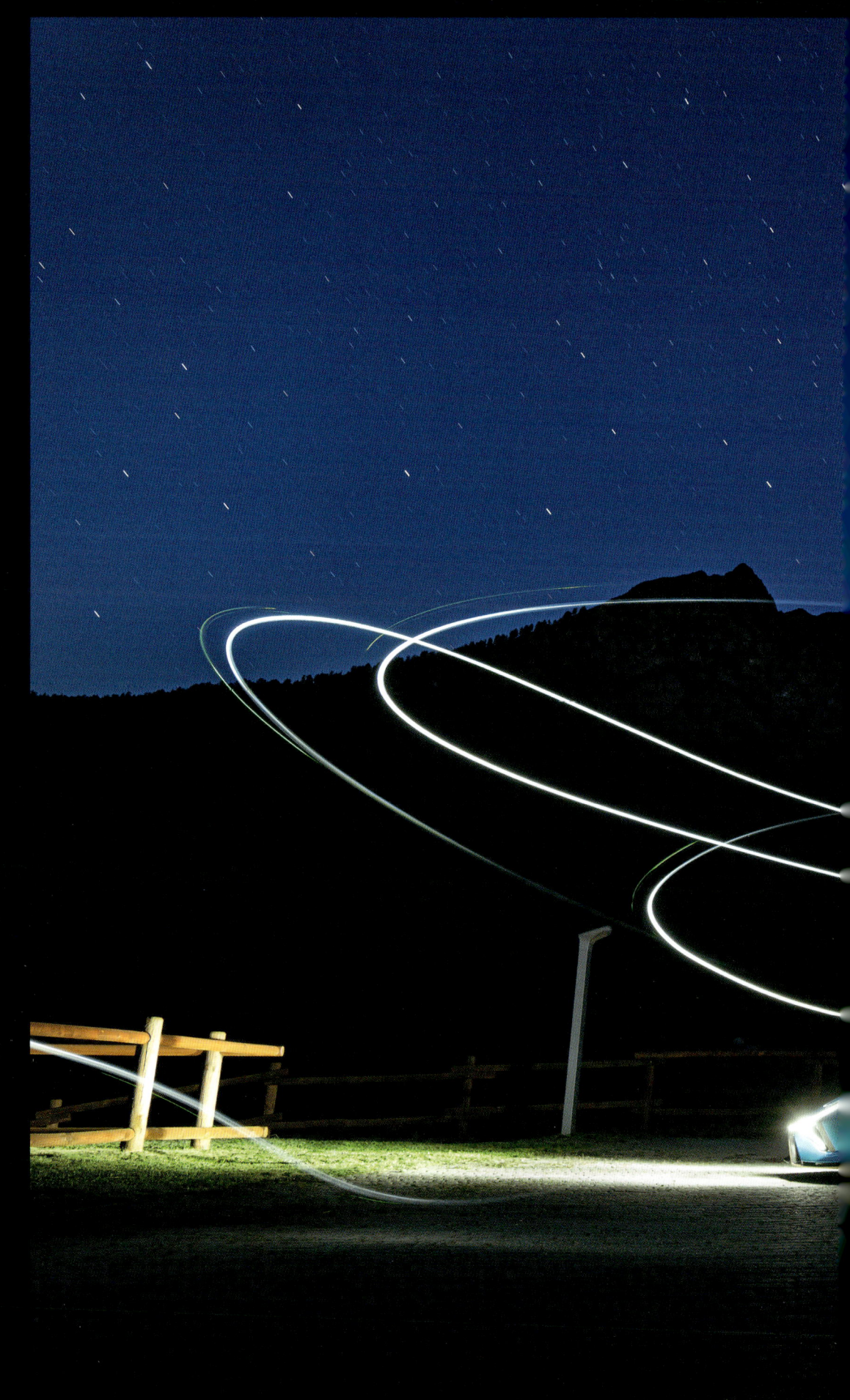

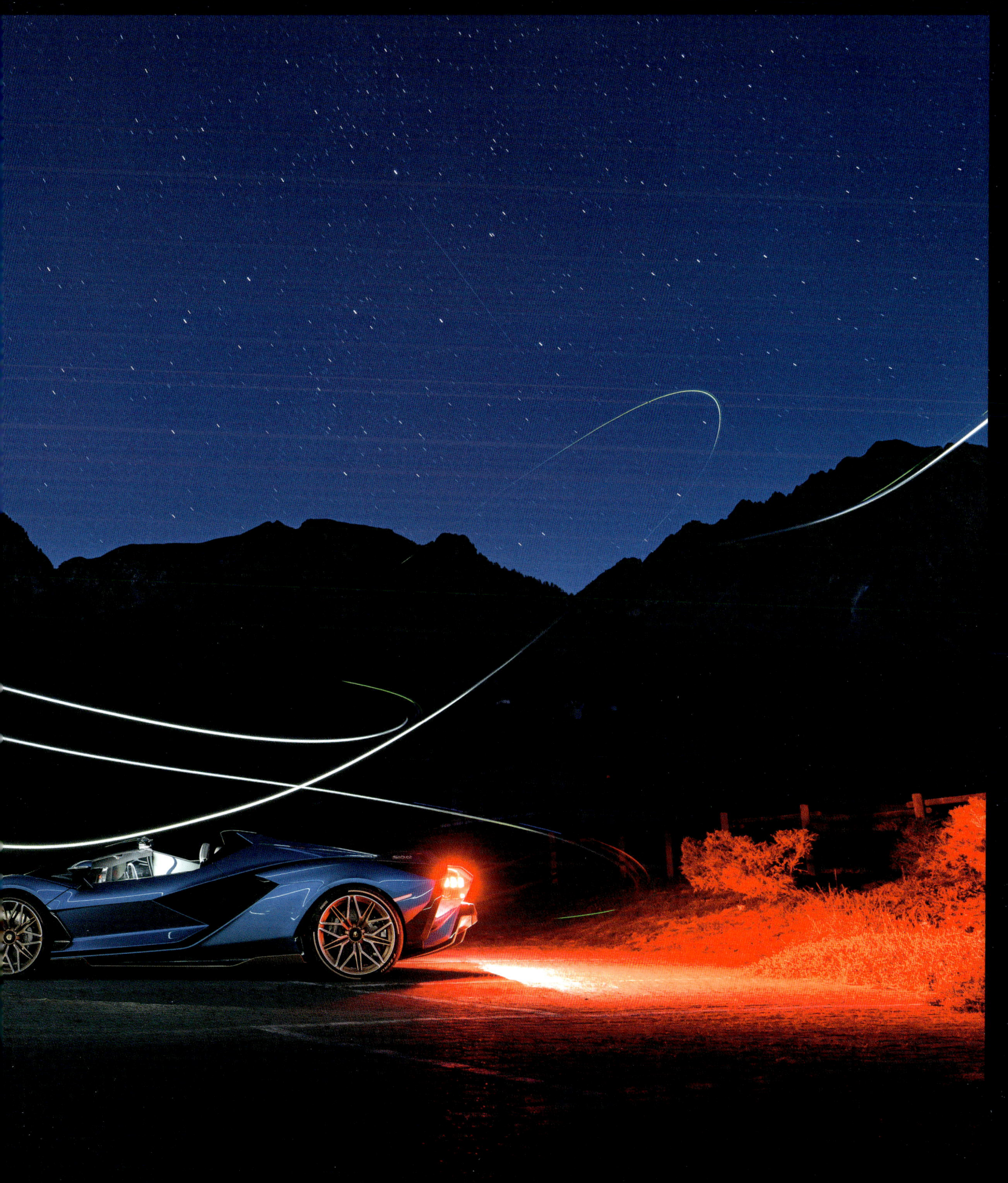

THE PHOTOGRAPHERS

Mattia Balsamini
FRIULI VENEZIA GIULIA
Mattia Balsamini was born on 24 September 1987 in Pordenone and moved to Los Angeles in 2008, enrolling at the Brooks Institute of California. In 2010, he interned at David LaChapelle's studio as an assistant and archivist. In 2011, upon receiving his Bachelor's degree with honours, he returned to Italy. Since 2012 he has been teaching photography at Venice's IUAV University, in the Master in Interactive Media for Interior Design programme. A member of the Contrasto photo agency, he focuses mainly on themes related to his region of origin, as well as on the idea of home and work as a crucial factor in human identity. His images have been displayed at the Milan Triennale, the MAXXI, the Sandretto Re Rebaudengo Foundation, Villa Manin and the Italian Cultural Institute of San Francisco. He collaborates with prestigious newspapers and clients, including Apple, Banca Nazionale del Lavoro, *Domus*, Enel, Eni, *Esquire*, Fendi, *Financial Times*, The Prada Foundation, *GQ*, *GEO*, *Icon*, *Internazionale*, *Libération*, *M Le Magazine du Monde*, Mercedes Benz, Nike, *The New York Times*, *The Observer*, Politecnico di Milano, *The Guardian*, *Vogue*, *Wallpaper**, *WIRED*.

Letizia Battaglia
PALERMO
Letizia Battaglia was born in Palermo in 1935. She started working as a photographer in Milan in 1971. Between 1974 and 1991, she headed a team of photographers at the Palermo-based daily newspaper *L'ORA*, including Franco Zecchin, Ernesto Battaglia, Shobha, Filippo La Mantia and others. Her photographs have featured in leading international publications as well as exhibited in galleries and museums. Her most recent exhibitions include *Fotografia come scelta di vita* at Casa dei Tre Oci in Venice and *Storie di strada* at Palazzo Reale in Milan.
She has been the recipient of the most prestigious international awards for social photography. She is not only a photographer, but also a film director, an environmentalist, a member of the local council, a city councillor for the Green party with Leoluca Orlando's administration during the period known as "Primavera Siciliana", a member of the Sicilian Regional Assembly, editor of *Grandevù* magazine and editor and director of Edizioni della Battaglia. She was a co-founder of the Giuseppe Impastato Centre of Documentation. She opened her first photography gallery in Southern Italy, Palermo, in 1978: the Laboratorio d'IF. In 1985 she became the first European woman to receive the Eugene Smith Award for social photography in New York and, in 1999, The Mother Johnson Achievement for Life in San Francisco. In 2007, the German Photography Society presented her with the Erich Salomon Prize in Germany. In May 2009, she won the Cornell Capa Infinity Award in New York. She was included in the long-list of 1,000 Women for the Nobel Peace Prize, nominated by PeaceWomen Across the Globe. She was the only Italian woman to be nominated by *The New York Times* as one of the 11 most representative women of 2017. She accepts invitations to hold lessons and workshops for museums and institutions in Italy and abroad. In 2017, she fulfilled her dream by inaugurating the International Photography Centre at the Cantieri Culturali alla Zisa in Palermo, where she is the director and curator of a selection of exhibitions and meetings dedicated to historical and contemporary photography.
Numerous books have been published and several documentaries made about her life and work, such as *Battaglia* (2004) by Daniela Zanzotto, *Amore Amaro* (2012) by Francesco Raganato in collaboration with Sky Arte, *La Mia Battaglia* (2016) by Franco Maresco, *Shooting the Mafia* (2019) by Kim Longinotto, *La mafia non è più quella di una volta* (2019) by Franco Maresco. She has been the director of *Mezzocielo*, a bimonthly political culture and environment magazine run by women since 1991.
She is a mother to three daughters – Cinzia, Shobha and Patrizia – a grandmother and a great-grandmother.

Simone Bramante
TRENTINO-ALTO ADIGE
Born in 1978, Simone Bramante is a creative director and photographer. Well known as Brahmino in the Instagram community, Simone strives to capture emotions, style and colours through his stories, running projects based on his unique narrative for personal and commercial purposes.
He believes stories are an incisive way to talk about life thanks to their authenticity, messages and emotions. His creative series have been featured, among others, on *Forbes*, *Huffington Post*, *GQ Portugal*, *Glamour Spain*, *Corriere della Sera*, *WithNews Japan* and *Buzzfeed*. He has shown his work in Los Angeles, San Francisco, Paris, Minsk and recently in Milan, where he exhibited his work from the Arctic Sea.

Fulvio Bugani
VALLE D'AOSTA
Born in Bologna in 1974, Fulvio Bugani is a freelance documentary photographer specializing in photojournalism. His work mainly explores socio-cultural issues and the interrelation of society and people, with this vision leading to both socially-themed projects and advertising work for major brands. He has collaborated with Doctors Without Borders and Amnesty International, and in 2012 was hired by Universal Music to produce a photo reportage for Zucchero Fornaciari's album *La Sesión Cubana*. On 8 December 2012, he was the only official photographer at Zucchero's historic concert in Havana. In 2015, he received an award at the World Press Photo contest and, that same year, began working with Leica Camera, which in 2017 chose him as one of its three International Brand Ambassadors. Between 2016 and 2017, he contributed to developing the new visual identity for Juventus Football Club. His photojournalism work has been exhibited at festivals and galleries throughout the world and published in numerous international magazines, including *TIME*, *The Guardian*, *LFI – Leica Fotografie International*, *Marie Claire*, *6Mois*, *Daily Mail* and *Cubadebate*.

Marco Casino
CAMPANIA
Born in 1986, Marco Casino is a multimedia photographer and filmmaker specializing in social reportage. He began pursuing photography as a profession in 2010, focusing on the phenomenon of neo-melodic music as a tool for the Camorra to reach the poorest demographics in the Campania region. Over the years he has completed numerous medium- and long-format projects in Europe, Africa and the Americas. He was the winner of Leica Talent 24x36 for 2011–12. In 2014, he was awarded first prize in the Short Feature category at the World Press Photo Multimedia Contest. He won the PDN Photo Annual awards back-to-back in 2014 and 2015, as well as the 30 under 30 competition sponsored by the Magnum Photos agency. In 2015 he was

also granted a scholarship by the Lucie Foundation in Los Angeles. From 2016 to 2018, he served as an expert consultant for the first three series of the *Master of Photography* TV show, co-produced by the Sky Art channels in Italy, Germany and the UK. Casino has also worked on marketing campaigns for major companies. His photographic and video work has appeared in international media and been displayed, both in solo and collective exhibitions, in such venues as the Pulitzer Hall of New York's Brown Institute for Media Innovation and the Royal Albert Hall in London.

Alessandro Cinque
UMBRIA
Born in 1988, Alessandro Cinque is a photojournalist currently based in Lima. His work investigates the social and environmental issues affecting minorities, focusing in particular on the devastating impact of mining on indigenous communities and their lands. Specifically, Alessandro has been documenting environmental contamination and public health concerns among the communities of Campesinos living along Peru's mining corridor. In 2017, he documented gold mining in Senegal and Kolbars smuggling goods at the border between Iraq and Iran. In 2019, while studying at the ICP in New York, he portrayed Williamsburg's Italian-American community and travelled to Arizona to photograph the abandoned uranium mines in the Navajo territories. His photographs have featured in *The New York Times*, the *NYT*'s *Lens* blog, *Marie Claire*, *Libèration*, *Internazionale*, *L'Espresso*, and others. In 2019, his work on Peru won first place at the POYi's Issue Reporting Picture Story. In the same year, he was selected as a finalist for the Eugene Smith Grant and Alexia Foundation Grant. In December 2019, Alessandro started a collaboration with Reuters on the coverage of Latin America, while expanding his project on the impact of Peru's mining industry on the Quechua populations.

Davide De Martis
SARDEGNA
Born in Sardinia in 1979, Davide De Martis graduated from the European Institute of Design in 2010. A multifaceted artist and photographer, he has always shown a keen interest in portraits: of people, but also of automobiles. He is a co-founder of the multidisciplinary Good Life Studio in Turin, a Certified by Leica photographer and, since 2017, a professor at the prestigious Leica Akademie.
He is known for his *auteur* approach to automotive photography, especially in his shots of classic cars. He works regularly in the field of corporate and advertising photography. His photographs have been displayed at Turin's PHOS Centre and at several contemporary art fairs, including The Others Fair and Photissima. They have also appeared in various specialized publications in Italy and abroad, as well as online on both company websites and social media.

Anna Di Prospero
LAZIO
Anna Di Prospero was born in Rome in 1987. She studied photography at Rome's European Institute of Design and New York's School of Visual Arts. Her photographic research stands out for the introspective style through which she explores the relation of man and space. Her works have been displayed in numerous solo and collective exhibitions in Italy and the US, including the Month of Photography Los Angeles, Milan's Triennale and the Palazzo delle Esposizioni in Rome. She has won several prestigious international prizes, such as the Sony World Photography award in the Portraiture category, the People Photographer of the Year at the International Photography Awards and Discovery of the Year at the 2011 Lucie Awards.

Camilla Ferrari
BASILICATA
Camilla Ferrari was born in 1992. Following a degree in communications, she studied at the Italian Institute of Photography in Milan. Her multimedia approach blends static and moving images in order to dissect the physical and emotional relation between humans and their surrounding environment, reflecting on the perception – and the power – of silence. Her works have been published in *National Geographic*, *NPR*, *US News*, *D – La Repubblica*, *CNN*, *6Mois*, *InsideOver* and *Elle Decor Italia*, among others. She is a member of Women Photograph. In 2018, she attended the Canon Student Development Programme at Visa Pour l'Image, the Nikon NOOR Masterclass in Turin, and was a finalist for the WMA Hong Kong Commission Grant. In 2019, she was selected by PDN as one of the thirty emerging talents in the international photography scene and by Artsy among the 20 Rising Female Photojournalists, while in 2020 she was nominated for the Joop Swart Masterclass of the World Press Photo.

Gabriele Galimberti
TOSCANA
Born in 1977, Gabriele Galimberti is a documentary photographer who lives mostly on airplanes and occasionally in Val di Chiana, Tuscany, where he was born and raised. He has spent the last few years working on long-format documentary photography projects around the world, some of which have become books, such as *Toy Stories*, *In Her Kitchen*, *My Couch Is Your Couch* and *The Heavens*. Gabriele's job mainly consists of telling stories, through portraits and short narratives, of people around the world, exploring their peculiarities and differences. Currently, he is travelling the globe, working on both solo and shared projects, as well as on assignments for international magazines and newspapers, such as *National Geographic*, *The Sunday Times*, *Stern*, *Geo*, *Le Monde*, *La Repubblica* and *Marie Claire*. His pictures have been exhibited in shows worldwide, such as the well-known Festival Images in Vevey, Switzerland, Les Rencontres de la Photographie in Arles and the prestigious Victoria and Albert Museum in London; they have won the Fotoleggendo Festival award in Rome and the Best In Show prize at the New York Photography Festival. Gabriele recently became a *National Geographic* photographer and has been contributing regularly to the magazine.

Piero Gemelli
EMILIA-ROMAGNA
Born in 1952, Piero Gemelli is an architect and photographer best known for his work in fashion, glamour and still life. He has created advertising campaigns and images for prestigious global jewellery, cosmetics and fashion brands, including Tiffany, Gucci, Ferré, Lancôme, Estée Lauder, Revlon, Shiseido, and many others. A long-time leading contributor to the Italian and several foreign editions of Vogue, he has also actively collaborated with many others of the world's most important fashion magazines. Regarded as one of the most significant Italian photographers worldwide, he has shown his work in numerous solo and collective exhibitions, starting from his first one-man show *Idea Progettata - fotografie 1983-1993* (Milan, 1994) and including *20 anni di Vogue Italia 1964-1984* (Milan, 1985), *À propos de la photographie italienne* (Musée de l'Elysée, Lausanne, 1992), *Lo sguardo Italiano. Fotografie di moda dal 1951 a oggi* (Rotonda della Besana, Milan, 2005) and the solo exhibit *W(H)O-MAN* (MyOwnGallery, Milan, 2010). Currently, he also works as an architect and art director, enriching his various projects with his own photography and graphic design work.

Vincenzo Grillo
LOMBARDIA
Vincenzo Grillo was born in Vibo Valentia, Calabria, in 1985. His love for photography dates to his childhood, following in the footsteps of his father, a well-known photographer from Southern Italy. At the age of eighteen he moved to Milan, where he began collaborating with the renowned Imaxtree agency, one of the big names in the Italian and international fashion scene. Over time, he specialized in street-style fashion photography, working with some of the world's most famous models and celebrities. Subsequently, he developed an interest in the digital advertising world, working for such brands as Fendi, Versace and Tod's among others. A tireless globetrotter, he divides his time between New York, Paris, Milan and London, always in search of the perfect shot.

Stefano Guindani
SICILIA
Born in 1969, Stefano Guindani is an artist who combines a love for fashion and lifestyle photography with an undying commitment to reportage. Over the years, he has worked with major companies, such as Condé Nast and Rai Cinema, contributing to the web and TV series *Ricette e Ritratti d'Attore*, which grew into an exhibition in Los Angeles and a book titled *Sguardi d'Attore*. In parallel, Guindani has cultivated a passion for reportage which, in partnership with the Francesca Rava Foundation, led to the publication of *Haiti Through the Eye of Stefano Guindani* and *Do You Know*, as well as the photography exhibitions *Ey You!* at Milan's Microsoft House in 2017, and *My Dream Home*, held in 2018 as part of FuoriSalone. For the past two years, Guindani has been a brand ambassador for Huawei, contributing to various projects in the realm of communications and mobile photography. He founded the studio SGP, which has become a benchmark in the photography world.

Gabriele Micalizzi
PUGLIA
Born in 1984, Gabriele Micalizzi is a photojournalist. From 2004 to 2005, he worked as a photo reporter for Milan's Newspress agency. In 2008 he co-founded the independent photography collective CESURA, with Magnum photographer Alex Majoli as the art director. In 2010, he began working as a photo reporter in conflict-affected areas, such as Bangkok during the Red Shirt riots, the Middle East during the Arab Spring and Greece during the economic crisis. On 11 February 2019, while documenting the US-led coalition's advance against ISIS forces in south-eastern Syria, he survived an RPG attack from the Islamic State militia. A Leica testimonial since 2016 and the winner of the first edition of European talent show *Master of Photography*, he contributes to numerous Italian and international publications, such as *The New York Times*, *Herald Tribune*, *The New Yorker*, *Newsweek*, *Stern*, *Corriere della Sera*, *L'Espresso*, *La Repubblica*, *Internazionale*, *Panorama*, *Ruptly*, *Sportweek*, *Wall Street Journal*.

Chiara Mirelli
PIEMONTE
Chiara Mirelli was born in Milan in 1976. In 2001 she received her diploma in photography from the Riccardo Bauer school, to which she added a degree in photo editing in 2006. In 2011, she completed a programme in documentary filmmaking at Milan's School of Cinema, Television and New Media. She has worked extensively in the music world, creating portraits, booklets, videos and reportages for various Italian artists. In addition to musicians, she has also portrayed numerous athletes and celebrities for leading Italian newspapers.

Roselena Ramistella
MOLISE
Born in 1982, Roselena Ramistella is a Sicilian photographer with a background in political science. Social issues, portraiture, and the interaction between the natural world and mankind are at the centre of her photographic research. Roselena investigates her subjects with a personal and intimate vision, always experimenting with new techniques to create a highly personal narrative and artistic style. She is a brand ambassador and teacher for Leica. Her work has been showcased and recognized across Europe and beyond, and in 2018 she won the Sony Award in the Natural World & Wild Life category. Roselena collaborates with important publications, such as *L'Uomo Vogue*, *Repubblica*, *Internazionale*, *Wordt Vervolgd – Amnesty International*, *Io Donna*, with her work also appearing in *The Guardian*, *BBC*, *The Times* and the *British Journal of Photography*.

Alberto Selvestrel
LIGURIA
Alberto Selvestrel was born in Turin on 29 November 1996. He first approached photography in 2014, aged seventeen, displaying a keen interest in landscapes from the start. A self-taught artist, he later furthered his research focusing mainly on the anthropic landscape and its modifications, crafting a distinctive style characterized by minimalist geometric compositions. Selvestrel's aesthetic principles involve a synthesis of maximum conceptual expression and minimal form. In 2017 he became the youngest member of the inQuadra collective. On 1 October 2017 he released his first book, titled *IMAGES*. His work has received enthusiastic reviews and was published by leading magazines and newspapers around the world. Selvestrel showcased his work in Italy in 2017, as well as in London and Brussels in 2018. In that same year he began a collaboration with the P3 Project Pixel Paper workshop, touring Italy's major cities alongside photographer Alex Liverani. On 18 March 2019 he released his second book *Link*, which included a foreword by the famous Italian photographer Giovanni Gastel. In 2019, he was chosen by Fujifilm as the only Italian spokesperson for the launch of their X-Pro3 camera.

Valentina Sommariva
ABRUZZO
Valentina Sommariva was born in 1986 in Milan, where she lives and works to this day. A freelance journalist, she specializes in portrait, travel and interior photography. Following a degree in architecture from the Politecnico di Milano, she studied design in London and then completed a Masters in contemporary photography and video art in Modena. She works with some of Italy's leading design firms, contributing to advertising campaigns and catalogues for major magazines, mainly in Italy, France, the UK and the US. At the same time, she pursues independent photographic and video projects focusing mainly on identity and the relationship between people and their living environments. Her works have been displayed in various exhibitions.

Wolfango Spaccarelli
MARCHE
Born in Milan in 1964, Wolfango Spaccarelli has a degree in political sciences with a focus on international law. He briefly considered a diplomatic career but eventually decided to professionally pursue his true passion, photography. In 1984 he began a long and fruitful collaboration with Carrstudio, a photography studio specializing, though not exclusively, in automotive photography. With extensive experience in both studio and outdoor photography, he has personally managed important accounts for clients in the automotive and publishing sector (RCS magazines, Rusconi editions, Hachette, and many others). He opened his own studio in 2005, further expanding his professional and creative collaborations with other top brands in the car world, including Lamborghini, Bugatti, Audi, Infiniti and Kia. He is also a regular contributor to important projects with both national and international publications, which have allowed him to broaden his photographic explorations at the intersection of the car universe and the human dimension.

Guido Taroni
CALABRIA
Born in Milan in 1987, Guido Taroni spent most of his youth at the family villa on Lake Como, where he developed his fascination with beauty, colours and shape. Guido got off to an early start, and by the age of seventeen he was already honing his craft at the Sancassani Studio of interior photography, eventually becoming an assistant to his uncle, the acclaimed Italian photographer Giovanni Gastel. He had his first solo show in Milan when he was just twenty-one. Titled *Sogni Sospesi* (Suspended Dreams), the exhibition was received enthusiastically by both audience and critics, being selected by the renowned art critic Vittorio Sgarbi to be part of the prestigious Festival dei Due Mondi. His good looks and refined appearance have also gained him numerous requests to work as a model, including for a recent Tod's campaign, and TV personality starring in the four-episode Sky Arte documentary *Grand Tour* about the best Italian properties of the National Trust FAI. He was the only photographer featured in the book *The Interiors and Architecture of Renzo Mongiardino: A Painterly Vision*, published by Rizzoli and presented in October 2017 in London, New York and Milan. The volume explores the sublime work of the architect Renzo Mongiardino (1916–98), reinforcing his place as a legend in the field. Another coffee table book was released in 2019, *Inside Tangier*. Published by Vendome Press, it explores a selection of exceptional properties, houses and gardens as well as the city's eccentric inhabitants.

Marco Valmarana
VENETO
Marco Valmarana was born in Venice in 1990. From his Venetian father, he inherited a love for water, the lagoon and Venetian traditions. And from his mother, a former PR specialist from Mexico City, he picked up a passion for communication, which led to his interest in social media. Fascinated with travelling and photography since he was a teenager, Marco got his start by publishing intriguing visual tales of Venice on his Instagram page in 2012. Through his love for cameras, travel and his hometown, he quickly succeeded in turning his hobby into a profession. Currently, he is an established content creator whose portfolio of clients includes major brands in the hotel, food & beverage and automotive industries as well as in the arts.

		Lamborghini	Production	Chassis number	Colour
Letizia Battaglia	Palermo	Aventador SVJ	2019	ZHWUM6ZD9KLA07782	Giallo Tenerife
Stefano Guindani	Sicilia	Urus	2018	ZPBEA1ZL8KLA00327	Rosso Anteros
Davide De Martis	Sardegna	Miura P400 S	1970	4644	Oro Metallizzato
Guido Taroni	Calabria	Urus	2020	ZPBEA1ZLXLLA08513	Blu Eleos
Gabriele Micalizzi	Puglia	Aventador SVJ Roadster	2019	ZHWEN6ZDXLLA09009	Bronzo Zenas
Camilla Ferrari	Basilicata	Aventador S	2019	ZHWEG4ZD5KLA08699	Nero Maia (Ad Personam colour)
Marco Casino	Campania	Huracán EVO Spyder	2020	ZHWET4ZF4LLA13455	Bianco Asopo
Roselena Ramistella	Molise	Urus	2019	ZPBEA1ZL6KLA04134	Giallo Auge
Valentina Sommariva	Abruzzo	Huracán EVO Spyder	2020	ZHWET4ZF7LLA13448	Azzurro Arione matt (Ad Personam colour)
Anna Di Prospero	Lazio	Aventador S	2019	ZHWEG4ZD7KLA08736	Rosso Nestor (Ad Personam colour)
Wolfango Spaccarelli	Marche	Diablo 6.0 SE	2001	ZA9DE01A01LA12898	Oro Elios
Alessandro Cinque	Umbria	Aventador SVJ	2018	ZHWEM6ZD8KLA07566	Verde Alceo
Gabriele Galimberti	Toscana	Huracán EVO RWD Spyder	2020	ZHWET5ZF1LLA15055	Arancio Anthaeus
Piero Gemelli	Emilia-Romagna	Aventador S Roadster	2019	ZHWEV4ZD5KLA08594	Blu Vathys (Ad Personam colour)
Marco Valmarana	Veneto	Countach 25th Anniversary	1990	ZA9C005A0KLA12085	Argento Luna
Mattia Balsamini	Friuli Venezia Giulia	Urus	2018	ZPBEA1ZL5KLA00253	Grigio Nimbus
Simone Bramante	Trentino-Alto Adige	Huracán EVO RWD	2020	ZHWEF5ZF4LLA14635	Giallo Belenus
Vincenzo Grillo	Lombardia	Huracán EVO	2020	ZHWEF4ZF9LLA14964	Verde Hector (Ad Personam colour)
Chiara Mirelli	Piemonte	Huracán EVO Spyder	2019	ZHWET4ZF0LLA13226	Verde Selvans
Alberto Selvestrel	Liguria	Huracán EVO	2020	ZHWEF4ZF6LLA15103	Blu Aegeus
Fulvio Bugani	Valle d'Aosta	Sián Roadster	2020	ZHWEK7ZD1LLA09494	Blu Uranus

Automobili Lamborghini wishes to thank all the photographers who took part in this project for their art and enthusiasm. We would also like to thank the local authorities, regional governments, provinces and municipalities for their support during the photo shoots.

For Palermo, Letizia Battaglia wishes to thank
Mayor Leoluca Orlando.

Stefano Guindani wishes to thank
Katia Bassi, Chief Marketing & Communication Officer at Automobili Lamborghini, for immediately supporting this project, which was conceived and developed with Clara Magnanini (Brand & Corporate Communication Manager at Lamborghini). For the operational work in Sicily, Andrea Lo Cicero, an excellent guide, Johnny Dalla Libera, Andreana Patti, Luca Ferlito, Pucci Piccione, Caroline Costa Aniceto, Antonio D'Alì Staiti, Francesco Alparone, Saretto Bambara, Mayor of Ortigia Francesco Italia, and Stefano d'Alessandro.

For Sardegna, Davide De Martis wishes to thank
The Fondazione Sardegna Film Commission for the logistical support, and Nüadventure for support and project production back-up. I would like to dedicate my photos to my beloved island, and to my parents, Sandro and Angelamaria.

For Calabria, Guido Taroni wishes to thank
Carmine Lupia, President of Cammino Basiliano, for being my guide during the shoot; Dr. Aversa and the archaeological site of Capo Colonna; the Municipalities that gave us such a warm welcome (Cropani, Cerva, Zagarise, Sant'Andrea Apostolo dello Ionio), Gino Fuoco, and all the people who agreed to be portrayed, including Maria, Carmela, Tota, Orsola and Pietro, and the Sersale Pacchiane. My gratitude also goes to Annalia Paravati Capogreco and Virginia Taroni.

For Puglia, Gabriele Micalizzi wishes to thank
Ester, Riccardo, Tecla, Guenda, Gianna, the Cagnetta family, the Negruzzi family, the fishermen at Bari's old port and Vincenzo's *barreto*, the town of Gravina (Egidio), the B&B on the bridge, the Jolly Bari Car Wash, Cristian's Paninoteca and Alessia Bari. Special thanks to Puglia, *aka* Italy's California.

For Basilicata, Camilla Ferrari wishes to thank
ALSIA Basilicata (especially Sergio Rossi); the Maneggio San Nicola di Matera (especially Berta and Alessia Nicoletti); the Municipality of Matera, and my assistant Raoul Ven.

 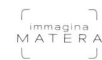

For Campania, Marco Casino wishes to thank
Carmine, an enthusiastic travelling companion, and his family; the mayors of the Municipalities of Procida and Amalfi for their welcome and support; all the friends who, through gestures large and small, made this project even more special. Thanks to my family, the foundation of my journey.

For Molise, Roselena Ramistella wishes to thank
All the people who helped me discover a wonderful place like Molise, Mrs. Renata in Roccasicura, Marilena in Rocchetta Alta, all the teenagers I met, who welcomed me with their energy and positivity; finally, my dear friend Giuseppe, for all his patience.

For Abruzzo, Valentina Sommariva wishes to thank
Alessia Interlandi, Maddalena Scarzella, Antonio Delle Monache, Matteo Benedetti, Peter Naguschiewski, Franca Pera of the Fucino Space Centre, and Fabio Santavicca, Mayor of Santo Stefano di Sessanio.

For Lazio, Anna Di Prospero wishes to thank
Eleonora Materazzo, Lorenzo Alfonso Volino, and Vincenzo Di Prospero.

For the Marche, Wolfango Spaccarelli wishes to thank
Lamborghini's technicians, Francesco Falcone and Mirko Petronelli, two wonderful guys who were both professional and helpful; my assistant Francesco Cottatellucci, tireless and a source of excellent suggestions, and the Municipalities of Urbino, Senigallia and Ascoli Piceno.

For Umbria, Alessandro Cinque wishes to thank
The Umbria Region and the mayors who took part in the project, the Municipality of Orvieto, Agriturismo Titignano, Sibillini Ranch, Gabriele Giuliani and Giovannino Cinque.

For Toscana, Gabriele Galimberti wishes to thank
My assistants Diego Scalet, Juri De Luca, Veronica Strazzari, the Chianini biker group for their leaning turns, and the kids Matteo Pallini, Jeremy Zecchini, and Pietro Butini who were portrayed in the photos.

For Emilia-Romagna, Piero Gemelli wishes to thank
The Municipalities of Dozza, Ravenna, Tresignana, Parma, and Salsomaggiore Terme, for the support offered with permits and their assistance in creating these images.

For Veneto, Marco Valmarana wishes to thank
The staff at the Canoviano Temple of Possagno, the Monumental Garden of Valsanzibio, Villa Barbarigo, and the Temple of Diana.

For Friuli Venezia Giulia, Mattia Balsamini wishes to thank
Edoardo Pedrotti and Teo Zanin.

For Trentino-Alto Adige, Simone Bramante wishes to thank
Visit Trentino, for their help in accessing Val San Nicolò and Val Venegia, where I shot the whole series.

For Lombardia, Vincenzo Grillo wishes to thank
Armando Grillo for his invaluable help during the shoot.

For Piemonte, Chiara Mirelli wishes to thank
Friends and folk involved in the shots: Giorginess, Liede, the skater Davide, Mr Fijodor, Eugenio Gege Odasso, and the La Maglianese band.

For Liguria, Alberto Selvestrel wishes to thank
Irene Volpiano, my mom Serena, my dad Roberto, and Massimo Delbo.

For Valle d'Aosta, Fulvio Bugani wishes to thank
Daniela Damonte for overall organization. On the operational side, heartfelt thanks go to Adele Grotti, Luca Vittorio Toffolon, Sylvie Cheney, Marco Armienti, Francesco Catanese, and Luca Finotello. My thanks, too, to Leica Camera Italia and Maurizio Beucci, as well as to all the wonderful people we met in the beautiful Aosta Valley, without whom it would have been impossible to make this project a reality. Special thanks to Lamborghini's technical staff who were there for the shoot, Mario Fasanetto and Marco Risi, together with carrier Federico Azzolini: a truly special team.